The Handbook of Environmental Chemistry

Volume 2 Reactions, Processes Part O

O. Hutzinger
Editor-in-Chief

Oxidants and Antioxidant Defense Systems

Volume Editor: Tilman Grune

With contributions by
G. Bartosz · H. K. Biesalski · A. Boveris · E. Cadenas
K. J. A. Davies · L. Gille · H. Griffiths · T. Grune
L.-O. Klotz · A. V. Kozlov · J. Krutmann · H. Nohl
F. V. Pallardo · D. Poppek · H. E. Poulsen · B. Riis
J. Sastre · P. Schröder · K. Staniek · J. Viña

Volume Editor

Dr. Tilman Grune
Research Institute of Environmental Medicine
Heinrich Heine University
Auf'm Hennekamp 50
40225 Düsseldorf, Germany
Tilman.Grune@uni-duesseldorf.de

Library of Congress Control Number: 2004116735

ISSN 1433-6839
ISBN 3-540-22423-8
DOI 10.1007/b98750
Springer Berlin Heidelberg New York

Springer-Verlag is a part of Springer Science+Business Media
springeronline.com

Printed in Germany

Cover Design: E. Kirchner, Springer-Verlag
Production Editor: Christiane Messerschmidt, Rheinau
Typesetting: Fotosatz-Service Köhler GmbH, Würzburg
Printed on acid-free paper 52/3141me – 5 4 3 2 1 0

The Handbook of Environmental Chemistry Also Available Electronically

For all customers who have a standing order to The Handbook of Environmental Chemistry, we offer the electronic version via SpringerLink free of charge. Please contact your librarian who can receive a password for free access to the full articles by registering at:

springerlink.com

If you do not have a subscription, you can still view the tables of contents of the volumes and the abstract of each article by going to the SpringerLink Homepage, clicking on "Browse by Online Libraries", then "Chemical Sciences", and finally choose The Handbook of Environmental Chemistry.

You will find information about the

- Editorial Bord
- Aims and Scope
- Instructions for Authors
- Sample Contribution

at springeronline.com using the search function.

Preface

Environmental Chemistry is a relatively young science. Interest in this subject, however, is growing very rapidly and, although no agreement has been reached as yet about the exact content and limits of this interdisciplinary discipline, there appears to be increasing interest in seeing environmental topics which are based on chemistry embodied in this subject. One of the first objectives of Environmental Chemistry must be the study of the environment and of natural chemical processes which occur in the environment. A major purpose of this series on Environmental Chemistry, therefore, is to present a reasonably uniform view of various aspects of the chemistry of the environment and chemical reactions occurring in the environment.

The industrial activities of man have given a new dimension to Environmental Chemistry. We have now synthesized and described over five million chemical compounds and chemical industry produces about hundred and fifty million tons of synthetic chemicals annually. We ship billions of tons of oil per year and through mining operations and other geophysical modifications, large quantities of inorganic and organic materials are released from their natural deposits. Cities and metropolitan areas of up to 15 million inhabitants produce large quantities of waste in relatively small and confined areas. Much of the chemical products and waste products of modern society are released into the environment either during production, storage, transport, use or ultimate disposal. These released materials participate in natural cycles and reactions and frequently lead to interference and disturbance of natural systems.

Environmental Chemistry is concerned with reactions in the environment. It is about distribution and equilibria between environmental compartments. It is about reactions, pathways, thermodynamics and kinetics. An important purpose of this Handbook, is to aid understanding of the basic distribution and chemical reaction processes which occur in the environment.

Laws regulating toxic substances in various countries are designed to assess and control risk of chemicals to man and his environment. Science can contribute in two areas to this assessment; firstly in the area of toxicology and secondly in the area of chemical exposure. The available concentration ("environmental exposure concentration") depends on the fate of chemical compounds in the environment and thus their distribution and reaction behaviour in the environment. One very important contribution of Environmental Chemistry to the above mentioned toxic substances laws is to develop laboratory test methods, or mathematical correlations and models that predict the environmental fate of new chemical compounds. The third purpose of this Handbook is to help in the basic understanding and development of such test methods and models.

The last explicit purpose of the Handbook is to present, in concise form, the most important properties relating to environmental chemistry and hazard assessment for the most important series of chemical compounds.

At the moment three volumes of the Handbook are planned. Volume 1 deals with the natural environment and the biogeochemical cycles therein, including some background information such as energetics and ecology. Volume 2 is concerned with reactions and processes in the environment and deals with physical factors such as transport and adsorption, and chemical, photochemical and biochemical reactions in the environment, as well as some aspects of pharmacokinetics and metabolism within organisms. Volume 3 deals with anthropogenic compounds, their chemical backgrounds, production methods and information about their use, their environmental behaviour, analytical methodology and some important aspects of their toxic effects. The material for volume 1, 2 and 3 was each more than could easily be fitted into a single volume, and for this reason, as well as for the purpose of rapid publication of available manuscripts, all three volumes were divided in the parts A and B. Part A of all three volumes is now being published and the second part of each of these volumes should appear about six months thereafter. Publisher and editor hope to keep materials of the volumes one to three up to date and to extend coverage in the subject areas by publishing further parts in the future. Plans also exist for volumes dealing with different subject matter such as analysis, chemical technology and toxicology, and readers are encouraged to offer suggestions and advice as to future editions of "The Handbook of Environmental Chemistry".

Most chapters in the Handbook are written to a fairly advanced level and should be of interest to the graduate student and practising scientist. I also hope that the subject matter treated will be of interest to people outside chemistry and to scientists in industry as well as government and regulatory bodies. It would be very satisfying for me to see the books used as a basis for developing graduate courses in Environmental Chemistry.

Due to the breadth of the subject matter, it was not easy to edit this Handbook. Specialists had to be found in quite different areas of science who were willing to contribute a chapter within the prescribed schedule. It is with great satisfaction that I thank all 52 authors from 8 countries for their understanding and for devoting their time to this effort. Special thanks are due to Dr. F. Boschke of Springer for his advice and discussions throughout all stages of preparation of the Handbook. Mrs. A. Heinrich of Springer has significantly contributed to the technical development of the book through her conscientious and efficient work. Finally I like to thank my family, students and colleagues for being so patient with me during several critical phases of preparation for the Handbook, and to some colleagues and the secretaries for technical help.

I consider it a privilege to see my chosen subject grow. My interest in Environmental Chemistry dates back to my early college days in Vienna. I received significant impulses during my postdoctoral period at the University of California and my interest slowly developed during my time with the National Research Council of Canada, before I could devote my full time of Environmental Chemistry, here in Amsterdam. I hope this Handbook may help deepen the interest of other scientists in this subject.

Amsterdam, May 1980 *O. Hutzinger*

Twentyone years have now passed since the appearance of the first volumes of the Handbook. Although the basic concept has remained the same changes and adjustments were necessary.

Some years ago publishers and editors agreed to expand the Handbook by two new open-end volume series: Air Pollution and Water Pollution. These broad topics could not be fitted easily into the headings of the first three volumes. All five volume series are integrated through the choice of topics and by a system of cross referencing.

The outline of the Handbook is thus as follows:

1. The Natural Environment and the Biochemical Cycles,
2. Reaction and Processes,
3. Anthropogenic Compounds,
4. Air Pollution,
5. Water Pollution.

Rapid developments in Environmental Chemistry and the increasing breadth of the subject matter covered made it necessary to establish volume-editors. Each subject is now supervised by specialists in their respective fields.

A recent development is the accessibility of all new volumes of the Handbook from 1990 onwards, available via the Springer Homepage springeronline.com or springerlink.com.

During the last 5 to 10 years there was a growing tendency to include subject matters of societal relevance into a broad view of Environmental Chemistry. Topics include LCA (Life Cycle Analysis), Environmental Management, Sustainable Development and others. Whilst these topics are of great importance for the development and acceptance of Environmental Chemistry Publishers and Editors have decided to keep the Handbook essentially a source of information on "hard sciences".

With books in press and in preparation we have now well over 40 volumes available. Authors, volume-editors and editor-in-chief are rewarded by the broad acceptance of the "Handbook" in the scientific community.

Bayreuth, July 2001 *Otto Hutzinger*

Contents

Foreword

The fact that oxygen free radicals and other oxidants are important in normal cellular metabolism has been recognized by many scientists over the last few decades. This has been accomplished by excellent research contributions by many scientists during this time.

This book is aimed at giving an insight into the basic mechanisms that are involved in the formation of and the defence against oxidants and free radicals. The chapters describe the most prominent endogenous and environmental sources of oxidizing species, the effects of oxidative stress on biological systems, and the cellular strategies for preventing and repairing oxidant-induced damage.

I would like to thank all the authors who have contributed to this book. I also want to thank many colleagues and friends for all the fruitful discussions and generous help over the years.

I hope the reader will enjoy reading this book and the content will help scientists, physicians, students, and post-docs gain an overview and a better understanding of this interdisciplinary field.

Düsseldorf, October 2004 Tilman Grune

Part A
Oxidants

The Handbook of Environmental Chemistry Vol. 2, Part O (2005): 1–18
DOI 10.1007/b101143

Endogenous Oxidant-Generating Systems

Hans Nohl[1] (✉) · Andrey V. Kozlov[2] · Lars Gille[1] · Katrin Staniek[2]

[1] Basic Research Institute for Pharmacology and Toxicology, Veterinary University of Vienna, Veterinärplatz 1, 1210 Vienna, Austria
Hans.Nohl@vu-wien.ac.at

[2] Ludwig-Boltzmann-Institute for Clinical and Experimental Traumatology, Donaueschingen Strasse 13, 1200 Vienna, Austria

Abstract Although organisms respiring air oxygen use their energy sources in an optimal way they are threatened by the compulsory formation of reactive oxygen species (ROS). A great variety of ROS sources have to be considered. Mitochondria, which are assumed to be mainly involved, produce ROS under pathophysiological conditions. Other cell organelles such as lysosomes and endoplasmatic reticulum may contribute to cellular ROS formation. Numerous oxidases and non-enzymatic compounds generate ROS. Nitric monoxide is also an important biomolecule with radical character. It is involved in a great variety of physiological and pathophysiological events.

Keywords Oxidases · Oxygen radicals · Mitochondria · Nitric oxide

Abbreviations

ROS Reactive oxygen species
ESR Electron spin resonance
SOD Superoxide dismutase
NOS Nitric monoxide synthase
UQH_2 Ubiquinol
UQ Ubiquinone
$UQ^{\bullet -}$ Ubisemiquinone
FeS_R Rieske iron-sulfur protein

1
Chemistry of Oxygen

Two oxygen atoms, each containing six valence electrons can be expected to form a double bonded structure with one δ and one π bond according to the valence bond theory. This, however, does not reflect the biradical character of molecular oxygen ($^3O_2$). In molecular orbital theory, the two binding electrons with highest energy occupy the two degenerated (identical energy levels) antibonding orbitals $2p\pi^*$ separately giving a formal explanation of the biradical character of $^3O_2$. The latter can be experimentally demonstrated using electron spin resonance (ESR).

According to the Pauli principle a pair of electrons must have opposite values of the spin quantum number. Therefore an energy-requiring spin inversion of one of the antibonding odd electrons is required to allow the completion of the vacant positions by the transfer of a pair of electrons (Fig. 1). The use of dioxygen as terminal electron acceptor in aerobic organisms, which is a prerequisite for optimal energy release, requires enzymes which reduce oxygen by mediating the transfer of single electrons. The thermodynamic restraint compelling single electron transfer to oxygen, instead of a pair of electrons, necessarily leads to odd electrons in the outer orbitals of this essential air constituent. Pairing of one electron only in the two orbitals leads to a paramagnetic species (superoxide radical) unless the second orbital remains occupied with an odd electron. This fact in combination with the existence of enzymes removing reactive oxygen species (ROS) in aerobic tissues suggests that oxygen radicals compulsorily appear in aerobic metabolism.

The parent oxygen radical built in aerobic organisms is the superoxide radical anion ($O_2^{\bullet-}$), which can be considered as the starter radical from which all other oxygen radicals as well as carbon-centered radicals are derived (Fig. 2). Some of the oxygen-reducing enzymes form stable complexes with oxygen while subsequently transferring two or even four electrons unless they release their reduction products. Examples in this respect are a couple of oxidases giving rise to the existence of H_2O_2 in aerobic systems, while cytochrome oxidase of mitochondria releases water after tetravalent reduction of oxygen.

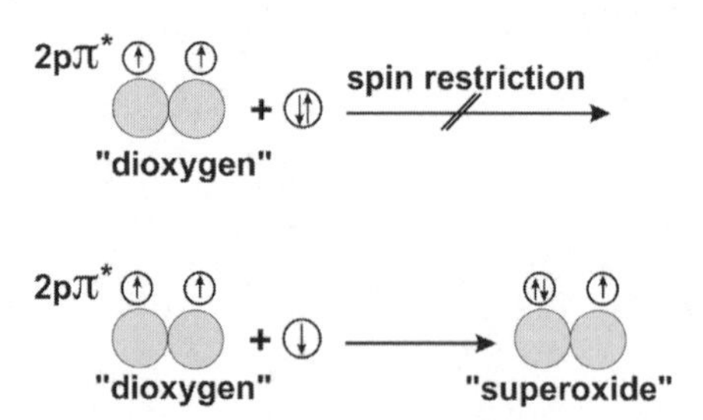

Fig. 1 Occupation of antibonding $2p\pi^*$ molecular orbitals is a prerequisite for single electron transfer to oxygen

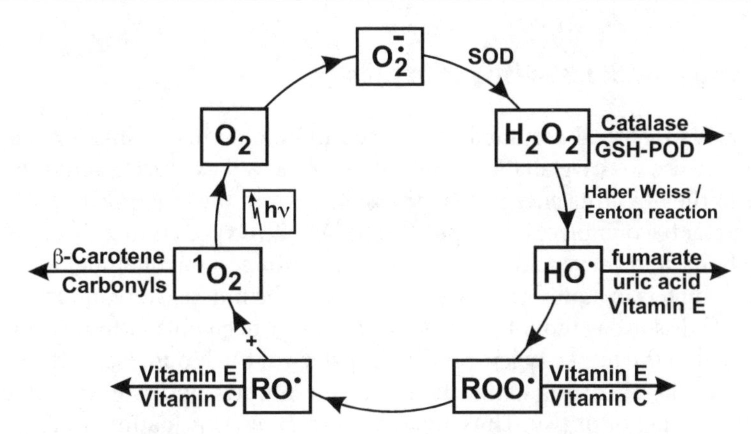

Fig. 2 Formation of various reactive oxygen species in biological systems derived from univalent reduction of molecular oxygen. *GSH-POD* glutathione peroxidase

Transition metals such as iron or copper are operating in the active center of these enzymes facilitating single electron transfer to oxygen by orbital-overlapping [1]. The highest density of cooperating enzymes with iron or copper in the active center exists in the respiratory chain of mitochondria. The great majority of these redox-cycling enzymes have the capability to pass on single electrons to the respective redox partner. In addition, the high rate of oxygen consumption by these mitochondrial electron carriers suggest mitochondria as the main intracellular source of oxygen radicals (Fig. 3).

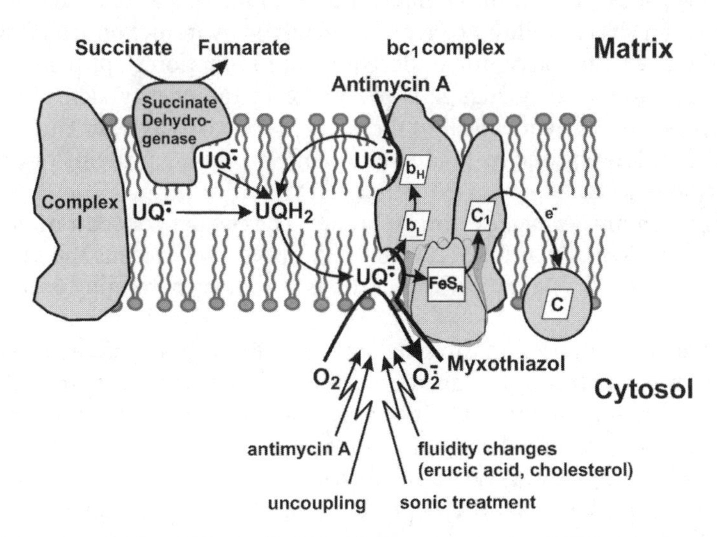

Fig. 3 Impact on the integral coordination of thermodynamics and kinetics is an explanation of ROS formation triggered by a great diversity of factors in the mitochondrial respiratory chain

2
Physiology and Biochemistry of Oxygen

Oxygen is in general assumed to be used in aerobic organisms for maximal energy release to drive anabolic and catabolic as well as muscle activities. This is due to the high free energy change resulting from the high positive standard redox potential of molecular oxygen (+0.815 V) [2]. Oxygen is mainly distributed in the body upon reversible pH-dependent binding to ferrohemoglobin (Bohr effect), which is in equilibrium with myoglobin the muscle store of oxygen. The physically dissolved amount of oxygen in plasma is negligible (0.13×10^{-3} mol L^{-1}) compared to the total oxygen-binding capacity of the blood. Doubling oxygen tension in the air of inhalation increases the physically dissolved amount of oxygen by a factor of two. Thus, upregulation of oxygen-loading in this way is insignificant while the risk of damaging the alveolar surfactant factor increases with increasing oxygen tension. In addition, other susceptible tissues such as the central nervous system, the heart and the eye will be subject to damage. Apart from a stimulation of oxygen radical formation, when increasing oxygen tension blood vessels reduce blood flow. A reasonable way to increase the oxygen-binding capacity of blood is the reactivation of the breathing reserve, which may account for more than 20% of the total oxygen capacity.

The bulbar respiratory center and the carotid bulbus regulating the rhythmic neuromuscular process of breathing have chemoreceptors sensitive to oxygen and CO_2. Lowering pO_2 selectively and reversibly reduces the outward voltage-dependent K^+ current of these chemoreceptors. Cheyne-Stokes breathing of dying persons indicates the alternate decrease of pO_2 and increase of pCO_2 due to the lower sensitivity of the chemoreceptors of the bulbar respiratory center.

More than 90% of cellular oxygen is consumed by mitochondria. Oxygen is the terminal electron acceptor, undergoing four reduction steps prior to water release. The free energy change is preserved in the energy-rich phosphate-binding of ATP while about 50% of the energy is released as heat. This ratio can be modulated if required by the induction of uncoupling proteins responsible for thermogenesis.

Oxygen readily permeates biomembranes. The cellular concentration of oxygen varies between 30 and 50×10^{-6} mol L^{-1}. This by far exceeds the Michaelis-Menten constant (k_M) of cytochrome oxidase for oxygen, ranging around 0.3×10^{-6} mol L^{-1}.

Oxygen concentration in the cell controls ATP synthesis via the respiratory chain of mitochondria and competes for cytochrome oxidase with the small biomolecule nitric monoxide (NO). It was reported that nitric monoxide removes oxygen from this terminal enzyme of the respiratory chain, thereby inhibiting energy-linked respiration. Increasing oxygen tension reactivates mitochondrial respiration associated with ATP production. The nitric monoxide synthase system (NOS) requires oxygen for NO synthesis from L-arginine. NOS introduces oxygen atoms from dioxygen into L-arginine thereby hydroxylating the guanidinium part of the latter and converts L-arginine to citrulline. Both cytochrome P_{450} together with tetrahydrobiopterin are involved in an overall five-electron oxidation of guanidinium nitrogen followed by the conversion into NO.

Under hypoxic conditions this pathway is hindered and NO production decreases.

This has severe consequences on a great variety of physiological functions. While NO increases blood flow via vascular relaxation and downregulates aggregation of platelets, tissues are less supplied with blood and the risk of embolus pathology increases when NO cannot be generated. Superoxide radicals transform nitric monoxide to peroxynitrite by a diffusion-limited reaction. This fatal reaction not only eliminates the bioregulating role of nitric monoxide but the outcoming reaction product has toxicological properties.

3
Superoxide Radicals Trigger Radical Chain Reactions

Two of the outer orbitals of dioxygen are occupied by single electrons only (Fig. 2). Completion of the vacant positions by a pair of electrons is not possible on thermodynamic grounds. Thus, oxygen is reduced step by step by the transfer of single electrons. This is performed in biological systems by special enzymes, the oxidases. The first reduction product is a superoxide radical, which in turn triggers the formation of secondary oxygen-centered radicals. Superoxide radicals are rapidly converted to hydrogen peroxide catalyzed by superoxide dismutase (SOD). Hydrogen peroxide is not totally removed due to the compartmentation of the metabolizing enzyme catalase. Non-metabolized H_2O_2 can undergo reductive homolytic cleavage if in contact with transition metals such as iron. The reaction product is the hydroxyl radical ($HO^{\bullet}$), which is the most potent oxidant in aerobic metabolism (Eq. 1).

$$H_2O_2 + Fe^{2+} \rightarrow HO^- + HO^{\bullet} + Fe^{3+} \quad (1)$$

Nohl et al. have shown that this homolytic cleavage is also possible without transition metals [3, 4]. The life time of the latter is extremely short indicating its high reactivity. Hydroxyl radicals have the highest oxidation potential of all aerobic metabolites. In order to pair the odd electron with a second electron, hydroxyl radicals attack all biomolecules and remove a further electron for completion of the orbital. Since hydroxyl radicals are often generated in or around biomembranes, the phospholipid bilayer is affected where $HO^{\bullet}$ radicals trigger hydrogen abstraction of polyunsaturated fatty acids, followed by autocatalytic degradation of the respective membrane. The first lipid radical product is an organic peroxide species which has a life time of about 10 s. However, it may destabilize with a second peroxide resulting in the formation of another powerful oxidant (alkoxyl-radical) and singlet oxygen, which differs from the triplet ground state of the dioxygen molecule by a spin inversion. Singlet oxygen may transfer its energy onto other biomolecules or destabilize by light emission.

4
From Where Are Reactive Oxygen Species Derived?

It is generally believed that mitochondria are the main cellular source of ROS during homeostasis. In fact mitochondria have many biochemical properties that favor these organelles as permanent producers of ROS.

In the early seventies Britton Chance and colleagues observed that isolated mitochondria release H_2O_2 when the respiratory chain was blocked with antimycin A [5]. The presence of SOD in the matrix of mitochondria rapidly converts $O_2^{\bullet-}$ radicals, generated in the respiratory chain, to H_2O_2. Later on Chance and his group reported that mitochondria also generate $O_2^{\bullet-}$-derived H_2O_2 in the absence of this respiratory poison. The authors observed H_2O_2 production under conditions of maximal membrane potential (state IV respiration) while down-regulation of the transmembraneous potential (state III respiration) stopped mitochondrial transfer of single electrons to oxygen out of sequence. From these observations it was assumed that the mitochondrial membrane potential at its highest level (state IV) triggers the deviation of single electrons to dioxygen, which results in the generation of superoxide radicals [6–8]. These results were obtained in the presence of oligomycin to keep and preserve membrane potential at high levels [6]. There are also some opposing articles in the literature that provide evidence that a low membrane potential initiates $O_2^{\bullet-}$ radical generation (Table 1). For example, aging is reported to stimulate $O_2^{\bullet-}$ radical formation while the membrane potential declines [9]. Bruce Ames and coworkers reported that oxidation of unsaturated fatty acids of cardiolipin causes a decrease of the activity of various electron carriers, which ultimately results in the uncontrolled leakage of single electrons to dioxygen out of sequence [10].

Two different sites of the mitochondrial respiratory chain were considered to contribute to the univalent reduction of dioxygen. An iron-sulfur protein of complex I was reported to undergo auto-oxidation independent of the height of the membrane potential [8, 11, 12]. The identity of the complex I constituent was concluded from the use of rotenone, which did not affect H_2O_2 generation from complex I substrates, in combination with selective inhibitors of iron centers. The second center suggested to be involved in mitochondrial ROS for-

Table 1 Mitochondrial ROS formation is reported to be a consequence of an increase (↑) or a decrease (↓) of the mitochondrial membrane potential ($\Delta\Psi$)

$\Delta\Psi$	ROS	References
↑	↑	Liu SS [7]
↑	↑	Korshunov SS et al. [6]
↓	↑	Tanaka T et al. [53]
↓	↑	Zamzami N et al. [54]

mation was reported to be activated when the membrane potential is high, e.g., under conditions of state IV respiration [6–8]. The reductant of oxygen was sensitive to myxothiazol, indicating the ubiquinol/bc_1 redox couple as the source of ROS formation (Fig. 3).

Apart from the proton-motive force considered by P. Mitchell to control mitochondrial ATP synthesis [13], Kadenbach et al. have recently described a second regulator of mitochondrial energy conservation [14]. These authors made the observation that cytochrome oxidase has various binding centers for ATP, which can be phosphorylated and dephosphorylated by hormones. Dephosphorylation inhibits energy-linked respiration associated with an increase of the membrane potential [14, 15]. Although evidence is still missing it is believed that dephosphorylation causes deviation of single electrons from the respiratory chain leading to mitochondrial ROS formation.

The fact that reductants of dioxygen must be one-electron carriers with a more negative redox potential than the $O_2/O_2^{\bullet-}$ couple favors mitochondria as the major ROS source in the cell [16, 17]. A variety of other bioenergetic properties favor mitochondria as the most active generators of reactive oxygen species in vivo. However, evidence about the capability of mitochondria for ROS formation came exclusively from in vitro experiments with mitochondria removed from their natural surrounding in the cell. The mechanical stress required to obtain mitochondria isolated from the cell essentially affects bioenergetic parameters reported to trigger ROS formation. Some reports suggest a high membrane potential to trigger mitochondrial ROS formation. Other reports in the literature claim that only low membrane potentials initiate ROS formation (Table 1). Depending on the isolation procedure, the transmembraneous potential of mitochondria may change in both directions. Another problem is the inevitable contamination with fragmented mitochondria where electron flow is out of control. Furthermore, detection methods routinely applied may interfere in the regular electron transfer simulating generation of ROS from mitochondria. Applying a novel non-invasive detection system with higher sensitivity Staniek et al. were unable to confirm mitochondrial ROS formation under "physiological" conditions [18]. However, we observed that any change of the physical membrane state caused a release of single electrons from the ubiquinol/bc_1 complex. In that case oxygen was reduced to $O_2^{\bullet-}$ radical from a nonphysiological oxygen reductant, which was identified to be a ubisemiquinone ($UQ^{\bullet-}$) operating on the cytosolic face of the inner mitochondrial membrane (Fig. 3). The regular transfer of a pair of reducing equivalents from ubiquinol (UQH_2) is split up in contact with the bc_1 complex into two separate steps. One reducing equivalent goes to the Rieske iron-sulfur protein (FeS_R) while the second reduces the two b-type cytochromes. Ubiquinone (UQ) at the matrix side is reduced and, after the transfer of the second electron, ubiquinol is formed which starts a new cycle (Q-cycle). This highly susceptible redox couple requires an integral coordination of thermodynamics and kinetics, which is normally governed by free conformational changes of the oxidants of ubiquinol. Any impediment of the Q-cycle causes a leakage of single electrons to dioxygen out of sequence [19].

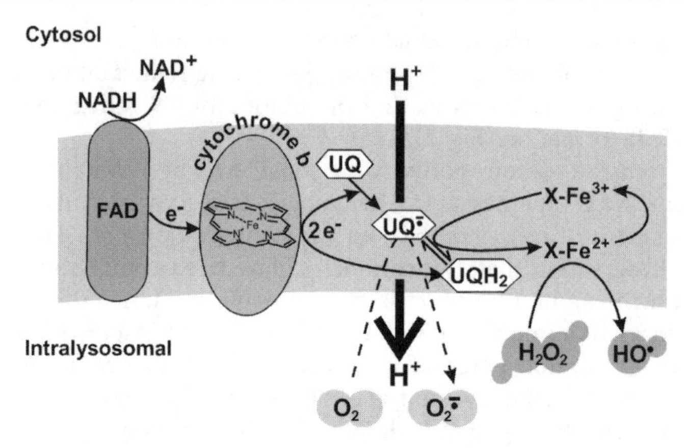

Fig. 4 A lysosomal redox chain drives H^+ translocation by redox-cycling ubisemiquinones

We have recently observed that under hypoxic conditions lysosomes activate a respiratory chain that produce ROS due to an autoxidizing ubisemiquinone. Using ESR spin-trapping with 5,5′-dimethyl-1-pyrroline *N*-oxide we observed a quartet indicating the formation of hydroxyl radicals. Redox-cycling ubiquinone not only compensates for translocation required to acidify the interior of lysosomes but it also requires oxygen as the terminal electron acceptor. Spontaneous dismutation of $O_2^{\bullet-}$ radicals derived from $UQ^{\bullet-}$ is enhanced due to the low pH inside the lysosomes. The dismutation product H_2O_2 rapidly undergoes homolytic cleavage to $HO^{\bullet}$ radical catalyzed by the high amount of iron in lysosomes (Fig. 4).

Apart from mitochondrial one-electron carriers, a variety of oxidases exist which transfer single electrons onto dioxygen (Table 2). These oxidases are constituents of cell membranes, cytosol, and organelles.

Since oxygen can only be reduced by the stepwise transfer of single electrons, special enzymes have been developed in aerobics that catalyze the reduction of dioxygen. All of these enzymes have transition metals in their active centers. These transition metals, which are normally copper or iron ions, facilitate electron transfer onto oxygen via orbital-overlapping. Another advantage of transition metal-binding to proteins is the great span of redox properties overcoming thermodynamic restraints for electron transfer to the oxygen molecule. While the majority of oxygen-activating enzymes are copper and iron proteins xanthine oxidase, which is widely distributed in all tissues, contains two atoms of molybdenum and eight atoms of iron. Xanthine oxidase is converted from xanthine dehydrogenase during the initial phase of hypoxia or anoxia. While both enzymes catalyze the oxidation of xanthine or hypoxanthine, the dehydrogenase reduces NAD^+ to NADH while the oxidase reduces dioxygen to superoxide radicals. Under conditions of ischemia or anoxia cytosolic calcium is elevated, activating a Ca^{2+}-dependent protease that catalyses the conversion from xanthine dehydrogenase to xanthine oxidase (Fig. 5). This process can be inhibited by a soybean trypsin inhibitor [20].

Table 2 Enzymatic sources of superoxide radicals

Enzyme	Location	Comment
Xanthine oxidase (XOD)	Intestine, endothelial cell (ischemia)	Forms both $O_2^{\bullet-}$ and H_2O_2 [55–58]
Indoleamine-2,3-dioxygenase (IDO)	Most animal tissues, especially small intestine (not in liver); virus infection and endotoxins induce IDO activities	Cleaves the indole ring of tryptophan and related compounds; $O_2^{\bullet-}$ is released and used for the cleavage process [59–61]
Aldehyde oxidase	Liver	Contains like XOD iron and molybdenum and has a broad substrate specificity [62, 63]
Dihydroorotate dehydrogenase	Mitochondria (liver)	Biosynthesis of UTP and CTP (pyridine nucleotides); forms $O_2^{\bullet-}$ and H_2O_2 [64]
NADPH oxidase	Plasma membrane of polymorphonuclear neutrophils	Involved in phagocytosis and rheumatoid arthritis; releases $O_2^{\bullet-}$ [65–67]
Exogenous NADH dehydrogenase	Mitochondria (heart)	Involved in adriamycin cardiotoxicity and reperfusion injury [68–70]

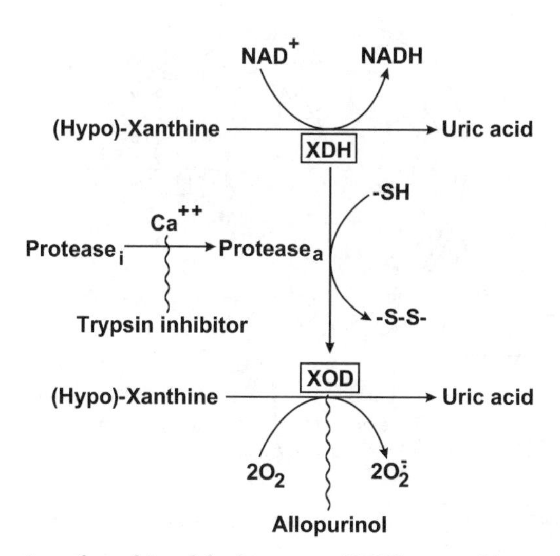

Fig. 5 Transformation of xanthine dehydrogenase (*XDH*) to xanthine oxidase (*XOD*) as a consequence of ischemia

Some conversion also results from sulfhydryl oxidation [21]. The shift of oxidase content in ischemic tissues triggers ROS formation by replenishing normal oxygen tension. Evidence for the role of xanthine oxidase as the major $O_2^{\bullet-}$ radical source responsible for oxidative injury is the inhibition of the insult in the presence of SOD or inhibition of the enzyme activity by allopurinol.

Since small amounts of xanthine dehydrogenase also seem to be converted to oxidases under normoxic conditions one can expect that these oxidases are a permanent but insignificant source of ROS under physiological conditions.

Great physiological significance can be attributed to the NADH oxidase of plasma membranes. These enzymes may indirectly contribute to oxygen activation by means of protein-bound iron reduction; however, they are more important for transmembraneous electron transfer coupled to proton translocation and regulation of cell growth [22–24].

Of pathophysiological significance are NADPH oxidases involved in phagocytosis and chronic inflammation (Fig. 6). Phagocytic leukocytes following activation respond with a burst of oxygen uptake. Oxygen is one substrate of the membrane-associated NADPH oxidase while the reductant is NADPH [25]. Activation of the NADPH oxidase is a complex concerted action of the enzyme involving flavoproteins and b-type cytochromes. The reducing equivalents for the univalent reduction of dioxygen are supplied by NADPH ultimately derived from the pentose-phosphate shunt. The first reduction product is the $O_2^{\bullet-}$ radical which may, however, undergo dismutation to H_2O_2 and subsequent

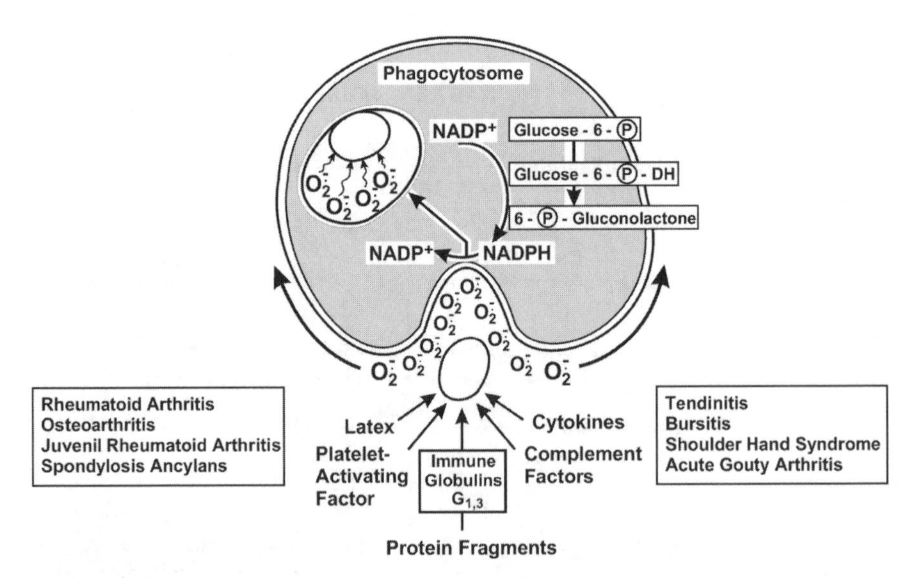

Fig. 6 Induction of superoxide radical release during phagocytosis. The onset of ROS formation is linked to the uptake of opsonized proteins and proinflammatoric factors derived from various diseases as listed in the *boxes*. The chronic inflammation associated with various diseases leads to serious damage of the surrounding tissue by ROS

homolytic cleavage to form $HO^{\bullet}$ radicals. These reactive oxygen species will be released into the extracellular fluid, which has a rather low antioxidative capacity. Reactive oxygen species released do not only activate further NADPH oxidases, which stimulate ROS release up to critical amounts, the uncontrolled ROS accumulation affects the whole tissue and releases leukotactic components. Ultimately this ends up in the establishment of a chronic inflammation.

Another oxidase of physiological and pathophysiological significance is the cytochrome P_{450} monooxygenase of the endoplasmatic reticulum. This endoplasmatic redox chain includes various redox couples that have the potential capability to directly reduce dioxygen to superoxide radicals (Fig. 7) [26]. Apart from autooxidizing semiquinones of flavoproteins, which are prosthetic groups of NAD(P)H dehydrogenases, cytochrome P_{450} monooxygenase may form unstable oxenoid complexes where electrons are shifted to the liganded oxygen molecule via heme-iron from the NADPH/cytochrome b_5 redox shuttle. The oxy-ferrous P_{450} is liganded to the electron-rich thiolate anion of cystein. The sulfur ligand places the electron density to the ferric-heme iron, which facilitates the release of superoxide radicals. While cytochrome P_{450} in that case is the direct $O_2^{\bullet -}$ radical source, more often the coupled electron carriers of microsomal respiration are indirectly involved in microsomal radical formation.

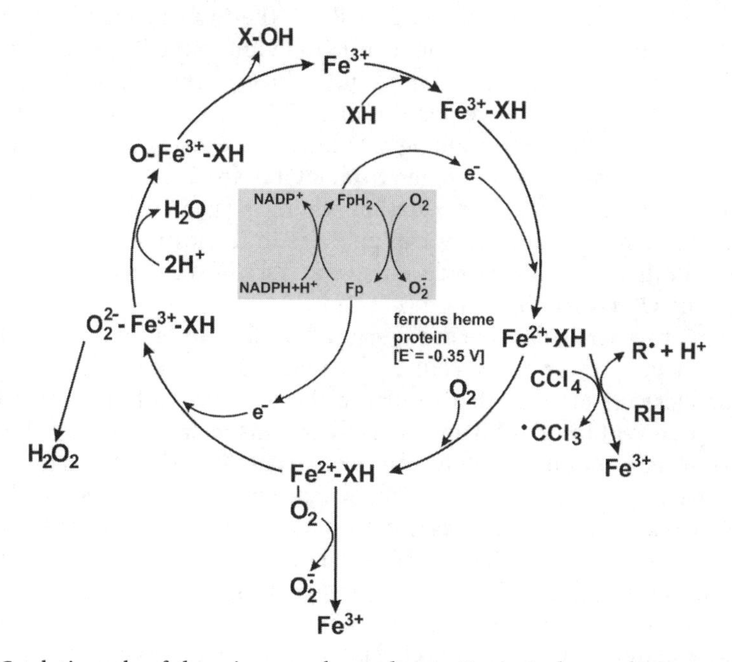

Fig. 7 Catalytic cycle of the microsomal cytochrome P_{450}/cytochrome b_5 system that is involved in the metabolism of endogenous and exogenous substrates. Under certain metabolic conditions these electron carriers can produce ROS by either direct autoxidation or indirectly via the formation of radicals from metabolized xenobiotics. *Fp* flavoprotein, *X* xenobiotic, e.g., CCl_4, *R* biomolecule

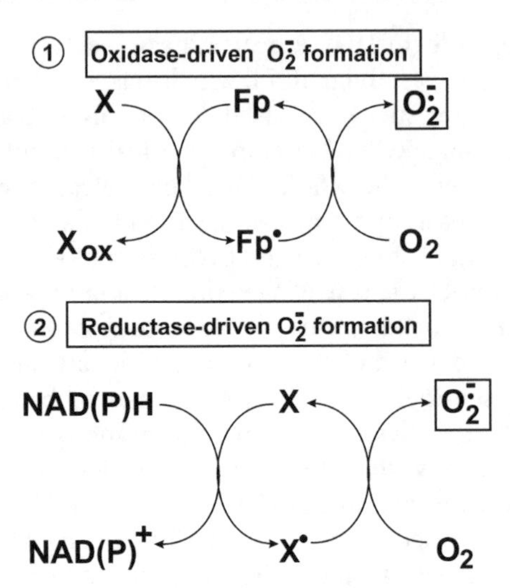

Fig. 8 Drug-mediated univalent reduction of O_2 by reductases and oxidases. *Fp* flavoprotein, *X* xenobiotic

Carbon tetrachloride (tetrachloromethane) is extremely toxic when metabolized in the liver. As demonstrated in Fig. 7 the toxic metabolite is formed when CCl_4 binds to the ferrous-heme iron of cytochrome P_{450}, which removes one electron by a complex reaction and releases the trichloromethyl radical$\cdot CCl_3$ which is a carbon-centered radical [27]. The trichloromethyl radical might combine directly with biological molecules, causing covalent modification as well as abstracting hydrogen from membrane lipids, thereby initiating a chain reaction of lipid peroxidation where oxygen-centered radicals are formed. In general, many drugs can give rise to the formation of superoxide radicals. Both are possible: oxidase-driven $O_2^{\bullet-}$ radical generation and reductase-driven $O_2^{\bullet-}$ radical formation (Fig. 8).

Iron plays a particular role in oxygen radical formation (Fig. 9) [1, 28]. Iron is not only part of the active center of oxidases, it is also involved in the homolytic cleavage of H_2O_2 that results in the formation of $HO^{\bullet}$ radicals (see Eq. 1). To prevent this deleterious role of iron, this metal is normally bound to storage proteins in the less dangerous ferric state [29]. However, a decrease of pH occurring during ischemia or the presence of reducing agents such as $O_2^{\bullet-}$ radical or ascorbic acid removes iron from the safe storage proteins [30]. Free iron, irrespective of its reduction state, can directly or after reduction trigger $HO^{\bullet}$ radical formation by homolytic cleavage of H_2O_2.

A variety of other biological components besides iron contribute to monovalent reduction of dioxygen (Table 3). Three percent of oxygen bound to the ferrous-heme center of hemoglobin separates a single electron from the latter and is released as $O_2^{\bullet}$ radical [31]. As a consequence, hemoglobin is oxidized to methemoglobin and is subsequently reduced again to oxygen-binding hemoglobin by an NADPH-dependent enzyme (diaphorase).

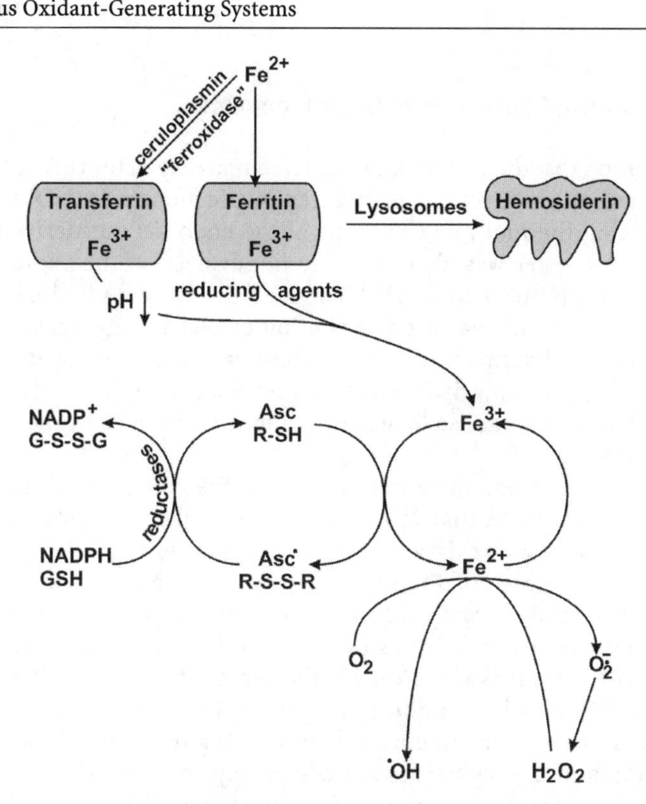

Fig. 9 Role of free iron in the catalysis of oxidative stress. Pathophysiological events can trigger the release of redox-active iron from transport and storage proteins. Ferric iron can be reduced by cellular reductants, resulting in autoxidation and subsequent HO• radical formation

Table 3 Biological components involved in the univalent reduction of oxygen

Biological Components	Metabolites
Catecholamines (+traces of transition metals)	*o*-Semiquinones/Quinones+$O_2^{\bullet-}$
Tetrahydropterin (+traces of transition metals)	Dihydropterin+$O_2^{\bullet-}$
Ascorbic acid (+traces of transition metals)	Ascorbyl radical+$O_2^{\bullet-}$
Iron (II)	Iron (III)+$O_2^{\bullet-}$
Cooper (I)	Cooper (II)+$O_2^{\bullet-}$
Hemoglobin	Methemoglobin+$O_2^{\bullet-}$
Myoglobin	Metmyoglobin+$O_2^{\bullet-}$
Flavins ($FADH_2$, $FMNH_2$)	FAD, FMN+$O_2^{\bullet-}$
Thiols	Disulfides
Semiquinones	Quinones

5 The Function and Generation of Nitric Monoxide

The recognition that the endothelium-derived relaxing factor (EDRF) is a small inorganic molecule was totally unexpected. Nitric monoxide (NO) was identified as exerting the multiple functions of the endothelium-derived relaxing factor [32]. The latter was also shown to be involved in the pathogenesis of ischemia-related tissue damage [33]. The great diversity of physiological functions such as vascular relaxation, inhibition of platelet aggregation, involvement in central and peripheral neurotransmission, utilization by macrophages and other cells, and a myriad of further physiological roles made it dubious that a small inorganic molecule like nitric monoxide can exert these activities [32, 34].

Scientific interest was therefore focused on the biological source of nitric monoxide. It was found that NO is a reaction product of an enzymatically catalyzed five-electron oxidation of a guanidinium nitrogen from L-arginine followed by the release of the free radical species nitric monoxide. Nitric monoxide formation is catalyzed by a single protein complex supported by cofactors and coenzymes such as cytochrome P_{450} and tetrahydrobiopterin. Two types of NO synthases are known, the constitutive NOS (cNOS) found in the brain (nNOS) and in endothelial cells (eNOS), and the inducible form (iNOS) which is expressed in macrophages. Recent data have shown that total NO production in the body does not adequately correspond to its biological effects. For instance, it is commonly accepted that NO produced by eNOS has mostly a regulatory and protective function [35, 36]. In contrast, the expression of iNOS often plays an important role in pathogenesis of a number of diseases [37, 38].

Cytochrome P_{450}, which is involved in the five-electron oxidation of nitrogen, is known to uncouple from substrate oxidation under certain conditions. In these cases reduced oxygen species such as superoxide and hydrogen-peroxide are formed [39, 40].

Some years ago it was reported that mitochondria have their own constitutive nitric monoxide synthase [41]. The existence of a NO-generating enzyme in mitochondria was called into question by others. Particularly the far-reaching bioenergetic function of this mitochondrial enzyme has provoked controversial discussions. The mitochondrial NO synthase is assumed to control energy-linked respiration, which means a regulation of ATP synthesis [42, 43]. Down-regulation of ATP synthesis is suggested to result from inhibition of cytochrome oxidase [44] following NO-binding to the ferric-heme center of this terminal mitochondrial electron carrier. Based on experimental data, it was argued that NO levels provided by mitochondrial NOS are not sufficient to inhibit energy-linked respiration via a blockade of cytochrome oxidase [45]. The life time of NO in biological systems is rather low, estimated to be in the millisecond range. Thermodynamically the reduction to the nitroxyl anion (NO^-) is favored. NO^- readily converts to nitrous oxide (N_2O). Other competing pathways transform

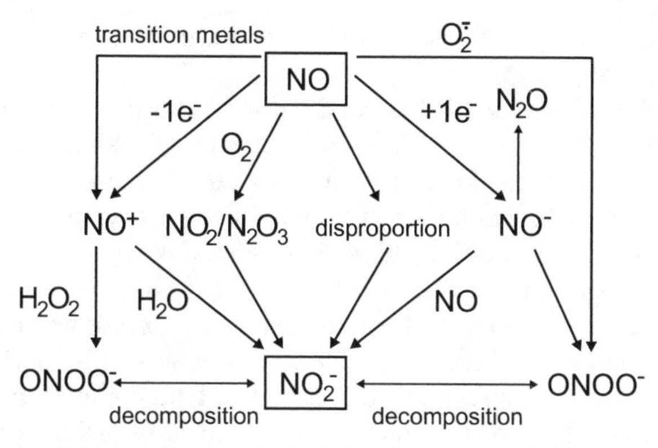

Fig. 10 Pathways of NO degradation yield nitrite via a variety of chemical reactions

NO^- to hydroxylamine, peroxynitrite ($ONOO^-$) or directly to NO_2^-. Peroxynitrite reacts with peroxynitrous acid giving rise to two NO_2^- and one O_2. Oxidation of nitric monoxide to the nitrosium ion (NO^+) requires transition metals or $O_2^{\bullet-}$ radicals. NO^+ reacts with nucleophilic partners or gives rise to NO_2^- and $ONOO^-$. Other pathways also lead directly or indirectly to NO_2^- (Fig. 10).

The first oxidation product is nitrite (NO_2^-), which was recently reported to undergo a reduction to its bioactive form via a particular segment of the mitochondrial respiratory chain (Fig. 11) [45, 46]. This could be of major signif-

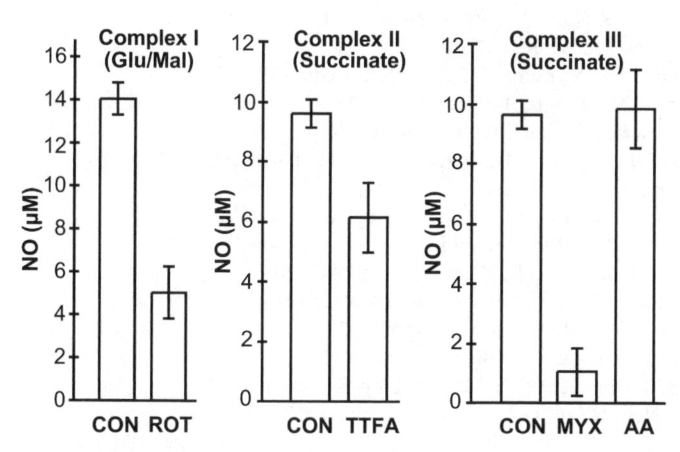

Fig. 11 Effect of specific inhibitors of the respiratory chain on the nitrite reductase activity of rat liver mitochondria. Mitochondria were incubated with nitrite for 2 h under argon in the presence of succinate or glutamate/malate (*Glu/Mal*). Nitric monoxide derived from nitrite was trapped with hemoglobin. The concentration of NO-hemoglobin complexes was measured using low temperature ESR spectroscopy. Other details are described in Kozlov et al. [46]. *CON* control, *ROT* rotenone, *TTFA* thenoyltrifluoroacetone, *MYX* myxothiazol, *AA* antimycin A

icance when cells are overflowed with NO_2^- in patients using NO donors such as trinitroglycerol [47, 48].

The radical nature of NO is best understood in terms of molecular orbital occupancy. The feature of the radical chemistry of NO can be derived from its capability for hydrogen atom abstraction. However, NO cannot be directly shown using ESR technique (in aqueous solutions) as the peak is widely broadened by spin delocalization. Instead NO can be detected using spin-trapping technique with deoxy-hemoglobin or dithiocarbamate [49, 50]. Nitric monoxide formed in blood reacts with hemoglobin to form nitrosyl complexes of hemoglobin. These complexes are very stable and may circulate in the blood for hours. Hemoglobin-NO complexes have a characteristic ESR signal, which allows qualitative and quantitative analysis of NO bound to hemoglobin. In order to measure NO levels in tissues a NO trap is required. The most common NO trap is diethyldithiocarbamate-iron complex [51], a water-insoluble NO trap, or its water-soluble analogue (*N*-methyl-*d*-glucamine dithiocarbamate) [52].

References

1. Koppenol WH (1994) Chemistry of iron and copper in radical reactions. In: Rice-Evans CA, Burdon RH (eds) Free radical damage and its control. Elsevier, p 3
2. Sawyer DT (1990) Int Rev Exp Pathol 31:109
3. Gille L, Nohl H (1997) Free Radic Biol Med 23:775
4. Nohl H, Jordan W (1983) Biochem Biophys Res Commun 114:197
5. Boveris A, Oshino N, Chance B (1972) Biochem J 128:617
6. Korshunov SS, Skulachev VP, Starkov AA (1997) FEBS Lett 416:15
7. Liu SS (1997) Biosci Rep 17:259
8. Votyakova TV, Reynolds IJ (2001) J Neurochem 79:266
9. Cavazzoni M, Barogi S, Baracca A, Castelli GP, Lenaz G (1999) FEBS Lett 449:53
10. Shigenaga MK, Hagen TM, Ames BN (1994) Proc Natl Acad Sci USA 91:10771
11. Genova ML, Ventura B, Giuliano G, Bovina C, Formiggini G, Parenti-Castelli G, Lenaz G (2001) FEBS Lett 505:364
12. Herrero A, Barja G (1997) Mech Ageing Dev 98:95
13. Mitchell P (1966) Biol Rev Camb Philos Soc 41:445
14. Kadenbach B, Arnold S (1999) FEBS Lett 447:131
15. Lee I, Bender E, Kadenbach B (2002) Mol Cell Biochem 234–235:63
16. Barja G (1999) J Bioenerg Biomembr 31:347
17. Hatefi Y (1985) Ann Rev Biochem 54:1015
18. Staniek K, Nohl H (1999) Biochim Biophys Acta 1413:70
19. Gille L, Nohl H (2001) Arch Biochem Biophys 388:34
20. Jones DP (1985) The role of oxygen concentration in oxidative stress: Hypoxic and hyperoxic models. In: Sies H (ed) Oxidative stress. Academic Press, London, p 151
21. Engerson TD, McKelvey TG, Rhyne DB, Boggio EB, Snyder SJ, Jones HP (1987) J Clin Invest 79:1564
22. Brightman AO, Wang J, Miu RK, Sun IL, Barr R, Crane FL, Morre DJ (1992) Biochim Biophys Acta 1105:109
23. Crane FL, Sun IL, Crowe RA, Alcain FJ, Low H (1994) Mol Aspects Med 15 Suppl:s1
24. Morre DJ, Morre DM (2003) Free Radic Res 37:795

25. Karlsson A, Dahlgren C (2002) Antioxid Redox Signal 4:49
26. Grover TA, Piette LH (1981) Arch Biochem Biophys 212:105
27. Kalyanaraman B, Mason RP, Perez-Reyes E, Chignell CF, Wolf CR, Philpot RM (1979) Biochem Biophys Res Commun 89:1065
28. Lieu PT, Heiskala M, Peterson PA, Yang Y (2001) Mol Aspects Med 22:1
29. Burkitt MJ, Mason R (1991) Proc Natl Acad Sci USA 88:8440
30. Gille L, Kleiter M, Willmann M, Nohl H (2002) Biochem Pharmacol 64:1737
31. Zhang L, Levy A, Rifkind JM (1991) J Biol Chem 266:24698
32. Palmer RM, Ferrige AG, Moncada S (1987) Nature 327:524
33. Moncada S, Palmer RM, Higgs EA (1991) Pharmacol Rev 43:109
34. Furchgott RF, Jothianandan D (1991) Blood Vessels 28:52
35. Aoki N, Johnson G, Lefer AM (1990) Am J Physiol 258:G275
36. Kubes P (2000) Gut 47:6
37. Moncada S, Higgs A (1993) N Engl J Med 329:2002
38. Nathan C (1992) FASEB J 6:3051
39. Kessova I, Cederbaum AI (2003) Curr Mol Med 3:509
40. Jaeschke H, Gores GJ, Cederbaum AI, Hinson JA, Pessayre D, Lemasters JJ (2002) Toxicol Sci 65:166
41. Ghafourifar P, Richter C (1997) FEBS Lett 418:291
42. Boveris A, Costa LE, Poderoso JJ, Carreras MC, Cadenas E (2000) Ann NY Acad Sci 899:121
43. Brookes PS, Bolanos JP, Heales SJ (1999) FEBS Lett 446:261
44. Brown GC (1995) FEBS Lett 369:136
45. Nohl H, Staniek K, Sobhian B, Bahrami S, Redl H, Kozlov AV (2000) Acta Biochim Pol 47:913
46. Kozlov AV, Staniek K, Nohl H (1999) FEBS Lett 454:127
47. Kozlov AV, Dietrich B, Nohl H (2003) Br J Pharmacol 139:989
48. Chen Z, Zhang J, Stamler JS (2002) Proc Natl Acad Sci USA 99:8306
49. Kozlov AV, Sobhian B, Costantino G, Nohl H, Redl H, Bahrami S (2001) Biochim Biophys Acta 1536:177
50. Kozlov AV, Sobhian B, Duvigneau C, Costantino G, Gemeiner M, Nohl H, Redl H, Bahrami S (1903) J Lab Clin Med 140:303
51. Mordvintcev P, Mulsch A, Busse R, Vanin A (1991) Anal Biochem 199:142
52. Komarov AM, Kramer JH, Mak IT, Weglicki WB (1997) Mol Cell Biochem 175:91
53. Tanaka T, Hakoda S, Takeyama N (1998) Free Radic Biol Med 25:26
54. Zamzami N, Hirsch T, Dallaporta B, Petit PX, Kroemer G (1997) J Bioenerg Biomembr 29:185
55. Landmesser U, Spiekermann S, Dikalov S, Tatge H, Wilke R, Kohler C, Harrison DG, Hornig B, Drexler H (2002) Circulation 106:3073
56. Vina J, Gimeno A, Sastre J, Desco C, Asensi M, Pallardo FV, Cuesta A, Ferrero JA, Terada LS, Repine JE (2000) IUBMB Life 49:539
57. Stockert AL, Shinde SS, Anderson RF, Hille R (2002) J Am Chem Soc 124:14554
58. Enroth C, Eger BT, Okamoto K, Nishino T, Pai EF (2000) Proc Natl Acad Sci USA 97:10723
59. Taniguchi T, Hirata F, Hayaishi O (1977) J Biol Chem 252:2774
60. Yoshida R, Urade Y, Tokuda M, Hayaishi O (1979) Proc Natl Acad Sci USA 76:4084
61. Morita T, Saito K, Takemura M, Maekawa N, Fujigaki S, Fujii H, Wada H, Takeuchi S, Noma A, Seishima M (2001) Ann Clin Biochem 38:242
62. Mira L, Maia L, Barreira L, Manso CF (1995) Arch Biochem Biophys 318:53
63. Wright RM, Riley MG, Weigel LK, Ginger LA, Costantino DA, McManaman JL (2000) DNA Cell Biol 19:459

64. Loffler M, Jockel J, Schuster G, Becker C (1997) Mol Cell Biochem 174:125
65. Dupuy C, Virion A, Ohayon R, Kaniewski J, Deme D, Pommier J (1991) J Biol Chem 266:3739
66. Vignais PV (2002) Cell Mol Life Sci 59:1428
67. Hiraoka W, Vazquez N, Nieves-Neira W, Chanock SJ, Pommier Y (1998) J Clin Invest 102:1961
68. Nohl H (1987) Eur J Biochem 169:585
69. Nohl H (1988) Biochem Pharmacol 37:2633
70. Nohl H, Gille L, Staniek K (1998) Z Naturforsch (C) 53:279

The Handbook of Environmental Chemistry Vol. 2, Part O (2005): 19–31
DOI 10.1007/b101144

Environmental Oxidative Stress – Environmental Sources of ROS

Peter Schröder (✉) · Jean Krutmann

Environmental Health Research Institute (IUF), Heinrich-Heine University Düsseldorf, Auf'm Hennekamp 50, 40225 Düsseldorf, Germany
Schroedp@uni-duesseldorf.de, Krutmann@uni-duesseldorf.de

Abstract Environmental factors are known sources for oxidative stress. In consequence of the numerous influences that define our environment, environmental oxidative stress can derive from several sources. Such sources can be categorised with respect to their mechanisms of action: Where are the reactive oxygen species generated? Where do they take effect? Are they generated chemically by a noxa, via a target biomolecule or are they physiologically generated by cells? Pollution, non-ionising (ultraviolet and infrared) and ionising radiations are known sources of environmental oxidative stress.

Keywords Reactive oxygen species · Ultraviolet · Infrared · Ionising radiation · Particles

Abbreviations
ROS Reactive oxygen species
UV Ultraviolet radiation
IR Infrared radiation

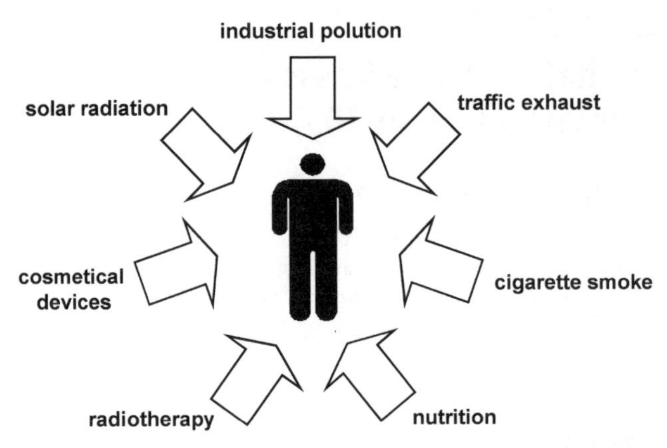

Fig. 1 Sources of environmental oxidative stress. The noxae shown represent environmental influences known to lead to an increased generation of ROS

1 Introduction

Reactive oxygen species (ROS), oxidative stress and the environment form a complex relationship with important influence on human health and disease. Environmental oxidative stress derives from different artificial and natural sources (Fig. 1) and is induced by several different mechanisms. To address the complexity of this field, this chapter will deal with the modes of action by which environmental factors can disturb the equilibrium between pro- and antioxidants favouring the former (for definition of oxidative stress see [1]). Adjacent, environmental sources of oxidative stress and their mode of action will be introduced.

2 Mechanisms of Action – How Do Environmental Factors Lead to Oxidative Stress?

When looking at how environmental sources induce oxidative stress several factors have to be taken into account. Where are the pro-oxidants generated? Where do they take effect? Are ROS generated chemically by a noxa, via a target biomolecule or are they physiologically generated by cells? Categorising the types of sources for environmental stress according to these points we would like to introduce three categories of noxae action (Fig. 2).

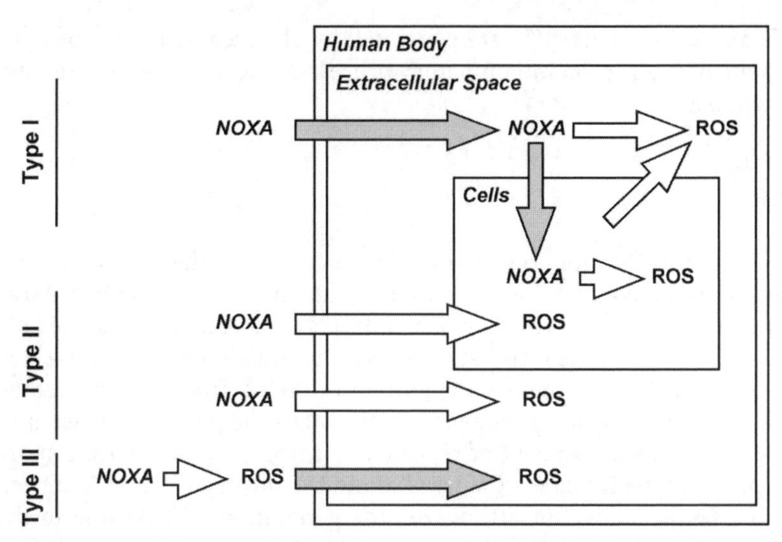

Fig. 2 Categories of environmental noxae inducing oxidative stress. A type I noxa is taken up actively or passively into the body (extracellular space and/or cells) where it generates ROS or induces the generation of ROS. The site of ROS generation is the extracellular space or the cell lumen. Type II noxae generate ROS by a noxa that does not reside inside the body. ROS generation happens via electronic excitation of an intracorporal substance induced by the noxa. Type III noxae produce ROS outside the body. The pro-oxidants enter the body in their reactive form and take effect there. For all kinds of noxae the damage caused by the noxa might lead to the generation of additional oxidative stress

2.1 Type I

A type I noxa is taken up actively or passively into the body (extracellular space and/or cells) where it generates ROS or induces the generation of ROS. The site of ROS generation is the extracellular space or the cell lumen. ROS generation can be a direct effect of the noxa (e.g. homolytic cleavage of atom bond within the noxa molecule), an interaction between the noxa and body components (e.g. one electron reduction, cytochrome P450), or a physiological reaction towards the noxa (e.g. a release of ROS by neutrophiles). The oxidative stress induced by type I noxae outlasts the period for which the noxa is present in the environment, which might lead to long term increased oxidative stress. As some ROS are able to pass through the cell membranes without reacting with membrane components, ROS might take effect in locations different to the place of ROS generation. Nevertheless the site of generation is of importance in terms of protection, effect and targets. Oxidants generated in the cytosol of a cell with a high concentration of cellular antioxidants like glutathione are less likely to cause severe damage. Additionally, repeated exposure can result in further increased and/or prolonged oxidative burden. Particulate matter is an example for type I noxae: particles of different composition are inhaled and enter body fluids and cells, where ROS are generated by chemical reactions of particle components and/or via inflammatory processes triggered by the presence of the particles.

2.2 Type II

Type II noxae differ from type I noxae in a major point: the generation of ROS is evoked by a noxa that does not reside inside the body. The ROS generation happens via electronic excitation of an intracorporal substance induced by the noxa. This results in the fact that direct ROS production due to a type II noxa only occurs as long as the noxa is present in the environment. Nevertheless, damage and biological consequences may outlast the period of exposure and may be enhanced as a result of repeated or chronic exposure. An important example of a type II noxa is UV irradiation. During exposure ROS are generated inside the body, but afterwards the generation of ROS due to direct exposure stops. Nevertheless the damage caused by the noxa might lead to the generation of additional oxidative stress, as is the case for all kinds of noxae.

2.3 Type III

Type III noxae produce ROS outside the body. The pro-oxidants enter the body in their reactive form and have an effect there. Type III noxae are rare and occur for example when the body comes into contact with ROS-containing fluids or aerosols, e.g. during direct exposure to high concentrations of ROS-generating chemicals.

3 Environmental Factors and Generation of Oxidative Stress

The environment is the sum of numerous influences on organic life, some of which are capable of promoting or generating oxidative stress. Such influences may be of natural origin but artificial, anthropogenic sources are gaining more and more importance. Their physical and chemical properties determine their mode of action and their degree of risk. The following environmental factors are known sources of environmental oxidative stress.

3.1 Pollution

Air pollution is an accompaniment of our modern industrial society. Although continuing efforts are undertaken to decrease the burden of exhausted particles we are confronted with several particulate substances, many of which are able to induce oxidative stress. Quartz, coalmine dust and asbestos are classical pathogenic particles, the most likely mechanism underlying their harmful and pro-inflammatory effects is oxidative stress [2, 3]. Increasing attention is being paid to other pollutants of great interest to environmental science, i.e. diesel exhaust particles [4, 5] and cigarette smoke [6–9]. Both noxae point out the complexity of the field as they are both not just single component systems but very complex mixtures of lots of components. Their composition varies not only between different sources but is liable to influences from the environment itself, leading to processes like aging of components and changes of composition. Even natural particles like pollen are thought to be contaminated with components that induce oxidative stress.

To interpret what effects inhaled particles might have, five factors must be evaluated: dose, deposition, dimension, durability and defence. First of all the dose that reaches the target tissue (e.g. the lung for particles) determines the oxidative burden created. Of course this dose is dependent on other factors, such as deposition in the target tissue and the dimension of the respective particles. The longer it takes the burdened organism to get rid of the particles, by mechanical, physiological or other means, the more the total amount of ROS created by the noxa is increased. Finally, how the tissue reacts to the particulate matter is important; an inflammatory defence reaction leads to the production of significant amounts of ROS.

A potent parameter driving the inflammation is the surface area of the particles. It is known that the ability of model ultrafine particulate matter to induce oxidative stress is not due to their composition but is related directly to their surface area [10–12].

Beside the oxidative stress created by the tissue's response to the presence of particulate matter, the oxidative activity of particulate matter is defined by a number of components that are able to generate oxidative stress. Presence of transition metals and organics within or attached to the particles play a major role in ROS generation. Electron spin resonance was applied to elucidate the participation of iron in generating ROS, namely hydroxyl radicals, via the Fenton reaction (Fig. 3) [13–17]. Besides iron, several other transition metals occurring in particulate matter (i.e. copper, chromium and vanadium) have been identified as generating ROS [13, 18]. Particles, especially when originating from combustion engines, often contain breakdown products and unburnt fuel. Among those mainly organic molecules are, for example, quinones capable of generating ROS [19] which inherit the ability to induce oxidative stress via the generation of free radicals. Due to their large functional surface, particles not only consist of their "core" components, but may be coated with several substances. Nevertheless, the ability to promote the generation of ROS

$$Fe^{2+} + H_2O_2 \longrightarrow Fe^{3+} + {}^{\bullet}OH + OH^-$$

Fig. 3 Fenton reaction. Hydrogen peroxide reacts with iron(II) to form iron(III), hydroxyl anion and the highly reactive hydroxyl radical

is not unique to either of these groups; the water-soluble "surface" fraction might be able to induce oxidative stress as well as the water-insoluble "core" fraction might.

The primary target tissue of pollution as a source of environmental oxidative stress is the respiratory tract [20], but inhaled particles have also been shown to be translocated to other organs, e.g. the liver [21, 22]. Increasing interest is also directed at alternative routes for how particles may enter the body, e.g. via the gastrointestinal tract [23–26] or across the skin [27, 28].

3.2 UV Radiation

UV radiation reaches living beings (e.g. the human body) from natural and artificial sources. The main source of UV is solar irradiation, but the use of artificial UV sources for therapeutic and increasingly also for lifestyle purposes is steadily rising. UV radiation is subdivided into three major bands of UV: UV-C of shortest wavelength (100–280 nm), UV-B (280–315 nm) and UV-A (315–400 nm). While UV-C originating from the sun is usually absorbed in the upper atmosphere, part of the UV-B and UV-A reach the earth's surface. Therefore they are of great importance as sources of environmental oxidative stress. As part of the natural solar radiation, together they account for 6.8% (UV-A 6.3% and UV-B 0.5%) of the total energy transferred. Their importance as a source of environmental stress is further enhanced by the fact that artificial UV-B and UV-A emitters are used increasingly, not only for therapeutic purposes but more and more for lifestyle reasons, e.g. for sun-beds or similar devices.

The dose of solar UV radiation reaching organisms is influenced by several factors: sun height, latitude, cloud cover, altitude, the ozone layer and ground reflection. The position of the sun in the sky is of importance because the UV radiation level rises with the height of the sun. The less atmosphere the UV rays pass through, the higher the dose that remains. Due to atmospheric transmission, the latitude of the exposed target is of interest, as is the altitude. Of course the composition of the atmosphere plays an important role, as well represented in the influence of cloud cover and the importance of the ozone layer. Finally not only the direct UV irradiation is of interest but indirect irradiation must be taken into account as well, mainly through UV reflected by the ground. The UV dose originating from artificial sources is determined by the power and emission spectrum of the UV source. Of course the time of exposure is a major factor concerning both natural and artificial UV sources.

UV-A, -B and -C reveal differences in their ability to penetrate the human skin: While UV-C barely reaches the upper epidermis, UV-B penetrates the epidermis and part of it reaches the dermis. UV-A, the radiation with the least photon energy and longest wavelengths within the UV band reaches the dermis and small amounts of it even reach the subcutaneous tissue [29, 30] (Fig. 4).

UV-A and UV-B radiation are known to cause several biological effects via different mechanisms, among which reactive oxygen species play an important

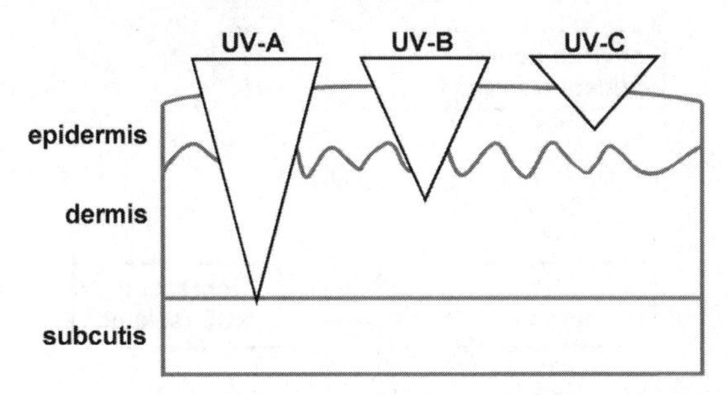

Fig. 4 Skin penetration by UV radiation. Subbands of the UV band penetrate the human skin with varying efficiency. UV-A penetrates the skin and reaches the dermis. UV-B hardly penetrates the epidermis and might reach the dermis. UV-C is totally absorbed within the epidermis

role. Both UV-A and UV-B have been shown to lead to oxidative damage and ROS production [31, 32]. In general, the action of UV radiation includes the following mechanisms [33]:

1. Direct interaction of photons with target molecules like DNA
2. Generation of reactive free radicals and/or ROS
3. Generation of proinflammatory molecules inducing the physiological production of ROS

Ultraviolet radiation is a potent inducer of ROS, namely superoxide radical ($O_2^{\bullet-}$), hydrogen peroxide (H_2O_2) and hydroxy radical ($HO^{\bullet}$). All three UV bands have been shown to be involved in generation of ROS and in formation of oxidative damage. Mechanistically, crucial involvement of chromophores in the UV-induced ROS generation is very likely. Porphyrin and porphyrin-like structures in the target tissue may be the mediators of ROS generation. The ability of such structures to mediate ROS formation has been extensively shown and is utilised by diagnostic and therapeutic methods, e.g. the photodynamic therapy of several cancer forms.

This underlines the fact that the biological effects of UV radiation may well be used for medical purposes, in spite of their detrimental side effects. These side effects drive continuing efforts to perform therapeutic efforts involving less, or sharper defined, UV radiation at minimal dosage or the use of less harmful radiation bands, such as visible light.

UV radiation makes an excellent example how a type II noxa might lead to secondary oxidative stress. UV irradiation is known to oxidatively cause a large scale deletion in the mitochondrial genome, termed "common deletion" [34, 35]. As the mitochondrial DNA (mtDNA) codes for several subunits of the mitochondrial respiratory chain and this electron transport chain is known as a major source of intrinsic ROS production, the UV-induced common deletion might lead to the enhanced leakage of ROS. This additional oxidative burden might then result in further damage to the mtDNA, again leading to further enhanced ROS leakage by the respiratory chain (Fig. 5).

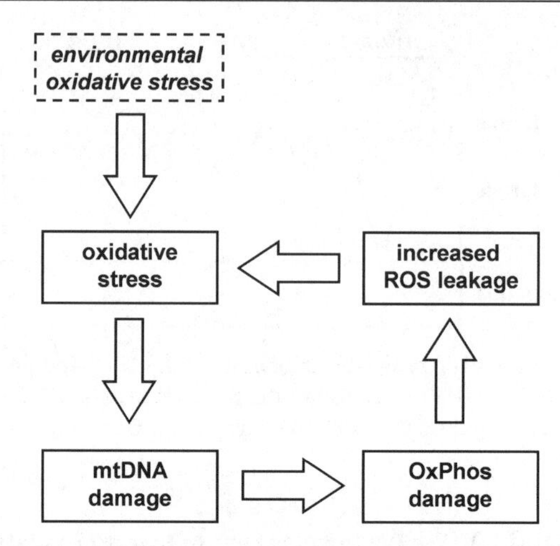

Fig. 5 Vicious cycle of oxidative damage in mitochondria. After an initial ROS generation due to an environmental noxa mtDNA damage occurs. This results in damage compounds of the mitochondrial respiratory chain (OxPhos), leading to the generation of additional ROS

The primary target tissue for UV irradiation is the skin, although for example the influence of ROS induced by non-ionising radiation like UV in degenerative ophtalmological diseases is receiving increased attention [36].

3.3 Infrared Radiation

Infrared radiation is another wave band of non-ionising radiation. The IR region is divided into IR-A (λ 760–1440 nm), IR-B (λ 1440–3000 nm) and IR-C (λ 3000 nm–1 mm). The main source of IR radiation is the sun, but artificial IR emitters are constantly gaining importance. They are used for therapeutic as well as for lifestyle purposes, e.g. as a "wellness" source.

Compared to UV radiation, only little research has been done in the field of IR-induced oxidative stress. While the photon energy of IR is lower than that of UV, the total amount of energy transferred by the sun contains approximately 54% IR while UV only accounts for 7% [37]. Most of the IR radiation lies within the IR-A band (approximately 30% of total solar energy). This is the most interesting part of IR, as it deeply penetrates tissues (e.g. human skin, see Fig. 6) [37–39]. The actual dose reaching the human skin, for example, is influenced by factors similar to those determining the UV dose, although the weightings of single factors (e.g. cloud cover) are different.

Although Kligman described in 1982 that skin damage similar to UV-induced damage could be generated by IR irradiation [40], it took some time before the first mechanistic insights into IR-induced skin damage were obtained [41]. Based on the knowledge of the oxidative cause of UV-induced skin damage [42] the oxidative properties of IR have come into focus [41, 43, 44].

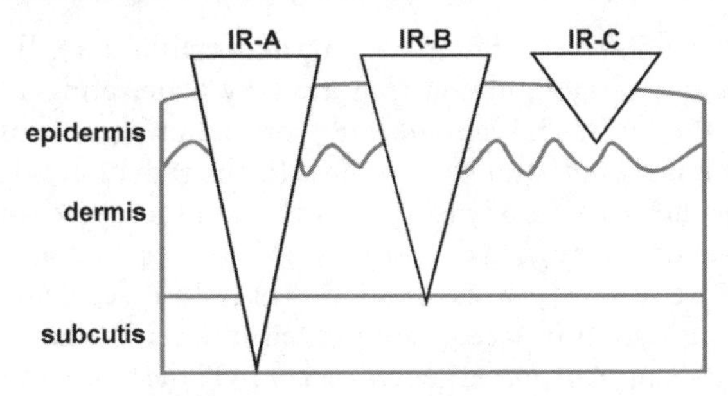

Fig. 6 Skin penetration by infrared radiation. Subbands of the IR band penetrate the human skin with varying efficiency. IR-A is able to reach epidermis, dermis and even the subcutis. IR-B penetrates into the dermis while IR-C is efficiently absorbed in the epidermis

Ongoing investigations underline that IR is able to induce significant ROS production (Schroeder and Krutmann, unpublished observations). The photobiological effect of IR is closely related to the respective chromophore(s). As such a chromophore must be able to absorb within the IR band it is necessary that molecules other than those discussed for UV must be considered [45]. While the endogenous chromophores for IR remain to be identified, exogenous chromophores for IR are used for therapeutic purposes [46, 47].

3.4 Ionising Radiation

Ionising radiation is highly energetic, extremely short waved radiation. Since Roentgen discovered x-rays at the end of the 19th century, and furthermore since the beginning of the nuclear era, the effects of ionising radiation have been subject to numerous investigations. Subjects of those investigations range from the side effects of diagnostically used X-rays to dealing with the consequences of a potential nuclear attack or accident.

Most of the biological effects of ionising radiation are mediated by reactive oxygen species. These ROS can either be generated primarily via radiolysis of water or they may be formed by secondary reactions [48], as the amount of primarily generated ROS could not account for the biological effects (e.g. activation of signal transduction) observed as a consequence of ionising radiation treatments [49, 50]. The primarily generated ROS is the hydroxyl radical, a highly reactive radical, usually short lived and fast reacting with a very limited diffusing range [51]. The secondary effects have been pinpointed to involve the mitochondria rather than representing a cytoplasmic phenomenon [52]. ROS produced due to secondary effects have been identified to include superoxide, hydroxyl radical and hydrogen peroxide [53, 54].

The main ionising radiations are alpha (α) particles, beta (β) particles, gamma (γ) rays, X-rays and neutrons. α-Particles have very little penetration power and can be stopped by a sheet of paper. β-Particles also have poor penetration and, on average, can travel only a few millimetres through biological tissue. γ-Rays, X-rays and neutrons are very penetrating and can readily pass through the body. The human body and other organisms can be confronted with ionising radiation in two ways: (i) the radiation is emitted from a source outside the target's body (e.g. X-ray or radiotherapy, cosmic radiation) or (ii) the radiation source is taken up into the body (e.g. radioisotopes inhaled or ingested). The oxidative stress created is in both cases determined by the dose of ionising radiation faced, represented in strength and quality of the irradiation. Upon intracorporeal generation of ROS the biological half-life time of the respective radioisotope is an important factor. Some radioisotopes (e.g. ^{32}P, ^{35}S) are excreted rather quickly, while others can be incorporated long term (e.g. ^{3}H, ^{14}C).

3.5 Nutrition

Nutrition is a factor gaining more and more importance in terms of oxidative stress. Besides the known protective effects of healthy nutrition (see below), it can also be responsible for arising oxidative stress. Nutrition is not strictly an environmental noxa itself but it represents a blend of several noxae (e.g. radioactive isotopes, particulate matter, nitric oxides). Pre-noxae, primarily less or not dangerous, can be modified by the organism (e.g. via cytochrome P450 systems) and represent a major factor in oxidative stress induced by nutrition. Such modification, as in the xenobiotic metabolism, might lead to potential sources for oxidative stress.

4 Targets Tissues of Environmental Oxidative Stress

The above mentioned examples illustrate that the tissue which has direct contact to the generated ROS is the first to be effected. As ROS are only able to travel short distances within the tissue or into neighbouring tissue before reacting with target molecules it is very likely that the primary oxidative damage is done close to the place of ROS generation. In type I noxae especially it does not have to be the tissue that first had contact with the noxa, but the noxa can be transported within the organism. In addition to the ROS that is generated directly by the noxa, secondary ROS can be generated, either through damaged molecules or as result of an immunoreaction. Damaged tissue is very likely to be subject to inflammation, during which additional ROS will be generated. The inflammation is not necessarily limited to the place where the primary oxidative damage took place.

Based on these considerations, the complexity of environmental science and medicine is obvious; source and effect can be connected via complicated mechanisms.

5 Protection

To think about protection against environmental oxidative stress it is first necessary to identify and characterise the respective source. If a source of environmental oxidative stress has been identified then the protective measures can be considered. Protection can generally take place on two levels: (i) physicochemical protection to lower the dose of exposure and (ii) physiological protection to increase the organism's antioxidative defense. (i) Physicochemical protection covers all means to lower the dose that the body is exposed to, including sunscreens against UV and IR radiation or exhaust filters to decrease particle emission. (ii) A more general approach is to strengthen the organism's defensive properties. Increased levels of endogenous (e.g. glutathione) or exogenous antioxidants (e.g. green tea polyphenols or carotinoids) have been proven to protect against oxidative stress [55–59]. The antioxidants may be applied via drug application, nutrition or through topical application.

References

1. Sies H (1985) Oxidative stress: introductory remarks. In: Sies H (ed) Oxidative stress. Academic Press, London, p 1
2. Banks DE, Parker JE (1999)Occupational lung disease, an international perspective. Oxford University Press, London
3. Kennedy TP, Dodson R, Rao NV, Ky H, Hopkins C, Baser M, Tolley E, Hoidal JR (1989) Arch Biochem Biophys 269:359
4. Hiura TS, Kaszubowski MP, Li N, Nel AE (1999) J Immunol 163:5582
5. Tsurudome Y, Hirano T, Yamato H, Tanaka I, Sagai M, Hirano H, Nagata N, Itoh H, Kasai H (1999) Carcinogenesis 20:1573
6. Chow CK (1993) Ann NY Acad Sci 686:289
7. Moller P, Wallin H, Knudsen LE (1996) Chem Biol Interact 102:17
8. Halliwell B, Cross CE (1994) Environ Health Perspect 102 Suppl 10:5
9. Epperlein MM, Nourooz-Zadeh J, Noronha-Dutra AA, Woolf N (1996) Int J Exp Pathol 77:197
10. Brown DM, Stone V, Findlay P, MacNee W, Donaldson K (2000) Occup Environ Med 57:685
11. Donaldson K, Stone V, Duffin R, Clouter A, Schins R, Borm P (2001) J Environ Pathol Toxicol Oncol 20 Suppl 1:109
12. Duffin R, Clouter A, Brown DM, Tran CL, MacNee W, Stone V, Donaldson K (2002) Ann Occup Hyg 46 Suppl 1:242
13. Ghio AJ, Pritchard RJ, Lehmann JR, Winsett DW, Hatch GE (1996) J Toxicol Environ Health 49:11
14. Gilmour PS, Brown DM, Beswick PH, MacNee W, Rahman I, Donaldson K (1997) Environ Health Perspect 105 Suppl 5:1313
15. Gilmour PS, Brown DM, Lindsay TG, Beswick PH, MacNee W, Donaldson K (1996) Occup Environ Med 53:817
16. Li XY, Gilmour PS, Donaldson K, MacNee W (1996) Thorax 51:1216
17. Li XY, Gilmour PS, Donaldson K, MacNee W (1997) Environ Health Perspect. 105 Suppl 5:1279

18. Lloyd DR, Carmichael PL, Phillips DH (1998) Chem Res Toxicol 11:420
19. Kendall M, Hamilton RS, Watt J, Williams ID (2001) Atmospheric Environment 35:2483
20. Menzel DB (1994) Toxicol Lett 72:269
21. Stuart BO (1976) Environ Health Perspect 16:41
22. Stuart BO (1984) Environ Health Perspect 55:369
23. Lomer MC, Harvey RS, Evans SM, Thompson RP, Powell JJ (2001) Eur J Gastroenterol Hepatol 13:101
24. Lomer MC, Thompson RP, Commisso J, Keen CL, Powell JJ (2000) Analyst 125:2339
25. Lomer MC, Thompson RP, Powell JJ (2002) Proc Nutr Soc 61:123
26. Pepelko WE (1991) Regul Toxicol Pharmacol 13:3
27. Pflucker F, Wendel V, Hohenberg H, Gartner E, Will T, Pfeiffer S, Wepf R, Gers-Barlag H (2001) Skin Pharmacol Appl Skin Physiol 14 Suppl 1:92
28. Sartorelli P, Cenni A, Matteucci G, Montomoli L, Novelli MT, Palmi S (1999) Int Arch Occup Environ Health 72:528
29. Health and Environmental Effects of Ultraviolet Radiation. WHO/EHG/95.16. 2002. WHO
30. Halliwell B, Gutteridge JMC (1999) Free Radicals in Biology and Medicine, 3rd edn. Thomas Press
31. Herrling T, Fuchs J, Rehberg J, Groth N (2003) Free Radic Biol Med 35:59
32. Herrling T, Zastrow L, Fuchs J, Groth N (2002) Skin Pharmacol Appl Skin Physiol 15:381
33. GonzalezS, Pathak MA (1996) Photodermatol Photoimmunol Photomed 12:45
34. Berneburg M, Gattermann N, Stege H, Grewe M, Vogelsang K, Ruzicka T, Krutmann J (1997) Photochem Photobiol 66:271
35. Berneburg M, Grether-Beck S, Kurten V, Ruzicka T, Briviba K, Sies H, Krutmann J (1999) J Biol Chem 274:15345
36. Glickman RD (2002) Int J Toxicol 21:473
37. Kochevar IE, Pathak MA, Parrish JA (1999) Photophysics, Photochemistry, and Photobiology. In: Freedberg IM, Eisen AZ, Wolff K, Austen KF, Goldsmith LA, Katz SI, Fitzpatrick TB (eds) Fitzpatrick's Dermatology In General Medicine. McGraw-Hill, New York, chap 16, p 220
38. Hellige G, Becker G, Hahn G (1995) Temperaturverteilung und Eindringtiefe wassergefilterter Infrarot-A-Strahlung. In: Vaupel P, Becker G (eds) Wärmetherapie mit wassergefilterter Infrarot-A-Strahlung. Hippokrates Verlag, Stuttgart, p 63
39. Cobarg CC (1995) Physikalische Grundlagen der wassergefilterten Infrarot-A-Strahlung. In: Vaupel P, Krüger W (eds) Wärmetherapie mit wassergefilterter Infrarot-A-Strahlung. Hippokrates Verlag, Stuttgart, p 19
40. Kligman LH (1982) Arch Dermatol Res 272:229
41. Schieke S, Stege H, Kurten V, Grether-Beck S, Sies H, Krutmann J (2002) J Invest Dermatol 119:1323
42. Ichihashi M, Ueda M, Budiyanto A, Bito T, Oka M, Fukunaga M, Tsuru K, Horikawa T (2003) Toxicol 189:21
43. Schieke SM (2003) Hautarzt 54:822
44. Schieke SM, Schroeder P, Krutmann J (2003) Photodermatol Photoimmunol Photomed 19:228
45. Karu T (1999) J Photochem Photobiol B 49:1
46. Koudinova NV, Pinthus JH, Brandis A, Brenner O, Bendel P, Ramon J, Eshhar Z, Scherz A, Salomon Y (2003) Int J Cancer 104:782
47. Tseng WW, Saxton RE, Deganutti A, Liu CD (2003) Pancreas 27:e42
48. Riley PA (1994) Int J Radiat Biol 65:27
49. Schmidt-Ullrich RK, Dent P, Grant S, Mikkelsen RB, Valerie K (2000) Radiat Res 153:245

50. Ward J (1994) Radiat Res 138 (Suppl):85s
51. Roots R, Okada S (1972) Int J Radiat Biol 21:329
52. Leach JK, Van Tuyle G, Lin PS, Schmidt-Ullrich R, Mikkelsen RB (2001) Cancer Res 61:3894
53. Cadet J, Douki T, Gasparutto D, Ravanat JL (2003) Mutat Res 531:5
54. Mikkelsen RB, Wardman P (2003) Oncogene 22:5734
55. Greul AK, Grundmann JU, Heinrich F, Pfitzner I, Bernhardt J, Ambach A, Biesalski HK, Gollnick H (2002) Skin Pharmacol Appl Skin Physiol 15:307
56. Liu GA, Zheng RL (2002) Pharmazie 57:852
57. Stab F, Wolber R, Blatt T, Keyhani R, Sauermann G (2000) Methods Enzymol 319:465
58. Stahl W, Sies H (2002) Skin Pharmacol Appl Skin Physiol 15:291
59. Stahl W, Sies H (2003) Mol Aspects Med 24:345

The Handbook of Environmental Chemistry Vol. 2, Part O (2005): 33–62
DOI 10.1007/b101145

Chemical Modifications of Biomolecules by Oxidants

Helen R. Griffiths (✉)

Life and Health Sciences, Aston University, Aston Triangle, Birmingham B4 7ET, UK
h.r.griffiths@aston.ac.uk

Abstract There is strong evidence in support of the oxidation of biomolecules during normal physiology and under conditions of environmental or pathological stress. Normally, the regulation of reduction/oxidation is vital for maintenance of cellular homeostasis. However, in conditions of excess free radicals, most intracellular substrates can be damaged, frequently leading to dysfunction and possibly disease. Whilst oxidants such as the hypohalous acids and reactive nitrogen species can cause a specific fingerprint of biomolecular damage, other commonly derived oxidants such as singlet oxygen, hydrogen peroxide and peroxy radicals cause generic types of damage. Herein, the chemistry of oxidation of DNA, lipids and proteins is considered in the context of our current knowledge on radical-specific oxidation, Furthermore, the formation of secondary oxidation products, arising from radical transfer between molecules and interaction between newly generated carbonyl groups and constituent amines, is examined.

Keywords DNA oxidation · Lipid peroxidation · Protein oxidation · Adduct · Hypohalous acid · Reactive oxygen species · Reactive nitrogen species · Singlet oxygen

Abbreviations

Ado Adenosine
cGMP Cyclic guanosine monophosphate
DNA Deoxyribonucleic acid
eNOS Endothelial nitric oxide synthase
Fapy Formamidopyrimidines
gC Glyoxylated cytosine
GSH Glutathione
Guo Guanosine
HNE Hydoxynonenal
iNOS Inducible nitric oxide synthase
IsoP Isoprostane
LDL Low density lipoprotein
M Metal
MDA Malondialdehyde
MPO Myeloperoxidase
NOS Nitric oxide synthase
PG Prostaglandin
PUFA Polyunsaturated fatty acid
RNA Ribonucleic acid
ROS Reactive oxygen species

1 Introduction

Molecules are exposed to a plethora of oxidant species in biological systems. Moreover, the interactions of these oxidants with the major classes of biomolecules, namely DNA, lipids and proteins, can yield multiple products. It has been suggested that the fingerprint of damage induced by oxidants may provide evidence for specific oxidant attack. Furthermore, the oxidative biomarker hypothesis propounds the theory that measurement of levels of oxidised biomolecules can provide an index of the levels of oxidant to which a biological system has been exposed [1].

Latterly, much interest has focussed on redox cell signalling, which involves the post-translational modification of signal transduction proteins by reactive oxygen and nitrogen species. Therefore, the purpose of this review is twofold: firstly, to review the nature of reaction of biologically relevant oxidants and secondly, yet most importantly, to consolidate our knowledge of the chemical modifications to biomolecules.

2
Free Radicals and Oxygen

Free radical species are inherently unstable by virtue of their single unpaired electron, thus tending to be highly reactive and transient, usually existing only at very low steady state concentrations (1 nM–100 μM). The single electron product of molecular oxygen, the superoxide anion radical ($O_2^{\bullet -}$) is one of the more stable radical species.

2.1
Superoxide, Peroxy Radicals and Peroxynitrite

Owing to its native electronic configuration as a diradical, molecular oxygen preferentially takes up electrons singly, where the first single electron product arising from this process is the superoxide anion radical ($O_2^{\bullet -}$). Whilst conventionally, chemical and physicochemical studies indicate that, for a radical species, $O_2^{\bullet -}$ is not very reactive per se, and is only damaging indirectly in aqueous solution through the formation of a more reactive radical species [2]. However, $O_2^{\bullet -}$ can act as both an oxidant and a reductant, depending on the environmental pH and the substrate with which it reacts [3]. When two molecules of $O_2^{\bullet -}$ react together, one is oxidised and the other is reduced in a dismutation reaction leading to the formation of hydrogen peroxide and oxygen, as shown in Eq. 1:

$$O_2^{\bullet -} + O_2^{\bullet -} + 2H^+ \rightarrow H_2O_2 + O_2^{\bullet -} \qquad (1)$$

where $k=10^5\ M^{-1}\ s^{-1}$ at pH 7.4.

This spontaneous dismutation reaction proceeds most rapidly at pH 4.8, the pKa for the equilibrium between $O_2^{\bullet -}$ and the hydroperoxy species ($HO_2^{\bullet}$). H_2O_2 is a stronger oxidant than $O_2^{\bullet -}$ and preferentially partitions into lipid and the hydrophobic cores of globular proteins and molecules where it exerts its oxidative effects [4]. Although $O_2^{\bullet -}$ is charged, it can cross the plasma membrane via anion channels and thus exhibit increased specificity in target selection.

At neutral pH, the enzyme-catalysed degradation of $O_2^{\bullet -}$ is more important for the generation of H_2O_2; the catalysed reaction proceeds at a rate of approximately $2\times10^8\ M^{-1}\ s^{-1}$. Although H_2O_2 is not a free radical, it is a reactive oxygen metabolite that may contribute further to free radical reactions. One of the first reactions of H_2O_2 to be described that involved radicals was the classical Haber–Weiss reaction. This involves the direct reduction of H_2O_2 by $O_2^{\bullet -}$ with the formation of the hydroxyl radical ($OH^{\bullet}$), O_2 and OH^- as shown in Eq. 2:

$$H_2O_2 + O_2^{\bullet -} \rightarrow O_2 + OH^- + OH^{\bullet} \qquad (2)$$

However, this reaction only proceeds very slowly under physiological conditions and is an unlikely source of significant quantities of $OH^{\bullet}$ in vivo, since $O_2^{\bullet -}$ would preferentially dismutate [5]. However, $O_2^{\bullet -}$ can also act as a powerful reducing agent, removing iron from saturated iron binding proteins such as lactoferrin and ferritin, and reducing it to the divalent ferrous state, see Eq. 3.

$$O_2^{\bullet -} + Fe(III) \rightarrow O_2 + Fe(II) \quad (3)$$

This is an important reaction since it provides a catalytic impetus for the metal-catalysed Haber–Weiss reaction. In this reaction, $O_2^{\bullet -}$ first reacts with an oxidised form of a trace metal ion, as previously described, or with low MW metal (M) ion chelates such as M-cysteine, M-histidine or M-EDTA. Transition metal ions are effective catalysts for this reaction owing to their labile "d" electron configuration. The reduced form of the metal ion generated by the reaction above subsequently reacts with hydrogen peroxide regenerating the oxidised form of the metal in a Fenton reaction and forming OH^- and significant fluxes of $OH^\bullet$. Once formed, the hydroxyl radical is highly reactive with all known biomolecules [6] by virtue of its large and positive redox potential, making it the most reactive of all radicals described.

Metal ions rarely exist in free solution but preferentially exist bound to carrier molecules such as the free amino acids cysteine, histidine or to these residues within a protein backbone. H_2O_2 is less reactive and more hydrophobic than $OH^\bullet$, permitting diffusion across great distances and through plasma membranes, away from its site of generation, and thus leading to $OH^\bullet$ formation in the presence of a catalytic metal ion and distant from the original source. Sites of metal binding on biomolecules are particularly susceptible to this sort of site-specific damage [7].

The hydroxyl radical reacts with most organic molecules at rates not far from diffusion controlled; consequently it has a very short half-life and is difficult to detect directly. It follows that a major determinant of the toxicity of $O_2^{\bullet -}$ and H_2O_2 in vivo is the availability and location of metal ions capable of catalysing $OH^\bullet$ formation.

2.2 Hypohalous Acids

Hypohalous acids, formed form the action of myeloperoxidase (MPO) on hydrogen peroxide in the presence of halide ions, are potent oxidising species, where HOCl predominates from monocytes and neutrophils owing to the relative concentration of chloride ions in plasma (100–140 mM Cl^-, 20–100 μM Br^-, and 100–500 nM I^-) [8]. In contrast, MPO from eosinophils preferentially oxidises Br^- to form hypobromous acid [9]. The pKa of the oxidants are 7.5 and 8.4, respectively; thus the major form of each at physiological pH will be HOBr and $HOCl/OCl^-$, where the anions might be expected to show lower reactivity with biomolecules. The most common reactions of these oxidising species are simple halogenation reactions, preferentially targeting unsaturated ring structures (see Scheme 1).

However, they can also elicit cleavage of carbon–carbon bonds, oxidative carboxylation or deamination reactions. Other reactions with amine-contain-

$$R{-}\underset{H}{C}{=}\underset{H}{C}{-}R + HOX \longrightarrow R{-}\overset{H}{\underset{X}{C}}{-}\overset{H}{\underset{OH}{C}}{-}R$$

Scheme 1 Halogenation of unsaturated bond

ing moieties include the formation of mono- and di-chloramine and bromoamines, respectively [10, 11]. The nitrogen halide bond is a labile structure, and allows further reaction with susceptible groups such as thiols, thereby regenerating the amine. Haloamines can also break down to yield either nitriles or aldehydes, as shown in Eqs. 4 and Eq. 5:

$$RCH_2NCl_2 \rightarrow RCN + 2Cl^- + 2H^+ \quad (4)$$

$$RCH_2NHCl + H_2O \rightarrow RCHO + NH_4^+ + Cl^- \quad (5)$$

2.3 Singlet Oxygen

A further important reaction involving the hypohalous acids is in the generation of singlet oxygen ($^1O_2$), which is characterised by the pairing of the two outer orbital electrons [12] as shown in Eq. 6:

$$HOCl + H_2O_2 \rightarrow {}^1O_2 + H_2O + H^+ + Cl^- \quad (6)$$

Whilst the occurrence of this reaction in physiological systems has been hotly debated the development of specific chemiluminescent probes has supported the notion that such a species can be generated during inflammation [12]. Other sources of singlet oxygen include phototherapy and simple UVA activation of endogenous photosensitisers such as porphyrins and flavins [13]. Recent studies using ^{18}O labelling have elegantly demonstrated that singlet oxygen can be formed from linoleic hydroperoxide. The study of the effects of singlet oxygen on biomolecules has been hampered by the lack of pure $^1O_2$ in high yield, and also the limited ability to be able to detected products of this reaction with high sensitivity. Recently the thermolysis of naphthalene endoperoxide as an endogenous generator has been applied to produce a pure chemical source of $^1O_2$ [14].

The rate constants for reaction of singlet oxygen vary significantly between biomolecules with greatest reactivity towards amino acid side chains, thereby affording some specificity of response [15]. Nevertheless, it can be considered as a powerful short-lived oxidant. The most frequent type of reaction initiated by this species is addition across unsaturated bonds yielding a cyclised product. This has been extensively reviewed by Davies [16] and will not be discussed further here.

2.4 Reactive Nitrogen Species

Nitric oxide (NO) is synthesised by one of three isoforms of nitric oxide synthase (NOS), which vary in their cellular localisation and the control of their activity. In general, NOS will catalyse the conversion of arginine to citrulline with the concomitant release of nitric oxide, a process that requires the presence of reduced nicotinamide adenine dinucleotide phosphate and oxygen. Giraldez et al. [17] reported the requirement for several other cofactors that play critical roles in the electron transfer process.

2.4.1 Reactions of Nitric Oxide and Formation of Other Reactive Nitrogen Species

In an environmental context, nitric oxide is known to be a toxic gas, and an important air pollutant. Nearly all cells produce NO as a freely diffusible gas that is preferentially soluble in hydrophobic compartments [18]. NO has a capacity to transfer easily across biological membranes and its coefficient of diffusion under physiological conditions exceeds that of molecular oxygen, rendering it an efficient signalling molecule. Indeed, it can migrate as far as 50 $\mu m\ s^{-1}$. Unlike other radical species, NO is relatively stable with a half-life in the order of 3–5 s in the extravascular space and is not toxic in its own right [19]. Indeed, it has been reported to have a cytoprotective effect under certain circumstances, where it may serve to terminate chain reactions in lipid peroxidation for example [20]. Instead, NO has an important role to play in modulation of biological activity, particularly in the brain. First discovered for its role as an endothelial derived relaxation factor [21], NO has subsequently been shown to react with metal ion clusters such as heme in guanylate cyclase, stimulating cGMP production and vasorelaxation in smooth muscle cells. It has subsequently been shown to react with many other metal ion centres, including zinc-cysteine motifs in zinc finger proteins and non-haem associated iron in glyceraldehyde-3-phosphate dehydrogenase, aconitase or lipoxygenase [22]. The direct effects of nitric oxide on metal ion centres is believed to predominate under normal physiological conditions, where the levels of NO are low. The reactivity of NO with iron is ultimately responsible for its degradation in biological systems, where red blood cell bound oxy-haemoglobin removes NO according to Eq. 7 [23]:

$$\text{Hb-Fe}^{2+}\text{O}_2 + {}^{\bullet}\text{NO} \rightarrow \text{Hb-Fe}^{3+} + \text{NO}_3^- \qquad (7)$$

Nitrate is frequently measured using the Griess reagent as an index of NO produced. NO itself participates in both reduction and oxidation reactions. Single electron reduction of NO yields the nitroxyl anion, whereas oxidation reactions yield the nitrosonium cation and $N_2O_3^+$, via autoxidation. Both the $^{\bullet}$NO radical and nitrosonium cation can be involved in direct reactions with biological systems, specifically with amino acid residues such as cysteine, with the resultant production of nitrosothiols (see later).

The oxidation of nitric oxide, firstly to nitrogen dioxide NO_2 and ultimately to $N_2O_3^+$, is now believed to be a rare event in biological systems, as it is dependent on the random interaction between three molecules described in Eq. 8 [24]:

$$2NO + O_2 \rightarrow 2NO_2 \quad (8)$$

Nevertheless, its rate of production via this pathway may be increased under exceptional circumstances where all three molecules are located in close proximity, for example in hydrophobic domains of proteins or within lipid bilayers [19]. Indeed, the majority of nitrogen dioxide may be found from the decomposition of peroxynitrite under physiological conditions.

Once formed, $^{\bullet}NO$ can exert indirect effects on biomolecules, including deamination, nitrosylation [25], nitration or proton abstraction reactions, as outlined in Schemes 2, 3 and 4 below.

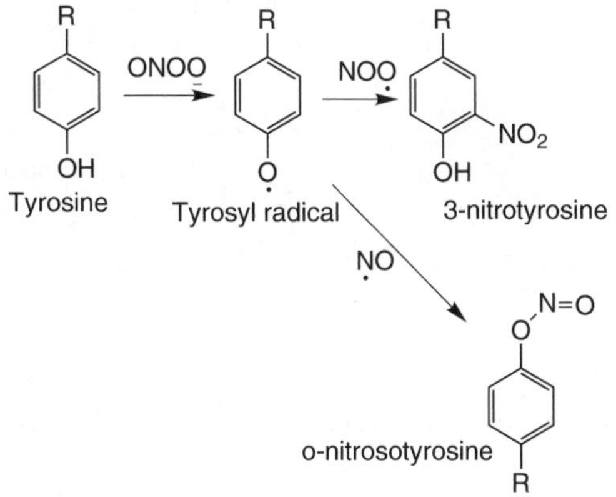

Scheme 2 Aromatic residues attacked by $ONOO^-$ or $^{\bullet}NO$ undergo nitration or nitrosation, respectively

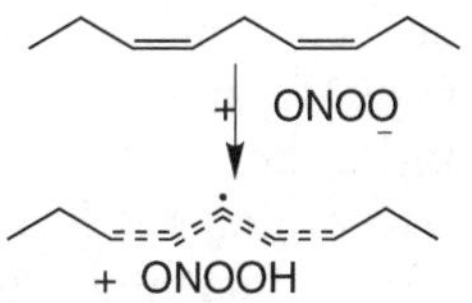

Scheme 3 Proton abstraction from unsaturated carbon chains by peroxynitrite

However, these indirect effects of nitric oxide derived products are far more prevalent under pathological conditions such as inflammation, where the production of both NO and $O_2^{\bullet-}$ by the professional phagocytic cell NADPH oxidase enzyme, and induction of iNOS yields the potent cytotoxic species peroxynitrite. Whilst nitric oxide will react with metal centres (as discussed above) at a rate of $\sim 5\times10^8\ M^{-1}\ s^{-1}$, and the superoxide anion can be dismutated by SOD at a rate of $2.3\times10^9\ M^{-1}\ s^{-1}$, the combined reaction below (Eq. 9), proceeds at a rate faster than either of these individual reactions:

$$^{\bullet}NO + O_2^{\bullet-} \rightarrow ONOO^- \quad (9)$$

where k=4.3–9×10^9 $M^{-1}\ s^{-1}$ [26].

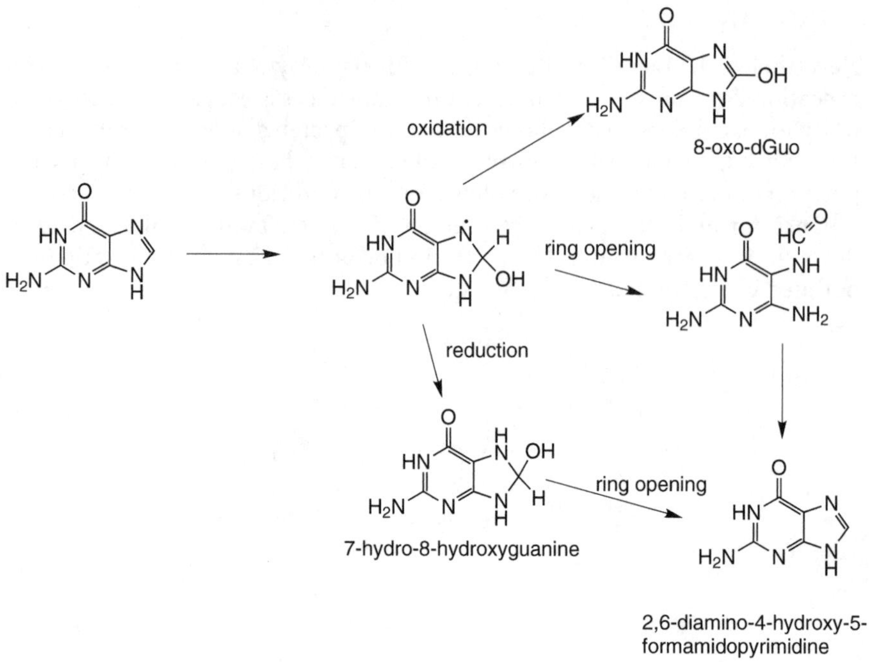

Scheme 4 Mechanism of formation of oxidation products of guanine

The sites of production of peroxynitrite are more likely to lie in closer proximity to the site of superoxide production as the latter is limited to transfer across the plasma membrane via anion channels, and has a shorter half-life (see above) [18]. It decomposes spontaneously at physiological pH, a process facilitated by the presence of Fe-EDTA and SOD, to yield nitrate (~65%) and nitrogen dioxide and the hydroxyl radical (~35%). Peroxynitrite is not a radical per se, and is less reactive than its protonated form. The pKa of the protonation reaction is 6.8, hence this proceeds efficiently at physiological pH yielding the potent oxidising species, peroxynitrous acid, which correspondingly decomposes to form the hydroxyl radical and nitrogen dioxide shown in Eq. 10 [27]:

$$ONOO^- \rightarrow ONOOH \rightarrow NO_2 + OH^{\bullet} \tag{10}$$

During inflammation, the rates of formation of peroxynitrite have been calculated in the order of up to 1 mM peroxynitrite per million cells every minute [28], making this an important source of oxidative and nitrosative stress in vivo.

In the next three sections, the effects of oxidants on three major classes of biomolecules will be examined, where consideration will be made to both radical specific products, and those generic products that arise following attack by many different species and mechanisms.

3 Damage of Nucleic Acids

3.1 By Superoxide, Peroxy Radicals and Peroxide

Whilst considerable efforts have been made to elucidate the chemical nature of nucleic acid base oxidation by reactive species it is important to remember that early observations from radiation biology demonstrated clearly that the sugar phosphate backbone, onto which the genetic information is encoded by sugar linked base structures, is susceptible to radical attack in the presence of oxygen. The major reactive species involved in this reaction is the hydroxyl radical, produced from the radiolysis of water. The abstraction of hydrogen atoms from any one of the carbon centres of deoxyribose by $OH^{\bullet}$ to form sugar radicals proceeds with second order kinetics and leads to the subsequent formation of many modified sugar products [29] or ring cleavage and eventual oxygen-dependent strand break formation. The involvement of peroxyl radicals in the process has been confirmed by Jones and O'Neill [30]. One unique reaction to the C5 centred radical is the formation of an intermediate capable of damaging both the base and the backbone [31]. Single strand breaks are rarely lethal to cells due to the associated induction of DNA repair processes. However, formation of SSBs at adjacent sites on parallel strands frequently results in double strand breaks, which are not subject to efficient or effective repair and therefore lead to cell death [32]. In the double helix, bases are sterically protected from ROS, however, over the last twenty years, it has been shown that the backbone offers little protection for bases from oxidants.

A second important point to note is that the reactions that will be subsequently described are not exclusively restricted to DNA bases. Indeed, RNA is also subject to oxidative damage where oxidised bases released into the cytoplasmic pool by repair processes may be subsequently incorporated into nucleic acids during de novo nucleic acid synthesis.

DNA is a particularly susceptible biomolecule to oxidative damage, since the transition metal ion, copper, plays an integral role in the maintenance of its quarternary structure [33, 34]. Divalent copper is bound to adenine residues on DNA where it plays an important role in stabilising the structure through cross-linking to histidine within the structural histone proteins. This condensation is important for both replication and RNA transcription. Indeed, sites of gene activation are commonly related to copper anchorage sites, and thus modification at these sites might be expected to affect transcription, as copper is integral for transciption factor activity. The freely diffusible species, hydrogen peroxide, can penetrate the hydrophobic core of the double helix where, in the presence of transition metal ions such as copper, it will produce the highly reactive hydroxyl radical. Hydroxyl radical mediated damage to DNA proceeds with second order kinetics, where the radical adds directly to one carbon of a double bond within the heterocyclic base yielding an OH adduct radical (reviewed by von Sonntag, [35]). In an oxygenated environment, peroxy radicals may also arise, which are potentially longer-lived species capable of further radical reactions [36]. Alternatively, for purine radical adducts at C4 or C5, a further dehydration and regeneration reaction may occur, restoring the parent base. In contrast, the C8 purine radical adduct can undergo a further single electron oxidation in the presence of oxygen (preferred) to yield 8-hydroxy-purines [37, 38], or under anoxic conditions, a reduction reaction is preferred to yield the corresponding formamidopyrimidine products. These reactions occur for both adenine and guanine, although the latter has received greater attention owing to the reported greater mutagenicity of 8-oxoGuo, as the syn conformation is able to mispair with adenine, resulting in a GC to TA transversion unless repaired prior to DNA replication [39].

In considering the reactivity of pyrimidine bases with the hydroxyl radical, it is evident that the simple ring structure favours the addition of oxygen after hydroxyl radical attack at either the C5 or C6 position [35, 40]. For example, from the C5 radical adduct, the resultant 5-hydroxy, 6-peroxyl radical adduct may subsequently eliminate $O_2^{\bullet -}$ after reaction with water to yield thymine glycol [35, 40]. In the absence of oxygen, reduction of either 5-OH or 6-OH adduct radicals, followed by protonation yields the corresponding hydroxy, hydropyrimidine shown below in Scheme 5. Reduction of the thymine-OH radical adduct also occurs to yield 5-hydroxy-5-methylhydantoin [37].

The methyl group of thymine is susceptible to hydrogen abstraction resulting in a carbon-centred allyl radical. Further oxidation of this radical intermediate can yield thymine-specific products, 5-hydroxymethyluracil and 5-formyluracil. Cytosine adducts also yield unique products under further

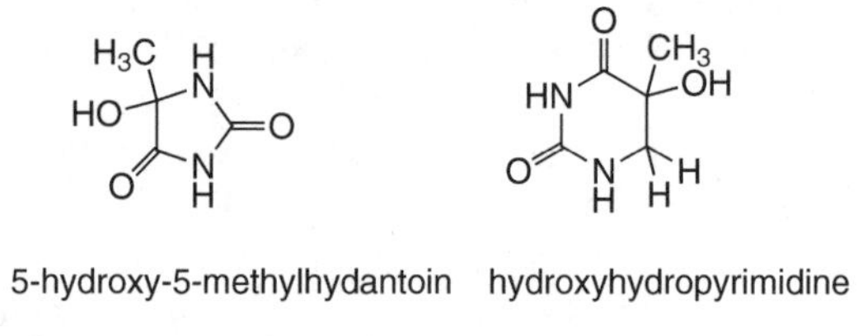

5-hydroxy-5-methylhydantoin hydroxyhydropyrimidine

Scheme 5 Chemical structures of 5-hydroxy-5-methylhydantoin and hydroxy, hydropyrimidine formed from hydroxyl radical attack on thymine

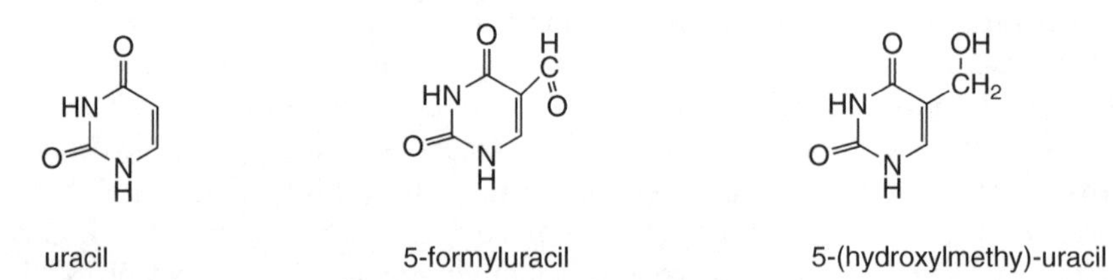

Scheme 6 Chemical structures of 5-hydoxymethyluracil, 5-formyluracil and uracil formed by hydroxyl radical attack on thymine and cytosine

oxidation, particularly related to their capacity to deaminate under certain conditions to yield uracil, as shown in Scheme 6.

Alloxan, another minor product of cytosine oxidation found in DNA, can also undergo decarboxylation to yield 5-hydroxyhydantoin [37, 40]. Similarly, thymine can be oxidised to 5-methyl-5-hydroxyhydantoin (Scheme 7).

5-hydroxyhydantoin

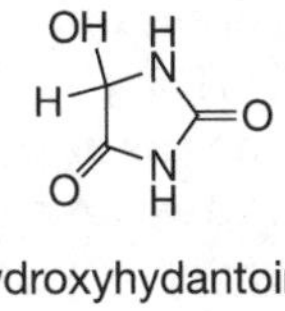

Scheme 7 Structure of the decarboxylation product of thymine oxidation, 5-hydroxyhydantoin

3.2 By Hypohalous Acids

Chlorinated nucleosides from both DNA and RNA have been described following attack by hypochlorous acid, including 5-chloro-2′deoxycytosine, 8-chloro-2′deoxyadenosine and 8-chloro-2′deoxyguanosine in DNA. These reactions are further enhanced by the presence of tertiary amines [41, 42]. Similarly, equivalent brominated species are produced following exposure to hydrogen peroxide and bromide under the catalysis of eosinophil peroxidase. The subsequent action of cytosine deaminase yields the potent mutagen, 5-bromo-2′deoxyuridine, which can subsequently be incorporated into genomic DNA [43, 44]. Recent studies have identified the formation of diamino-oxazalone from the action of HOCl on dGuo [45, 46]. A further interesting consequence of hypohalous acid production has been proposed: 8-oxodGuo is easily converted to further oxidation products (e.g. spiroiminodihydantoin, guanadinohydantoin and iminoallantoin) by various oxidants including HOCl, thus questioning the stability and validity of this biomarker in vivo [47–49] (Scheme 8).

HOCl may also indirectly damage DNA through interaction with nitrite to yield nitryl chloride at inflammatory sites. The major products from nitryl chloride attack on guanine or DNA are 8-nitroguanine and 8-nitroxanthine [50]. This reaction is predicted to be feasible in vivo, based on the known con-

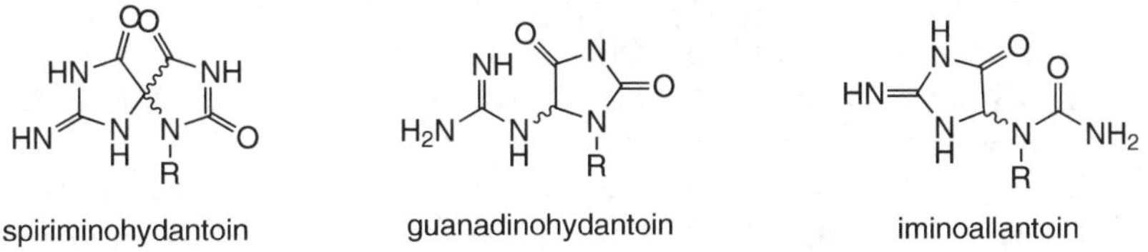

spiriminohydantoin guanadinohydantoin iminoallantoin

Scheme 8 HOCl mediated degradation of 8-oxodGuo yields spiroiminodihydantoin, guanadinohydatoin and iminoallantoin

centrations of HOCl and peroxynitrite formed at inflammatory sites; however, it has yet to be identified from isolated DNA.

3.3 By Reactive Nitrogen Species

Direct nitration of the C-8 position of guanine in both RNA and DNA can result from peroxynitrite attack [51, 52]. Additionally, the formation of 5-guanidino 4-nitroimidazole nucleoside has also been reported [53]. The combined action of peroxynitrite-induced base damage and DNA repair processes can result in depurination. Alternatively, peroxynitrite attack on guanine and adenine can yield the C-8 hydroxylation product, where guanine is preferentially oxidised over the bases. The resultant 8-oxoGuo is more reactive with peroxynitrite than its precursor, Guo; it is effectively degraded to yield secondary oxidation products such as cyanuric acid, oxaluric acid and oxazalone [54]. Therefore, the actual levels of 8-oxoGuo show only small increases in response to a peroxynitrite flux as a result of ring cleavage. Peroxynitrite has also been suggested to mediate single electron reduction and backbone breaks through an active intermediate similar to $^{\bullet}OH(NOOH^{*})$ [55].

N_2O_3 formed by a third order reaction, can deaminate DNA bases yielding uracil from cytosine, xanthine from guanine, methyl cytosine from thymine and hypoxanthine from adenine [56]. Furthermore, it can react with secondary amines to yield carcinogenic *N*-nitrosoamines, which can damage DNA by alkylation, [57].

3.4 By Singlet Oxygen

The products from the reaction of singlet oxygen with DNA bases vary according to the base environment. In free solutions, dGuo reacts with singlet oxygen to produce the 4S and 4R diastereoisomers of spiroiminodihydantoin, with a lesser yield of 8-oxodGuo [53, 58]. However, the rate of reaction of 8-oxodGuo with singlet oxygen is two orders of magnitude greater than that of the native base, resulting in a complex array of products from the free nucleoside. In contrast, the oxidation of DNA by singlet oxygen results in a largely specific reaction with guanine, yielding 8-oxodGuo, which is believed to proceed via the formation of an unstable C5-hydroperoxide [14]. Others have recently ascribed the reduction of an unstable 8-hydroperoxydGuo from the endoperoxide in the ultimate formation of 8-oxodGuo [59]. It is interesting to note that use of the Fpg protein only removes 35% of base modifications, where this raises the possibility that other non-Fpg sensitive lesions may be formed.

Single strand breaks have also been observed following exposure of DNA to singlet oxygen [60] but, in the absence of FapyG formation, this is unlikely to arise from a charge transfer reaction and may therefore result from direct damage to the sugar moiety.

3.5
By Secondary Oxidation

Base modifications can also arise as a secondary consequence of oxidation of other species, e.g. via lipid peroxidation products or through glycation. Many lipid peroxidation derived aldehydic products, e.g. malondialdehyde (MDA), hydroxynonenal (HNE), acrolein and crotonaldehyde, have been shown to form exocyclic adducts. Furthermore, strand breakage and abasic sites may arise from haem-catalysed lipid peroxide mediated attack [61, 62]. Aldehydes generate adducts through one of two major mechanisms: either direct reaction with DNA bases or through generation of more reactive electrophilic compounds such as bifunctional epoxides. In the case of the former, the products of reaction are 1,N^2-propano-2′-deoxyguanosine from α,β-unsaturated aldehydes, which are characterised by a six-member ring formation on the nucleobase see scheme 9. With MDA, cyclopyrimidinopurinone is formed from dA and an acyclic adduct from dC [63, 64]. Further oxidation of the α,β-unsaturated aldehydes yields epoxycarbonyl derivatives, which in turn will react with DNA to form substituted ethano, and etheno adducts.

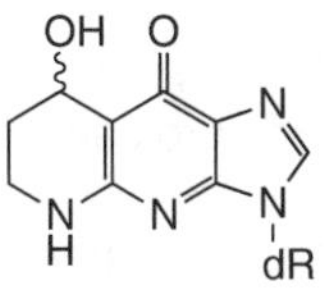

OH O
N
N N N
H dR

Scheme 9 Acrolein adducts to deoxyguanosine to yield 1,N^2-propano-2′-deoxyguanosine

This has been observed with HNE, which can react through an epoxide intermediate to form unsubstituted etheno (ε) adduct, unsaturated five-member extended ring formations on adenine, cytosine and guanine. Substituted ε-analogues have been characterised from reaction of epoxidised HNE with adenine and guanine. Epoxides of other α,β-unsaturated aldehydes such as acrolein, crotonaldehyde may yield similar adducts. The biological relevance of 1,N^2-εdGuo has been confirmed by in vitro misincorporation studies that show that this lesion strongly blocks replication at the lesion and beyond. In addition, dATP or dGTP are incorporated opposite the lesion [65]; however, it can be removed by the nucleotide excision repair pathway [66].

Indeed, one of the most toxic known breakdown products of lipid peroxidation to cells is *trans*, *trans*, 2,4, decadienal epoxide (DDE). This product reacts with both dAdo and dGuo, via the formation of a 4,5-epoxy-2(E)-decenal, to yield two highly mutagenic lesions, 1N^6-etheno-2′-dAdo and 1,N^2-etheno2-dGuo [54, 67, 68].

Many of the aldehydes produced from lipid peroxidation possess bifunctional electrophilic aldehyde groups allowing the formation of cross-links between self or adjacent molecules. Acrolein and related aldehydes are known to induce DNA-protein cross-links, where dGuo within a CpG sequence in duplex DNA is cross-linked to the N-terminal α-amine of the peptide structure [69].

A further secondary product of autoxidation is glyoxal, a two-carbon reactive ketoaldehyde generated during glycation and lipid peroxidation, which been shown to interact with DNA in vitro to form an ethano adduct with guanine residues. Glyoxal is also believed to be a product of deoxyribose degradation and ascorbate oxidation, and may be responsible for direct genotoxic damage. It readily forms DNA adducts, generating potential carcinogens such as glyoxalated deoxycytidine (gdC) [70]. Furthermore, dGuo-glyoxal-dC, dGuo-glyoxal-dA, cross-links, have been detected in glyoxal-treated DNA [71].

Extra-chain cross-links have been described following methylglyoxal treatment of DNA, where methylglyoxal cross-links a guanine residue of the substrate DNA and lysine and cysteine residues near the binding site of the DNA polymerase during DNA synthesis; replication is restricted at the cross-link [72].

4 Damage of Lipids and Membranes

4.1 By Superoxide, Peroxy Radicals and Peroxide

Lipid peroxidation of biological membranes is a destructive process, proceeding via an autocatalytic chain reaction mechanism [73]. Membrane phospholipids contain hydrogen atoms adjacent to unconjugated olefinic bonds, which make them highly susceptible to free radical oxidation. This is characterised by an initiation step, one or more propagation steps and a termination step [1], which may involve the combination of two radical species or interaction with an antioxidant molecule such as vitamin E. The products formed from such reactions include lipid peroxides, lipid alcohols and aldehydic by-products such as malondialdehyde and 4 hydroxynonenal [73].

The oxidation is initiated via allylic hydrogen atom abstraction from a polyunsaturated fatty acid with a methylene-interrupted double bond. This double bond weakens the CH bond of the methylene group, facilitating the elimination of the hydrogen atom. Initiation can be catalysed by several agents including hydrogen peroxide, hydroxyl radical, singlet oxygen, peroxynitrite, lipid hydroperoxides or a mixture of Fe(II) and Fe(III) salts, via an $OH^{\bullet}$-independent mechanism [74]. Whilst superoxide is generally not considered to be highly reactive, its interaction with its hydrated form $HO_2^{\bullet}$ may account for membrane damage in biological systems in the absence of iron [75]; however, the occurrence of this form may be limited at physiological pH owing to the pKa for formation of 4.2. In addition, the charged nature of superoxide will limit its capacity to penetrate the hydrophobic membrane compartment.

Once formed, polyunsaturated fatty acids (PUFA) can undergo double bond rearrangement to produce a diene conjugated molecule, which enhances stability of the radical intermediate. In the presence of oxygen, chain propagation can occur via lipid peroxy radical formation (LOO$^{\bullet}$), and initiation of an autocatalytic cycle as outlined below in Eqs. 11 to Eq. 13 [76]:

$$LH + Fe(III) \rightarrow L^{\bullet} + Fe(II) + H^{+} \quad (11)$$

$$L^{\bullet} + O_2 \rightarrow LOO^{\bullet} \quad (12)$$

$$LOO^{\bullet} + LH \rightarrow LOOH + L^{\bullet} \quad (13)$$

Lipid peroxy radicals will react further to yield cyclic peroxides, cyclic endoperoxide and finally to form aldehydes, including malondialdehyde [77] or other end-products such as isoprostanes or pentane and ethane [78, 79].

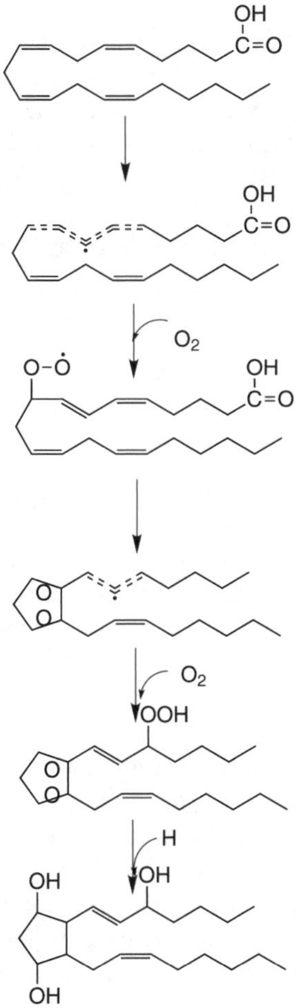

Scheme 10 Mechanism for formation of the 5-series F_2-isoprostanes

Isoprostanes (IsoPs) are derived as free radical oxidation products of arachidonic acid. Isoprostanes can be formed by two separate routes: direct peroxidation via an endoperoxide route and a dioxetane/endoperoxide mechanism. In neither case is the enzymic involvement of cyclooxygenase invoked. Again, the important initiation step in the formation of IsoPs is the abstraction of hydrogen from a bis-allylic methylene group, with the subsequent addition of molecular oxygen to form a peroxyl radical intermediate. This is followed by endocyclisation, and the further addition of oxygen to form parent IsoPs that resemble the parent prostaglandins, PGG_2 and PGH_2. The point of attachment of the hydroxyl group defines the regioisomer of IsoP as 5-, 12-, 8- or 15- IsoP, where the side chains are predominantly organised *cis*- to the prostane ring. Scheme 10 illustrates the mechanism of formation of 5-F_2-isoprostanes from arachidonic acid. These products are formed in situ within the membrane phospholipid, but may be released post-formation by phospholipase activity.

The PGH_2-like endoperoxides are central in the formation of IsoPs, but can also undergo rearrangement to yield E-ring, D-ring and thromboxane ring products [79]. Furthermore, in vivo, the E_2 and D_2 isoprostanes are susceptible to dehydration yielding cyclopentenone A_2 and J_2 isoprostanes. These cyclopentenones are reactive structures that undergo Michael addition reactions with thiols, as discussed later [80].

More recently, isoprostanes from the unsaturated fatty acid docosahexaenoic acid, which is enriched in the brain, have been described and termed neuroprostanes [81].

4.2 By Hypohalous Acids

As indicated above, the majority of reactive oxygen species are capable of eliciting general lipid peroxidation reactions, where the products described so far do not give any indication of the reactive species involved. In contrast, the hypohalous acids show reactivity with membrane phospholipids, with the generation of specific haloamines and halohydrins. Specifically, hypohalous acids can add across an unsaturated C=C bond within a fatty acid chain to yield alpha beta chlorohydrin and bromohydrin isomers [8] as shown in Scheme 1.

The carbon–halogen bond of the halohydrins is relatively stable, and these molecules are sufficiently bulky to cause membrane rupture if incorporated into red cell membranes. The addition reaction is favoured by the weaker oxidant, HOBr. Amine-containing phospholipid headgroups also have the potential to undergo N-halogenation reactions with the formation of haloamines. For hypochlorous acid, this product predominates under physiological conditions. In contrast, the higher reactivity of HOBr, yields both dibromoamine and bromohydrins under physiological conditions, where all products derived from HOBr are expected to exceed those from HOCl [10].

4.3
By Reactive Nitrogen Species

In contrast to the effects of the other species considered, nitric oxide has been reported to protect mammalian cells from oxidative killing, and also to inhibit lipid peroxidation in a variety of milieu [82]. Such effects have been attributed to the propensity for NO to partition into lipid environments and quench small organoperoxyl radicals. However, once formed, peroxynitrite itself is a potent initiator of lipid peroxidation cascades (see above). Furthermore, the combined product of nitrite with HOCl, nitryl chloride, has been reported to effectively oxidise lipids [83].

Lipid-derived peroxy radicals are important secondary species in inducing biomolecular damage since they are relatively long-lived species [84] enabling a greater specificity of reaction. PUFA peroxidation within membranes can also result in in situ damage, i.e. membrane disorganisation, alterations to ion transport and reduction in membrane fluidity [85]. This has a profound effect on the structure and functions of membranes, to such an extent as to cause cell dysfunction and death via apoptosis. In addition, the lipid peroxides once formed can interact with both proteins and DNA, yielding Schiff base products, and frequently dysfunction (vide infra).

5
Damage of Proteins

5.1
By Superoxide, Peroxy Radicals and Peroxide

Proteins constitute more than 50% of the dry weight of cells, and as such are predicted to be important targets for ROS. It has long been known that oxidative damage induced by selected radicals on lysozyme, α-1-antitrypsin, and apolipoprotein B in LDL inhibits their function [1]. However, the effects of oxidation are not always deleterious, where collagenase is released from a latent form by HOCl and also by the metal ion dependent product of superoxide anion [86]. Indeed, the chemical nature of oxidation and the biological consequences of this oxidation are dependent on: (1) the protein, (2) the oxidising species and (3) whether or not the oxidant can gain access to susceptible amino acid residues within that protein, i.e. three-dimensional structure constraints [87]. Finally, the dissipation of a radical intermediate is under the influence of "charge transfer": the movement of an unpaired electron along a peptide or protein to a susceptible amino acid of lower redox potential. This is most clearly illustrated when the azide radical reacts with a dipeptide of tryptophan and tyrosine: the tryptophanyl radical is soon formed but a rapid intramolecular transfer process takes place in which the free radical centre becomes located on the tyrosine residue [88]. Knowledge of the redox potentials of free radical couples enables prediction

Table 1 Biologically relevant amino acid redox couples

Couple		E/V
$RS^{\bullet}$	H^{+}/RSH	1.0
$Trp^{\bullet}$	H^{+}/Trp	0.98
$TyrO^{\bullet-}$	$H^{+}/TyrOH$	0.65

of the direction of charge transfer. Thus in any system where a protein is under radical attack, the original site of attack may not be the ultimate site of damage. Some examples of biologically relevant redox couples are shown in Table 1.

The complexity of protein structure, arising from the primary sequence and higher structures such as carbohydrate moieties for structure stabilisation and metal centres at catalytic sites, makes the study of protein oxidation complex. Examination of proteins that have undergone radical attack for the loss of the parent amino acid reveals that the amino acids most susceptible to oxidative attack are those capable of delocalising charge, such as amino acids containing aromatic and cyclical side chains and thiol-containing side chains where sulphur increases valency to stabilise the intermediate radical [89, 90]. Nonetheless, a large number of aliphatic residues are subject to oxidation, particularly with $OH^{\bullet}$ which rarely diffuses more than 12 Å [91]. Even the least reactive amino acids react with hydroxyl radical with a rate constant of between 10^{7} and $10^{8}\ M^{-1}\ s^{-1}$, with the result that most side chains are oxidised [6]. Given that the average size of an amino acid is 5 Å, $OH^{\bullet}$ is unlikely to diffuse further than two amino acids away from its site of generation.

The process of protein oxidation can cause backbone scission following hydroxyl radical attack and also frequently introduces new functional groups to side chains, such as hydroxyls by addition and carbonyls through hydrogen abstraction, which contribute to altered function and turnover. Carbonyls are ubiquitous products generated in response to a wide variety of oxidising stimuli, including hydroxyl, alkoxy and peroxy radicals arising on amino acid side-chains. They may also arise from the addition of oxidised sugars and lipids [92]. Both oxo-acids and aldehydes are formed following oxidative attack, having either the same number of carbon atoms or one less than the original amino acid [93]. Lysine can also undergo limited attack at the C6 carbon atom to yield the corresponding carbonyl product, aminoadipic semi-aldehyde. Scheme 11 shows the formation of glutamic semialdehyde, from either arginine or proline oxidation. These two products are the major products formed during metal-catalysed oxidation of proteins [94].

Histidine is frequently involved in the co-ordination of metal ions, and is thus a target for metal-catalysed oxidation [95], generating 2-oxo-histidine and its ring-ruptured products. It has also been postulated that the formation of Schiff bases between newly formed carbonyl moieties and epsilon amino groups on lysine residues may account for protein aggregation.

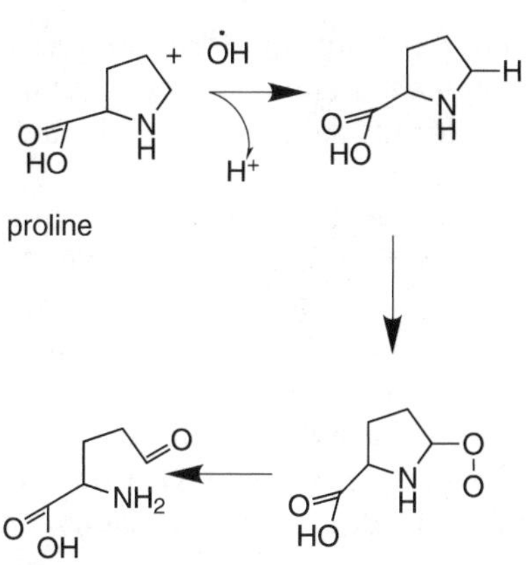

Scheme 11 Oxidation of proline causes ring opening with the resultant formation of glutamic semialdehyde

The chemistry of free radical oxidation of cysteine and its dimeric form cystine has been studied at length by radiation chemists [96]. Radiolytic generation of $OH^{\bullet}$ causes many different oxidative changes to cystine, which may or may not cause disulphide bond cleavage (Eqs. 14 to Eq. 18) [97, 98].

$$RSSR + OH^{\bullet} \rightarrow RSSOHR \quad (14)$$

$$RSSR + OH^{\bullet} \rightarrow RSSOH + R^{\bullet} \quad (15)$$

$$RSSR + OH^{\bullet} \rightarrow RSOH + RS^{\bullet} \quad (16)$$

$$RSSR + OH^{\bullet} \rightarrow RSSR\,(^{-}H) + H_2O \quad (17)$$

$$RSSR + OH^{\bullet} \rightarrow RSSR^{\bullet -} + OH^{-} \quad (18)$$

Hydrogen abstraction results in positive ion formation as shortlived species that may in turn act as oxidants [99] or may rearrange from thiols [98]. In addition, formation of a thiyl radical has been detected using pulse radiolysis [99]. These intermediates are not stable and decay rapidly via several pathways [96] that may result in the formation of oxy acids (cysteic acid or disulphides). However, the proportion of the product yielded is governed by many factors, including oxygen tension and substrate concentration.

The free thiol group of cysteine readily undergoes reversible oxidation to form a disulphide via a thiyl radical, which can be "repaired" in the presence of a thiol donor such as glutathione. This is important during reversible cellular signalling processes. Further oxidation leads irreversibly to cysteic acid [96]. Oxidation of the single amino acid cysteine is also complicated by the observation that the product may be a composite mixed disulphide formed between other thiol-containing molecules.

Methionine also represents an important and early target for oxidation, where end products include the sulphoxide and the sulphone. In addition to the hydroxyl radical, hypochlorous acid, nitric oxide and singlet oxygen are all capable of eliciting this change.

Previously, we have examined the formation of amino acid hydroperoxides following exposure to different radical species [100]. We observed that valine was most easily oxidised, but leucine and lysine are also prone to this modification in free solution. Scheme 12 illustrates the mechanism for formation of valine hydroperoxide. However, tertiary structure becomes an important predictor in proteins, where the hydrophobic residues are protected from bulk aqueous radicals, and lysine hydroperoxides are most readily oxidised. Hydroperoxide yield is poor from Fenton-derived oxidants as they are rapidly broken down in the presence of metal ions [101]. Like methionine sulphoxide, hydroperoxides are also subject to repair, in this case via glutathione peroxidase. They can also be effectively reduced to hydroxides, a reaction supported by the addition of hydroxyl radical in the presence of oxygen. Extensive characterisation of the three isomeric forms of valine and leucine hydroxides has been undertaken by Fu et al. [102, 103], and therefore will not be discussed further here.

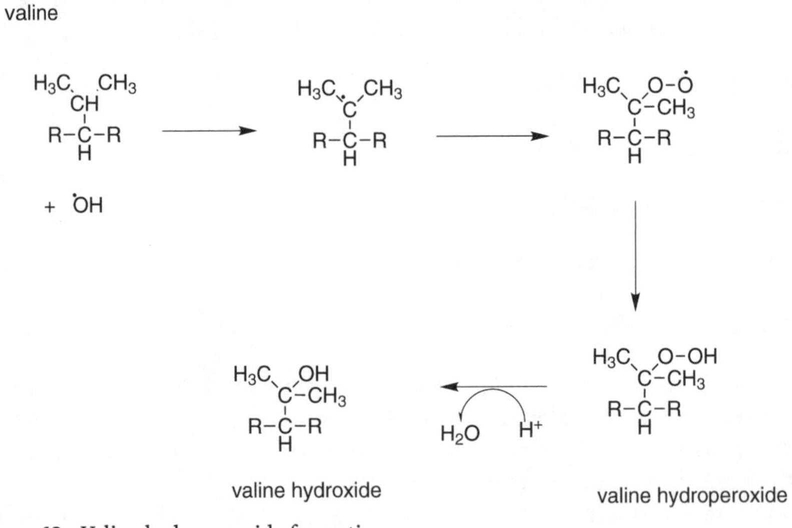

Scheme 12 Valine hydroperoxide formation

The most frequent oxidative changes to aromatic amino acids are ring addition reactions with the introduction of oxygen. In the case of tryptophan, this causes ring cleavage to yield *N*-formyl kynurenine (NFK), with a corresponding exponential decrease in indole fluorescence. Many of these changes are also associated with generation of a novel autofluorescence of longer wavelength that confers increased sensitivity and specificity in analysis [104]. NFK is the first major product to appear following oxidation by hydroxyl radical,

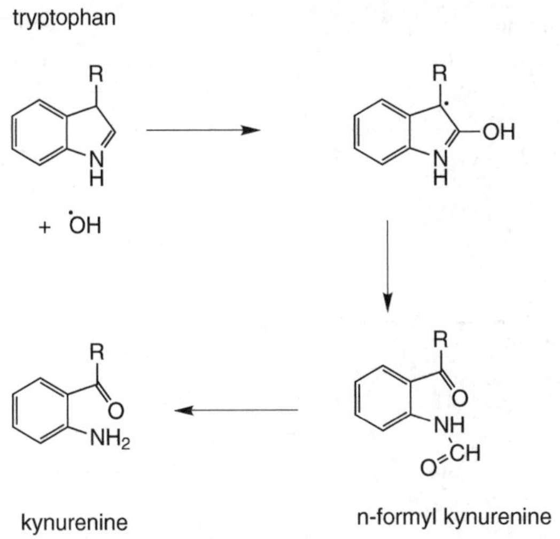

tryptophan

+ ˙OH

n-formyl kynurenine

kynurenine

Scheme 13 Mechanisms of tryptophan oxidation

as shown in Scheme 13, and is not produced by superoxide anion radicals [105]. This undergoes further degradation to yield kynurenine via elimination of CO_2, and hexahydropyrolloindole.

The light energy emitted by UV radiation can be directly absorbed by tryptophan, producing an excited intermediate. The tryptophanyl radical can be dissipated though transfer to dissolved oxygen (to yield $O_2^{\bullet-}$) by antioxidants or through charge transfer. In lysozyme, one tryptophan side chain is oxidised to indolyl free radical, which is produced quantitatively. The indolyl radical subsequently oxidises a tyrosine side chain via hydrogen abstraction to yield the phenoxyl radical in an intramolecular reaction. Other silent yet susceptible residues are bypassed as the charge can be stabilised through dimerisation and enolisation, resulting in dityrosine formation [106]. This fluorescent cross-link (excitation maximum 290 nm, emission maximum 410 nm) has been suggested to be important in mediating oxidation induced protein aggregation [107]. However, the yield of the biphenol is reduced to about 4% in the presence of oxygen [88]. This is consistent with the repair of phenoxyl by superoxide anion shown in Eq. 19:

$$PheO^{\bullet} + O_2^{\bullet-} + H^+ \rightarrow PheOH + O_2 \qquad (19)$$

Dityrosine may be a significant product in metalloproteins, such as globins, where the intermediate is stabilised through a change in redox status of the associated transition metal. Studies on the oxidation of tyrosine again show a preference for the electrophilic addition of hydroxyl radical to the unsaturated ring of tyrosine, yielding protein-bound L-DOPA. L-DOPA is a relatively long-lived species that can confer reducing activity in its environment [89, 90]. The possible routes for tyrosine oxidation are illustrated in Scheme 14.

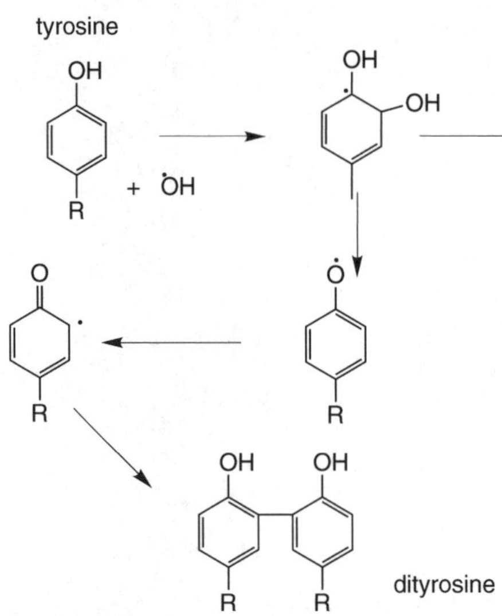

Scheme 14 Tyrosine oxidation by hydroxyl radicals

Hydroxylation of phenylalanine is a typical reaction of hydroxyl radicals, in which the resultant tyrosine is hydroxylated at the ortho- meta- or paraposition, and produces three isomeric tyrosines, *o*-tyrosine (2-hydroxyphenylalanine), *m*-tyrosine (3-hydroxyphenylalanine), and *p*-tyrosine (4-hydroxyphenylalanine) [108], shown in Scheme 15. This reaction has been used extensively to trap hydroxyl radicals in vivo, and thereby facilitate their measurement [108].

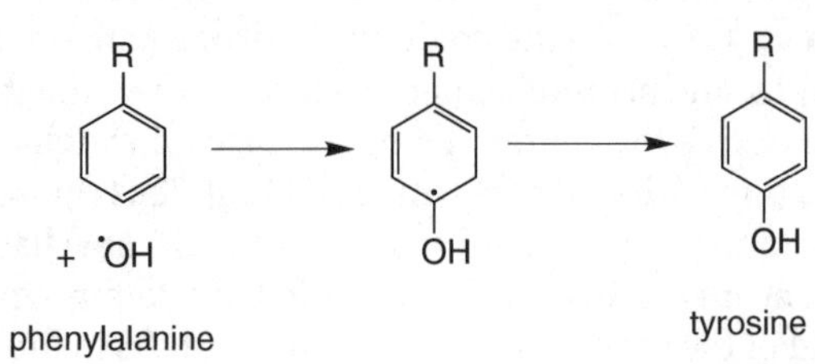

Scheme 15 Mechanism for tyrosine formation from phenylalanine following hydroxyl radical attack

5.2 By Hypohalous Acids

As previously discussed, the nucleophilic non-radical oxidant, HOCl, is particularly reactive and yields 3-chlorotyrosine as a specific and stable product from tyrosine. More recently, a dichlorinated species, 3,5,dichlorotyrosine has been described [109], which is formed four times less readily than the monochlorinated species at low fluxes of HOCl. Bromotyrosine may also be generated under the action of eosinophil peroxidase, and the interaction of HOBr with lysine produces *N*-bromoamine derivatives, although the yields tend to be low [9]. Corresponding chloramines are also unstable, preferentially breaking down to yield protein carbonyl residues [110] shown in Eq. 20:

$$RNH_2 + HOBr \rightarrow RNBr + H_2O \tag{20}$$

Lysine chloramines can transfer radicals, following primary attack of apolipoprotein B by hypochlorous acid, onto the lipid moiety in a second step involving homolytic fission reactions [110].

Reactivity is not limited to direct addition to the phenoxyl ring structure, and reactions with thiols, amines and olefinic compounds have been reported. Both HOBr and HOCl can elicit oxidation of cysteine residues, where oxidation of the thiol moiety to a sulphinic acid is associated with activation of the latent form of matrix metalloproteinase-7 [111]. Higher doses of HOCl cause inactivation of the enzyme through site-specific modification of tryptophan and the adjacent glycine to yield an unknown product that lacks four mass units [112].

5.3 By Reactive Nitrogen Species

Besides protein oxidation resulting in increased levels of protein carbonyls, the oxidative attack on protein by reactive nitrogen species primarily results in the addition of nitrate groups to the ortho position of amino acids, a process referred to as "protein nitration" [113]. This reaction occurs particularly with tyrosine (Scheme 2), tryptophan, phenylanine, methionine and cysteine. This is likely to proceed via a free radical mechanism where, for tyrosine, single electron oxidants derived from $ONOO^-$ initiate hydrogen abstraction from the ring to yield a phenoxyl intermediate that can undergo an addition reaction with $ONO_2^{\bullet}$ to yield 3-nitrotyrosine. Important catalysts for this reaction include transition metals, and myeloperoxidase, where the reaction proceeds via the formation of NO_2Cl [114]. MPO can also catalyse tyrosine nitration in proteins from nitrite and hydrogen peroxide [115].

Nitryl chloride, formed from peroxynitrite and HOCl is also capable of nitrating, chlorinating and hydroxylating tyrosine [116]. However, the reduction of nitrotyrosine by hypohalous acid has also been demonstrated. Several recent studies indicate that peroxynitrite can lead to generation of both nitrated and hydroxylated phenylalanine residues. The formation of 3-nitrotyrosine and 4-nitrophenylalanine was favoured over the hydroxylation reaction when phenylalanine was treated with peroxynitrite in vitro [117].

A recent study has described the inhibition of tyrosine nitration but not oxidation by tryptophan and tryptamine, indicating that nitration of the indole ring proceeds more favourably than nitration of the phenoxyl ring [118]. Methionine oxidation can be elicited by peroxynitrite and proceeds via two-step single electron reduction reactions yielding methionine sulphoxide as described previously [119].

$ONOO^-$ can react with cysteine residues via oxidative, nitrosative and nitrosylation to yield disulphides and nitrosothiols. For glutathione (GSH), the main NO mediated nitrosative pathway has been elucidated as described by Eqs. 21 to Eq. 23:

$$2^{\bullet}NO + O_2 \rightarrow 2^{\bullet}NO_2 \tag{21}$$

$$GS^- + {}^{\bullet}NO_2 \rightarrow GS^{\bullet} + NO_2^- \tag{22}$$

$${}^{\bullet}NO + GS^{\bullet} \rightarrow GSNO \tag{23}$$

In contrast, the mechanism of nitrosation by peroxynitrite does not proceed via a radical reaction but involves direct electrophilic attack [120] as shown in Eq. 24:

$$GS^- + ONOOH \rightarrow GSNO + HOO^- \tag{24}$$

Nitrosothiols appear to play important roles in the signalling function of NO, since addition to thiols interferes with the activation of signalling cascades such as MAPK, affecting the regulation of proliferation, growth and differentiation [121].

5.4 By Singlet Oxygen

As observed for amino acid oxidation elicited by other radical oxidants, the preferential targets for singlet oxygen appear to be tryptophan, tyrosine, methionine, histidine, cysteine and cystine. In this case, their chromogenic properties and absorption spectra are important determinants of their reactivity. The light energy emitted by UV radiation can be directly absorbed by tryptophan, producing an excited intermediate. The tryptophanyl radical can be dissipated though transfer to dissolved oxygen (to yield $^1O_2$) [122]. Alternatively, these excited radical species may serve to transfer electrons onto other biomolecules. The mechanisms of singlet oxygen mediated oxidative damage are poorly understood; however, much of the existing knowledge has been

Table 2 Products of amino acid oxidation elicited by singlet oxygen

Parent amino acid	Product
Tryptophan	Hydroxypyrolloindoles, NFK, kynurenine
Tyrosine	L-DOPA (minor), 3a-hydoxy-6-oxo-2,3,3a,6,7,7a-hexahydro-1H-indol-2-carboxylic acid, 3a-hydoxy-6-oxo-2,3,3a,6,7,7a-hexahydro-1H-indol-2-ketone
Histidine	Aspartic acid derivative, asparagine derivative, His–His cross-links
Methionine	Methionine sulphoxide
Cysteine	Cystine, cysteic acid
Cystine (RSSR)	RSS(=O)R

recently reviewed by Davies [16]. The end-products of singlet oxygen mediated oxygen are tabulated below (Table 2), where a common mechanism for formation appears to involve endoperoxide cyclisation.

5.5
By Secondary Protein Oxidation Products

Whilst glucose addition is a simple condensation reaction between the carbohydrate aldehyde group and protein amino group, the resultant Schiff base can undergo further reactions involving oxidative reactions, and these are collectively referred to as glycoxidation reactions. Protein-bound glucose can undergo autoxidative glycoxidation, whereby the autoxidation of sugar via dicarbonyl residues generates superoxide radicals adjacent to the protein backbone. Free sugars may also undergo autoxidation in the presence of catalytic metal ions. The major products of glycoxidation are collectively referred to as "advanced glycation end-products" (AGE) where *N*-ε-(carboxymethyl) lysine is the main structure produced [123].

Ascorbate (vitamin C) degradation products can undergo non-enzymatic glycation (Maillard reaction) with proteins to form highly cross-linked structures with brown pigmentation and characteristic fluorescence. Proteins in the body, especially the long-lived proteins, develop similar changes during aging and diabetes. Several studies have shown excessive degradation of ascorbate in plasma in diabetes, and in the ocular lens during aging and cataract formation. Recent studies have suggested that ascorbate degradation product-mediated glycation plays a role in lens pigmentation and cataract formation. One of the major degradation products of ascorbate, L-threose, promotes the formation of 2-acetamido-6-(3-(1,2-dihydroxyethyl)-2-formyl-4-hydroxymethyl-1-pyrrolyl)-hexanoic acid (formyl threosyl pyrrole), formed by the condensation of the epsilon-amino group of lysine with two molecules of threose [124].

Reactive aldehydes derived from lipid peroxidation, which are able to bind to several amino acid residues, are also capable of generating novel amino acid oxidation products. By means of specific polyclonal or monoclonal antibodies, the occurrence of malonaldehyde (MDA) and 4-hydroxynonenal (4-HNE) bound to cellular protein has been shown. Lysine modification by lipid peroxidation products (linoleic hydroperoxide) can yield neo-antigenic determinants such as *N*-ε-hexanoyl lysine. Both histidine and lysine are nucleophilic amino acids and therefore vulnerable to modification by lipid peroxidation-derived electrophiles, such as 2-alkenals, 4-hydroxy-2-alkenals, and ketoaldehydes, derived from lipid peroxidation. Histidine shows specific reactivity toward 2-alkenals and 4-hydroxy-2-alkenals, whereas lysine is an ubiquitous target of aldehydes, generating various types of adducts. Covalent binding of reactive aldehydes to histidine and lysine is associated with the appearance of carbonyl reactivity and antigenicity of proteins [125].

Isoprostane endoperoxide rearrangement yields gamma ketoaldehydes which cross-link with proteins through lysine residues. Cyclopentenones, derived from IsoPs, are highly reactive with thiols such as cysteine, where following Michael addition reaction, the thiol group is lost. Thiols are critical redox switches and sensors within cells, and modification by cyclopentenones is associated with a loss of function, as reported for the negative regulator of Nrf2, Keap1 [80].

Histone proteins play a central role in protecting and organising nuclear DNA; however, they are also susceptible to the effects of oxidative attack resulting in the formation of hydroperoxide species. Radicals generated during copper-catalysed degradation of hydroperoxides react with both pyrimidine and purine bases [126]. Similarly, protein-bound DOPA can promote further radical generating events, transferring damage to DNA [127]. Nevertheless, under certain circumstances, amino acids may serve to repair oxidative damage to other biomolecules such as DNA. Cysteine, methionine, tyrosine and particularly tryptophan derivatives have been shown to repair guanyl radicals (generated by radiolysis in the presence of thiocyanate) in plasmid DNA with rate constants in the region of approximately 10^5, 10^5, 10^6 and 10^7 M^{-1} s^{-1}, respectively. The implication is that amino acid residues in DNA-binding proteins such as histones might be able to repair DNA by an electron transfer reaction [128].

6 Conclusion

It is well established that oxidants are important modulators of biological processes through the post-synthetic modification of all the major classes of biomolecules. There is still much to be uncovered with respect to the chemical mechanisms of oxidation under biological conditions, and the consequences of oxidation, halogenation, nitration or nitrosation for regulation in normal physiology and for dysregulation during disease.

Acknowledgements HRG gratefully acknowledges financial support from the FSA, UK for development of biomarkers of oxidative damage to biomolecules.

References

1. Griffiths HR, Møller L, Bartosz G, Bast A, Bertoni-Freddari C, Collins A, Coolen S, Haenen G, Hoberg A-M, Loft S, Lunec J, Olinski R, Parry J, Pompella A, Poulsen H, Verhagen H, Astley SB (2002) Mol Aspects Med 23:101
2. Bielski BHJ, Richter HW (1977) J Am Chem Soc 99:3019
3. Fridovich I (1986) Arch Biochem Biophys 247:1
4. Lynch RG, Fridovich I (1978) J Biol Chem 253:4697
5. McCord J, Day ED (1978) FEBS Letts 86:139
6. Anbar M, Neta P (1967) Int J Appl Radiat Isotopes 18:493
7. Samuni A, Chevion M, Czapski G (1981) J Biol Chem 256:12632
8. Carr AC, Winterbourn CC, van den Berg JJM (1996) Arch Biochem Biophys 327:227

9. Weiss SJ, Test ST, Eckmann CM, Roos D, Regiani S (1986) Science 234:200
10. Carr AC, van den Berg JJM, Winterbourn CC (1998) Biochim Biophys Acta 1392:254
11. Thomas EL, Bozeman PM, Jefferson MM, King CC (1995) J Biol Chem 270:2906
12. Tatsuzawa H, Maruyama T, Misawa N, Fujimori K, Hori K, Sano Y, Kambayashi Y, Nakano M (1998) FEBS Lett 439:329
13. Tyrell RM (2000) Methods Enzymol 319:290
14. Martinez GR, Loueiro APM, Marques SA, Miyamoto S, Yamaguchi LF, Onuki J, Almeida EA, Garcia CCM, Barbosa LF, Medeiros MMG, diMascio P (2003) Mutat Res 544:115
15. Wilkinson F, Helman WP, Ross AB (1995) J Phys Chem Ref Data 24:663
16. Davies MJ (2003) Biochem Biophys Res Comms 305:761
17. Giraldez RR, Panda A, Zweiler JL (2000) Am J Physiol Heart Circ Phys 2778:2020
18. Liu X, Miller MJS, Joshi MS, Thomas DD, Lancaster JR (1998) Proc Natl Acad Sci USA 95:2175
19. Thomas DD, Liu X, Kantrow SP, Lancaster JR (2001) Proc Natl Acad Sci USA 98:355
20. Phoa N, Epe B (2002) Carcinogenesis 23:469
21. Palmer RMJ, Ferrige AG, Moncada S (1987) Nature 327:524
22. Cooper EE (1999) Biochim Biophys Acta 1411:290
23. Liu X, Miller MJS, Joshi MS, Sadowska LO, Krowicka H, Clark DA, Lancaster JR Jr (1998) J Biol Chem 273:18709
24. Augusto O, Bonini MG, Amanso AM, Linares E, Santos CX, Menezes SL (2002) Free Radic Biol Med 32:841
25. Goldstein S, Czapski G, Lind J, Merenyi G (2000) J Biol Chem 275:3031
26. Beckman JS (2001) Circ Res 89:295
27. Pryor WA, Squadrito GC (1995) Am J Physiol 268:L699
28. Kanazawa H, Shiraishi S, Okamoto T, Hirata K, Yoshikara J (1999) Am J Resp Crit Care Med 159:272
29. VonSonntag C (1980) Adv Carb Chem and Biochem 37:1
30. Jones GDD, O'Neill P (1990) Int J Radiat Biol 57:1123
31. Dizdaroglu M (1986) Biochem J 238:247
32. Mello Filho AC, Meneghini RD (1984) Biochim Biophys Acta 781:56
33. Lewis CD, Laemli UK (1982) Cell 29:171
34. George AM, Saboljev SA, Hart LE Cramp WA, Harris G, Hornsey S (1987) Brit J Cancer 55 (suppl VIII) 141
35. von Sonntag C (1987) Chemical basis of radiation biology. Taylor and Francis, New York, p 116
36. Steenken S (1989) Chem Rev 89:503
37. Dizdaroglu M (1992) Mutat Res 275:331
38. Dizdaroglu M, Jaruga P, Rodriguez H (2003) Oxidative damage to DNA mechanisms of product formation and measurement by mass spectrometry. In: Cutler R, Rodriguez H (eds) Critical reviews of oxidative stress and ageing. World Scientific, Danvers, p 165
39. Cheng KC, Cahill DS, Kasai H, Nishimura S, Loeb LA (1992) J Biol Chem 267:166
40. Dizdaroglu M (1993) FEBS Letts 315:1
41. Henderson JLP, Ryun J, Heinecke JW (1999) J Biol Chem 274:33440
42. Masuda M, Suzuki T, Friesen MD, Ravanat JL, Cadet J, Pignatelli B, Nishino H (2001) J Biol Chem 276(44):40486–40496
43. Shen Z, Mitra SN, Wu W, Chen Y, Yang Y, Qin J, Hazen SL (2001) Biochem 40:2041
44. Henderson JP, Ryun J, Williams MW, McCormick ML, Parks, Rolnaur LA, Heinecke JW (2001) Proc Nat Acad Sci 98:1631
45. Ravanat JL, Danki T, Duez P, Gremand E, Herbert K, Hofer T, Lasserre L, SaintPierre C, Favier A, Cadet J (2002) Carcinogenesis 23:1911

46. Suzuki T, Masuda M, Friesen MD, Fenet B, Ohshima H (2002) Nucleic Acids Res 30:2555
47. Niles JC, Wishnok JS, Tannenbaum SR (2001) Org lett 3:963
48. Suzuki T, Masuda M, Friesen MD, Ohshima H (2001) Chem Res Tox 14:1163
49. Burrows CJ, Muller JG, Kornyushyna O, Luo W, Duarte V, Leipold MD, David SS (2002) Env Health Persp 110:713
50. Chen H-JC, Chen Y-M, Wang TF, Wang KS, Shiea J (2001) Chem Res Tox 14:536
51. Masuda M, Nishino H, Ohshima H (2002) Chem Biol Interactions 139:187
52. Byun J, Mueller DM, Heinecke JW (1999) Biochem 38:2590
53. Niles JC, Wishnok JS, Tannenbaum SR (2001) J Am Chem Soc 123:12147
54. Lee SH, Oe T, Blair AI (2002) Chem Res Tox 15:300
55. Epe B, Ballmaier D, Roussyn I, Briviba K, Sies H (1996) Nucleic Acids Res 24:4105
56. Spencer JPE, Wong J, Jenner A, Aruoma OI, Cross CE, Halliwell B (1996) Chem Res Tox 9:1152
57. Oshima H, Bartsch H (1999) Methods Enzymol 301:40
58. Cadet J, Berger M, Douki T, Morin B, Raoul S, Ravanat JL, Spinelli S (1997) Biol Chem 378:1275
59. Ravanat JL, Martinez GR, Medeiros MMG, diMascio P, Cadet J (2004) Arch Biochem Biophys 423(1):23–30
60. DiMascio P, Wefers H, Do-Thy H-P, Lafleur MV, Sies H (1989) Biochim Biophys Acta 1007:151
61. Sawa T, Akaike T, Kida K, Fukushima Y, Takagi K, Maeda H (1998) Cancer Epidemiol Biomarkers Prev 7:1007
62. Kanazawa A, Sawa T, Akaik T, Maeda H (2000) Cancer Letts 156:51
63. Stone K, Uzieblo LJ, Marnett LJ (1990) Chem Res Tox 3:33
64. Stone K, Ksebati LJ, Marnett (1990) Chem Res Toxicol 3:467
65. Langouet S, Mican AN, Muller M, Fink SP, Marnett LJ, Muhle SA, Guengerich FP (1998) Biochem 37:5184
66. Marnett LJ (2002) Toxicol 181–182:219
67. Pandya GA, Moriya M (1996) Biochem 35:11487
68. Akasaka S, Guengerich FP (1999) Chem Res Tox 12:501
69. Kurtz AJ, Lloyd RS (2003) J Biol Chem 278:5970
70. Mistry N, Cooke MS, Griffiths HR, Lunec J (2003) Lab Invest 83:1
71. Kasai H, Iwamoto-Tanaka N, Fukada S (1998) Carcinogenesis 19:1459
72. Murata-Kamiya N, Kamiya H (2001) Nucleic Acids Res 29:3433
73. Braughler JM, Duncan LA, Chase RL (1986) J Biol Chem 261:10282
74. Halliwell B, Chirico S (1993) Am J Clin Nutr 57:715S and 724S
75. Bielski BMJ, Arudi RL, Sutherland MW (1983) J Biol Chem 258:4759
76. Parola M, Bellomo G, Robino G, Barrera G, Dianzani MU (1999) Antioxid Redox Signal 1:255
77. McCall MR, Frei B (1999) Free Radic Biol Med 26:1034
78. Meagher EA, Fitzgerald GA (2000) Free Radic Biol Med 28:1745
79. Roberts LJ, Morrow JD (2002) Cell Mol Life Sci 59:808
80. Levonen A-L, Landar A, Ramachandran A, Caeser EK, Dickinson DA, Zanoni G, Morrow JD, Darley-Usmar VM (2004) Biochem J 378:373–382
81. Reich EE, Montine TJ, Morrow JD (2002) Adv Exp Med Biol 507:519
82. Niziolek M, Korytowski W, Girotti AW (2003) Free Radic Biol Med 34:997
83. Schmitt D, Shen Z, Zhang R, Colles SM, Wu W, Salomom RG, Chen Y, Chisholm GM, Hazen SL (2001) Biochem 40:2041
84. Willson RL, Dunster CA, Forni LG (1985) Phil Trans Roy Soc Lond B311:545

85. Uchida K (2003) J Lip Prog 42:318
86. Weiss SJ, Peppin G, Ortiz X, Ragsdale C, Test ST (1985) Science 227:747
87. Greenacre S, Ischipoulos H (2001) Free Radic Res 34:541
88. Prutz WA, Siebert F, Butler J (1982) Biochim Biophys Acta 705:139
89. Davies MJ, Fu S, Wang H, Dean RT (1999) Free Radic Biol Med 27:1151
90. Griffiths HR (2000) Free Radic Res 33:S47
91. Berlett BS, Stadtman ER (1997) J Biol Chem 272:0313
92. Burchan PC, Kuhan YT (1996) Biochem Biophys Res Commun 220:996
93. Stadtman ER, Berlett BS (1991) J Biol Chem 266:17201
94. Stadtman ER, Levine RL (2003) Amino Acids 25:207
95. Stadtman ER (1993) Annu Rev Biochem 62:797
96. Creed D (1984) Photochem Photophys 39:577
97. Purdie JW (1969) Can J Chem 47:1037
98. Bonifacic M, Schafer K, Mockel H, Asmus KD (1975) J Phys Chem 71:1496
99. Bonifacic M, Asmus KD (1976) J Phys Chem 80:2426
100. Robinson S, Bevan R, Lunec J, Griffiths HR (1998) FEBS letts 430:297
101. Fu S, Gebicki S, Jessup W, Gebicki JM, Dean RT (1995) Biochem J 311:821
102. Fu SL, Dean RT (1997) Biochem J 324:41
103. Fu SL, Hick LA, Sheil MM, Dean RT (1995) Free Radic Biol Med 9:281
104. Singh A, Bell MJ, Korroll GW (1984) Radiolysis and photolysis of aqueous aerated tryptophan solutions. In: Bors W, Saran M, Tait D (eds) Oxygen radicals in chemistry and biology. de Gruyter, Berlin, p 491
105. Griffiths HR, Lunec J, Blake DR (1992) Amino Acids 3:183
106. Stuart-Audette M, Blouquit Y, Faraggi M, Sicard-Roselli C, Houee-Levin C, Jolles P (2003) Eur J Biochem 270:3565
107. Davies KJA (1997) J Biol Chem 262:9895
108. Kaur H, Fagerheim I, Grootveld M, Puppo A, Halliwell B (1998) Anal Biochem 172: 360
109. Chapman ALP, Senthihnohan R, Winterbourn CC, Kettle AJ (2000) Arch Biochem Biophys 377:95
110. Hazell LJ, Davies MJ, Stocker R (1999) Biochem J 339:489
111. Peppin G, Weiss SJ (1986) Proc Natl Acad Sci USA 83:4322
112. Fu X, Kassim SY, Parks WC, Heinecke JW (2003) J Biol Chem 278:28403
113. Yi D, Smythe GA, Blount BC, Duncan MW (1997) Arch Biochem Biophys 344:253
114. Dalen CJV, Winterbourn CC, Senthilomohan R, Kettle AJ (2000) J Biol Chem 275: 11638
115. Sampson JB, Ye Y, Rosen H, Beckman JS (1998) Arch Biochem Biophys 15:207
116. Eiserich JP, Cross CE, Jones AD, Halliwell B, van der Vliet A (1996) J Biol Chem 271: 19199
117. Ferger B, Themann C, Rose S, Halliwell B, Jenner P (2001) J Neurochem. 78:509
118. Nakagawa H, Takusagawa M, Arima H, Furukawa K, Kinoshita T, Ozawa T, Ikota N (2004) Chem Pharm Bull 52:146
119. Scott LJ, Russell GI, Nixon NB, Dawes PT, Mattey DL (1999) Biochem Biophys Res Comm 255:562
120. Schrammel A, Gorren AC, Schmidt K, Pfeiffer S, Mayer B (2003) Free Radic Biol Med 2003 34:1078
121. Klotz LO, Schroeder P, Sies H (2002) Free Radic Biol Med 33:737
122. Redmond RW, Gamlin JN (1999) Photochem Photobiol 70:391
123. Ikeda K, Nagai R, Sakamoto T, Sano H, Araki T, Sakata N, Nakayama H, Yoshida M, Ueda S, Horiuchi S (1998) J Immunol Methods 215:95

124. Nagaraj RH, Monnier VM (1995) Biochim Biophys Acta 1253:75
125. Uchida K (2003) Amino Acids 25:249
126. Luxford C, Morin B, Dean RT, Davies MJ (1999) Biochem J 344:125
127. Morin B, Davies MJ, Dean RT(1998) Biochem J 330:1059
128. Milligan JR, Aguilera JA, Ly A, Tran NQ, Hoang O, Ward JF (2003) Nucleic Acids Res 31:6258

The Handbook of Environmental Chemistry Vol. 2, Part O (2005): 63–75
DOI 10.1007/b101146

The Broad Spectrum of Responses to Oxidative Stress in Proliferating Cells

Kelvin J. A. Davies (✉)

Ethel Percy Andrus Gerontology Center and Division of Molecular & Computational Biology, The University of Southern California, 3715 McClintock Avenue, Los Angeles, CA 90089-0191, USA
kelvin@usc.edu

Abstract Proliferating mammalian cells exhibit a broad spectrum of responses to oxidative stress, depending on the stress level encountered. Very low levels of hydrogen peroxide, for example 3–15 μM, or 0.1–0.5 μmol 10^{-7} cells, cause a significant mitogenic response with 25–45% growth stimulation. Higher H_2O_2 concentrations of 120–150 μM, or 2–5 μmol 10^{-7} cells, cause a temporary growth-arrest that appears to protect cells from excess energy usage and DNA damage. After 2–4 h of temporary growth-arrest many cells begin to exhibit up to a 40-fold transient adaptive response in which genes for oxidant protection and damage repair are preferentially expressed: This transient adaptation is maximal at approximately 18 h after peroxide addition. The H_2O_2 originally added is metabolized within 30–40 min and if no more is added the cells will gradually de-adapt, so that at 36 h after the initial H_2O_2 stimulus they have returned to their original level of H_2O_2 sensitivity. At levels of H_2O_2 of 250–400 μM, or 9–14 μmol 10^{-7} cells, mammalian fibroblasts are not able to adapt, but instead enter a permanently growth-arrested state in which they appear to perform most normal cell functions, but never divide again. This state of permanent growth-arrest has often been confused with "cell death" in toxicity studies that have relied solely on cell proliferation assays as measures of viability. If the oxidative stress level is further increased to 0.5–1.0 mM H_2O_2, or 15–30 μmol 10^{-7} cells, apoptosis results. This oxidative stress-induced apoptosis involves nuclear condensation, loss of mitochondrial transmembrane potential, degradation/down regulation of mitochondrial mRNAs and rRNAs, and degradation/laddering of both nuclear and mitochondrial DNA. At very high H_2O_2 concentrations of 5.0–10.0 mM, or 150–300 µmol 10^{-7} cells and above, cell membranes disintegrate, proteins and nucleic acids denature, and necrosis swiftly follows. Cultured cells grown in 20% oxygen are essentially pre-adapted or pre-selected to survive under conditions of oxidative stress. If cells are instead grown in 3% oxygen, much closer to physiological cellular levels, they are more sensitive to an oxidative challenge but exhibit far less accumulated oxidant damage.

Keywords Oxidative Stress · Free radicals · Cell proliferation · Aging · Protein degradation · Apoptosis · Growth arrest

1
Introduction to Oxidative Stress

It has been said that, "a disturbance in the pro-oxidant/anti-oxidant systems in favor of the former may be denoted as an oxidative stress" [1]. Oxidative stress can result from increased exposure to oxidants or from decreased protection against oxidants; both problems may even occur simultaneously. This view of oxidative stress as an imbalance between oxidant exposure and oxidant protection is strongly supported by an extensive literature, e.g., [1–6]. Although oxygen is by no means the only oxidizing agent to which cells or organisms are exposed, it is certainly the most ubiquitous.

Ground-state (or triplet) molecular oxygen is the major form of element number 8 encountered environmentally, and it contains two unpaired electrons. In other words, molecular oxygen is a bi-radical [2, 3]. A small percentage of the oxygen in our environment actually exists in a singlet state, caused by absorption of electromagnetic energy. The two unpaired electrons in triplet molecular oxygen have parallel spins. Absorption of energy (e.g., from UV light) transiently "flips" one of these spins producing the short-lived, spin-paired, singlet oxygen species. Ground state molecular oxygen is only a moderately active oxidizing agent because the "spin restriction rule" prevents its full reduction by two spin-paired electrons [2, 3]. Since singlet oxygen can directly accept two spin-paired electrons it is a more powerful oxidizing agent than ground-state oxygen [2, 3]. Singlet oxygen is thought to be a significant mediator of photo-oxidative stress.

The spin-restriction rule causes most oxygen reduction reactions to proceed one-electron at a time [2, 3]. This series of one-electron oxygen reduction steps has become known [1–6] as the univalent pathway for oxygen reduction (Scheme 1). In the univalent pathway, oxygen is first reduced to the superoxide anion radical and then to hydrogen peroxide. A third electron reduction (with water elimination) produces the extremely powerful oxidant, hydroxyl radical ($HO^{\bullet}$). This dangerous univalent pathway finally grinds to a halt with the production of water, following the fourth-electron reduction. Ground state oxygen itself is a mild oxidizing agent; whereas singlet oxygen, hydrogen peroxide, and (especially) the hydroxyl radical are stronger oxidants. Superoxide is actually

The Univalent Pathway for Oxygen Reduction

$$O_2 \xrightarrow{e^-} O_2^{\cdot -} \xrightarrow[2H^+]{e^-} H_2O_2 \xrightarrow[OH^-]{e^-} {}^{\cdot}OH \xrightarrow[H^+]{e^-} H_2O$$

Scheme 1 $O_2^{\cdot -}$ = superoxide, H_2O_2 = hydrogen peroxide, $^{\cdot}OH$ = hydroxyl radical

a mild reducing agent that can donate an unpaired electron to selected cellular constituents, thus beginning a damaging free radical cascade [5].

Other, related, oxygen-based oxidants are also of biological importance [1–6]. These include the chlorinated oxygen products hypochlorous acid and hypochlorite (HOCl and OCl^-) produced by phagocytes as an antibacterial defense, and nitric oxide ($NO^{\cdot}$) used for blood vessel vasodilation; as well as the product of its reaction with superoxide, peroxynitrite ($ONOO^-$). Many of the initial products of cellular oxidation actually act as more powerful propagators of oxidative damage; these include lipid peroxides, oxidized proteins, and oxidized sugars. Finally, living organisms are exposed to many oxidizing environmental agents, foods, medications, and drugs. Significant environmental oxidants include ozone (O_3), various oxides of nitrogen (NO_X), numerous products of (industrial) partial combustion, and various pesticides and herbicides (e.g., paraquat). Some food items are direct oxidants produced by cooking, some are strong catalysts of oxidation (e.g., iron and copper), while others still may undergo slow oxidation by oxygen in the body (e.g., various sugars and sulfhydryls). Several medications and drugs are metabolized to form oxidizing agents (e.g., acetaminophen), while others actually catalyze oxygen radical production by mitochondria (e.g., doxorubicin or Adriamycin).

The oxidative stress inducing agents and conditions discussed above can cause damage to proteins, lipids, carbohydrates, and nucleic acids [1–6]. Thus cellular enzymes and structural proteins, membranes, simple and complex sugars, and DNA and RNA are all susceptible to oxidative damage [1–6]. Surviving an oxidizing environment is actually one of the greatest challenges faced by living organisms. The concept that over time we eventually surrender to a form of organic "rust" led to the free radical theory of aging [7, 8].

2 Antioxidant Defense and Repair Systems

The first-level of cellular responses to oxidative stress is to use antioxidant defense and repair system to minimize the damage that actually occurs, and to remove or repair whatever cellular components do get damaged [1–6, 8–11]. Although living organisms may eventually succumb to oxidant-induced aging, oxidatively damaged cellular constituents are actually maintained at very low levels for most of the life span. Minimal damage accumulation is achieved by multiple interacting systems of antioxidant compounds, antioxidant enzymes, damage removal enzymes, and repair enzymes. In addition, aerobic organisms maintain very tight control of both oxygen perfusion and cellular/tissue oxygen concentration.

Antioxidant compounds include the well-known vitamins C and E, ubiquinone, uric acid, and many others. Antioxidant compounds are "sacrificed" to oxidation in order to directly protect more important cellular components [1–4, 6]. Antioxidant enzymes, such as superoxide dismutases, glutathione peroxidases, and quinone reductases, act catalytically to convert oxidants to less reactive species [5, 6]. The essential cytoplasmic and nuclear proteolytic enzyme, proteasome recognizes and selectively degrades oxidized proteins [6, 8, 9]. Oxidized membrane lipids are recognized and selectively removed by lipases, particularly phospholipase A_2 [6, 8, 10]. Oxidatively modified DNA is subject to removal or excision repair by a wide series of DNA repair enzymes including endonucleases, glycosylases, polymerases, and ligases [6, 8, 11].

Last, but certainly not least, oxygen concentration is tightly controlled in multicellular organisms, directly limiting the possibility of oxidation. In human beings the respiratory and circulatory systems combine to perfuse tissues with oxygen, and remove carbon dioxide. Nevertheless, oxygen concentration falls from atmospheric levels (20%) in the lungs to only 2–5% in the tissues. In addition, the very low K_M of mitochondrial cytochrome oxidase for oxygen [12] assures that most oxygen in cells is actually bound by the cytochrome oxidase complex. Thus a major component of oxidative stress defenses is to keep actual cellular oxygen tension to a minimum. This fact is often overlooked by scientists performing cell culture experiments in which cells are directly exposed to atmospheric oxygen levels. By definition such cells have already been selected or adapted to be able to grow under conditions of very high oxidative stress.

3
Mitogenic Effects of Low Oxidant Concentrations

Exposure of dividing mammalian cells in culture to very low concentrations of oxidants actually stimulates cell growth and division (Scheme 2). This fascinating mitogenic effect is seen, for example, with exposure of fibroblasts to hydrogen peroxide at 3–15 μM, or 0.1–0.5 μmol 10^{-7} cells [13–15]. Presumably at such low concentrations H_2O_2 does not cause a true oxidative stress. In all probability a level of H_2O_2 in this range actually acts as a signaling agent for mitosis, although the mechanism is unknown. Interestingly, this growth-stimulatory effect of very low H_2O_2 concentrations is also seen with bacteria [16, 17] and yeast cells [18].

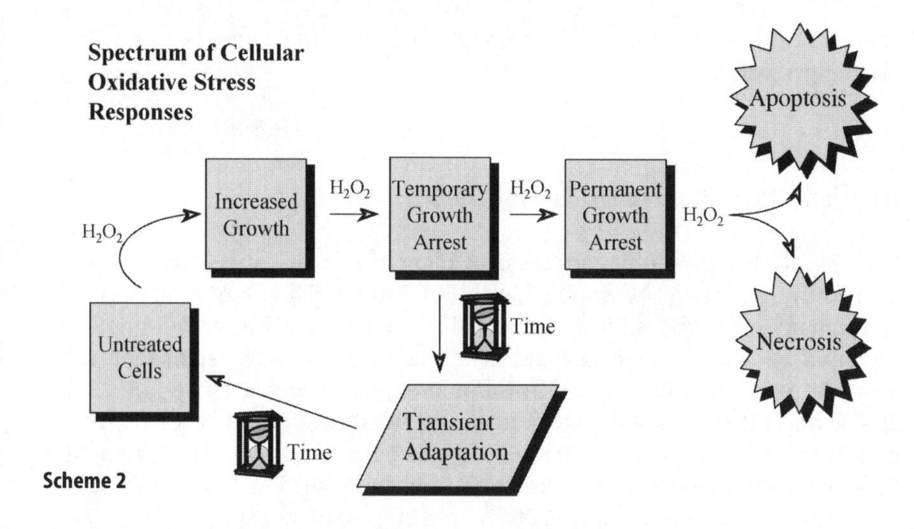

Scheme 2

4
Temporary Growth-Arrest as a Defense Against Oxidative Stress

Although 3 µM H_2O_2 is mitogenic, 120–150 µM H_2O_2, or 2–5 µmol 10^{-7} cells, typically causes a temporary growth-arrest (Scheme 2) in mammalian fibroblasts [15]. This temporary growth-arrest lasts for some 2–4 h, and appears to be caused by expression of the *gadd*45, *gadd153* [19–21], and *adapt*15 genes [22, 23]. Many investigators have treated all growth-arrest responses as toxic outcomes of oxidant exposure. I would like to suggest that temporary growth-arrest is actually a defense mechanism.

During peroxide-induced temporary growth-arrest, the expression of many housekeeping genes is halted, while expression of a select group of shock or stress genes is induced. This response can be viewed in analogy, as the "Medieval castle defense against toxicity." When attacked by an invading army the inhabitants of castles in medieval times would raise the drawbridge, lower the portcullis, conserve their resources, and hope that they could outlast their attackers. Similarly, I suggest, proliferating cells exposed to 10^{-6} M H_2O_2 shut-off expression of all but the most essential shock/stress genes, supercoil their DNA to protect it against oxidation, temporarily arrest divisional processes, and conserve resources for future use.

Interestingly, it does not matter which stage of the cell cycle fibroblasts are in, when a hydrogen peroxide stress is applied. At any point in the cell-cycle, application of 120–150 µM hydrogen peroxide, or 2–5 µmol 10^{-7} cells, simply stops cell-cycle progression for 2–4 h [22, 23]; after which the cells are at least temporarily synchronized. Low µM levels of H_2O_2 are easily metabolized (largely by glutathione peroxidases and catalases) during a 2–4 h temporary growth-arrest period so that when cells re-enter the growth cycle there is no longer a threat to counter. I propose that temporary growth-arrest is used as an effective counter measure against a wide variety of toxic agents, not just hydrogen peroxide.

5 Transient Adaptation to Oxidative Stress

Very important early studies on transient adaptation to oxidative stress were performed by Spitz et al. [24] and by Laval [25]. After 4–6 h of temporary growth-arrest many cells exposed to 120–150 µM hydrogen peroxide, or 2–5 µmol 10^{-7} cells, undergo further changes that can be characterized as transient adaptation to oxidative stress (Scheme 2). In the mammalian fibroblasts we [15, 22, 23, 26–31] and others [24, 25] have studied, maximal adaptation is seen approximately 18 h after initial exposure to hydrogen peroxide; i.e., some 12–14 h after they exit from temporary growth-arrest. In bacteria such as *E. Coli* and *salmonella* maximal adaptation is seen 20–30 min after oxidant exposure [16, 17, 26], whereas yeast cells require some 45 min for maximal adaptation [18, 26].

It is important to note that the adaptation referred to in this section simply means increased resistance to oxidative stress, as measured by cell proliferation capacity. Furthermore, the adaptation is transient, lasting some 18 h in mammalian fibroblasts, 90 min in yeast, and only 60 min in E. Coli. In our studies we have been especially careful to avoid selecting for pre-existing resistant cells in the population by repeatedly checking that transiently adapted cells can actually de-adapt.

In both procaryots and eucaryots, transient adaptation to oxidative stress depends upon transcription and translation. A large number of genes undergo altered expression during the adaptive response. Some genes are up-regulated, some are down-regulated, some are modulated early in the adaptation, while the expression of others is affected at later times. In mammalian fibroblasts we observe three broad "waves" of altered gene expression during adaptation; one at 0–4 h following H_2O_2 exposure, one at 4–8 h, and one at 8–12 h. Inhibiting either transcription or translation during the adaptive response greatly limits the development of increased H_2O_2 resistance. If both transcription and translation are inhibited, little or no adaptation will occur. Therefore, the transient adaptive response to oxidative stress depends largely on altered gene expression but partly on increased translation of pre-existing mRNAs. It further appears that message stabilization (for some mRNAs), increased message degradation (for other mRNAs), and altered precursor processing, are all involved in altered translational responses [6, 15, 18, 22, 23, 26–31].

Elegant studies in *E. Coli* and *salmonella* have shown that two particular regulons are responsible for many of the bacterial adaptive responses to oxidative stress: the oxyR regulon [32] and the soxRS regulon [33]. In mammalian cells no "master regulation molecules" have been found but at least 40 gene products are involved in the adaptive response. Several of the mammalian adaptive genes are involved in antioxidant defenses and others are damage removal or repair enzymes. Many classic shock or stress genes are involved very early in adaptive responses. As indicated in the section above, *gadd153*, *gadd45*, and *adapt15* play important roles in inducing temporary growth-arrest, which is a very important early portion of the adaptive response to oxidative stress [15, 19, 20–23]. The transcription factor, adapt66 (a *mafG* homologue) is probably responsible for inducing the expression of several other adaptive genes [28]. A number of other "adapt" genes have recently been discovered but their functions are not yet clear. One of these genes is the calcium-dependent *adapt33* [27], and another is *adapt73*, which appears to also be homologous to a cardiogenic shock gene called *PigHep3* [29]. *Adapt 78* has also been called *DSCR1* [30, 31] and, in addition to its induction during oxidative stress adaptation, now appears to also be involved in Down syndrome and Alzheimer's disease [34–37].

Numerous other genes have been shown to be inducible in mammalian cell lines following exposure to the relatively mild level of hydrogen peroxide oxidative stress that we find will cause transient adaptation. These include the protooncogenes c-*fos* and c-*myc* [36], c-*jun*, *egr*, and JE [37–39]. Similar oncogene induction has also been reported following exposure to *tert*-butylhydroperoxide [38, 39]. The induction of heme oxygenase by many oxidants, including mild peroxide stress, may have a strong protective effect, as proposed by Keyse and Tyrrell [40]. Other gene products that have been reported to be induced by relatively mild hydrogen peroxide stress in dividing mammalian cell cultures include: the CL100 phosphatase [41]; interleukin-8 [42]; catalase, glutathione peroxidase, and mitochondrial mangano-superoxide dismutase [43]; natural killer-enhancing factor-B [44]; mitogen-activated protein kinase [45]; and gamma-glutamyl transpeptidase [46]. Relatively low levels of nitric oxide have also been shown to induce expression of c-*jun* [47], c-*fos* [47,48], and zif/268 [48]. The list of oxidant stress-inducible genes is much longer than the space limitations of this review article will allow; apologies are extended to those investigators whose studies have not been cited here. It is, however, very important to note that many of the gene inductions reported in this paragraph have not actually been studied in an adaptive cell culture model. Thus, although many of the genes discussed in this paragraph may appear to be good candidates for involvement in transient adaptation to oxidative stress, their actual involvement in the adaptive process remains to be tested.

In concluding this section, a note must be made of important studies involving permanent (or stable) oxidative stress resistance. Investigators have chronically exposed cell lines to various levels of oxidative stress over several generations, and have selected for pre-existing or mutant phenotypes that confer oxidative stress resistance. Several such studies have reported dramatic increases in catalase activity (relative to the parent population), such as the 20-fold higher levels reported by Spitz et al. [49]. Stable oxidative stress resistance may tell us a great deal about the importance of individual genes to overall cellular survival, and the value of such cell lines should not be underestimated. It should be clear, however, that transient adaptive responses in gene expression, and stable stress resistance are quite different entities.

6 Permanent Growth-Arrest

If dividing mammalian cells are exposed to higher concentrations of H_2O_2 than those that cause temporary growth-arrest and transient adaptation, they can be forced into a permanently growth-arrested state (Scheme 2). Thus, cells exposed to H_2O_2 concentrations of 250–400 μM, or 9–14 μmol 10^{-7} cells, will never divide again [15].

Countless cytotoxicity studies have measured "cell death" by loss of proliferative capacity. These include studies with oxidizing agents, alkylating agents, heavy metals, various forms of radiation, etc. The common assumption of such investigations is that loss of divisional competence equals cell death. It is, indeed, true that many cells exposed to sufficient stress will both stop dividing and die (see next two sections). It is now clearly incorrect to conclude, however, that all permanently growth-arrested cells will die as a direct consequence of toxicant exposure. In other words, cells may survive an oxidative stress yet be permanently growth-arrested [15, 50].

Studies of both cell populations and individual cells have revealed that cultured mammalian cells (normal doubling time of 24–26 h) can survive for many weeks, or even months, following exposure to 100–200 μM H_2O_2 without dividing again [15]. In the past, studies estimating percent cell viability by growth curves or colony formation measurements alone have completely missed this permanent growth-arrest response. Arrested cells still exclude trypan blue, maintain membrane ionic gradients, utilize oxygen, and make ATP; in other words they are alive but do not divide. Interestingly, such permanently growth-arrested cells may make good cellular models for certain aging processes [50]. Whether the cessation of proliferation induced by oxidative stress (or other stressful exposures) in some way mimics the loss of divisional competence typical of terminally differentiated cells remains to be seen.

7 Cell Suicide or Apoptosis

A fraction of cells exposed to higher H_2O_2 concentrations of 0.5–1.0 mM, or 15–30 μmol 10^{-7} cells, will enter the apoptotic pathway (Scheme 2). The mechanism of oxidative stress-induced apoptosis appears to involve loss of mitochondrial transmembrane potential [51], release of cytochrome *c* to the cytoplasm [52], loss of bcl-2 [53], down-regulation and degradation of mitochondrially encoded mRNA, rRNA, and DNA [54–56], and diminished transcription of the mitochondrial genome [57]. Current thinking about toxicant-induced apoptosis suggests that, in multicellular organisms, the repair of severely damaged cells represent a major drain on available resources for the "host" organism. To avoid this difficulty, it is suggested, individual cells within organisms (or organs or tissues) will "sacrifice" themselves for the common good of the many. Apoptotic cells are characterized by "blebbing", nuclear condensation, and DNA laddering [58]. Such cells are engulfed by phagocytes, which prevent an immune reaction and recycle usable nutrients [51–60].

Certain toxicants, such as staurosporine, can induce widespread apoptosis in fibroblast cell cultures, with greater than 80% cell suicide [55, 57]. Even higher levels of apoptosis (98% or more) are routinely observed upon withdrawal of IL-2 from in vitro cultures of T-lymphocytes [55, 57]. In contrast, the highest levels of apoptosis we can induce in fibroblasts by H_2O_2 exposure never exceed 30–40% [55]. The cause of such disparity is not at all clear but it may suggest either slightly divergent pathways to apoptosis, or different efficiencies of repair processes for various toxicants. The apoptotic pathway may be very important in several age-related diseases such as Parkinson's, Alzheimer's, and sarcopenia. Importantly, many mitochondrial changes, including loss of membrane potential [51] and down-regulation and degradation of mitochondrial polynucleotides [54–56], are common to apoptosis directly induced by oxidants and to apoptosis induced by staurosporine or IL-2 withdrawal. Furthermore, over expression of the p53 gene has been seen to result in induction of multiple "redox-related" gene products, and initiation of apoptosis [59]. These observations support a strong involvement of oxidative stress mechanisms in general apoptotic pathways.

8 Cell Disintegration or Necrosis

At even higher concentrations of hydrogen peroxide, e.g., 5.0–10.0 mM, or 150–300 μmol 10^{-7} cells, most cells simply disintegrate or become necrotic (Scheme 2). Membrane integrity breaks down at such high oxidant stress levels and all is then lost [15, 61]. Studies that purport to examine cellular responses to 10.0 mM H_2O_2 in mammalian fibroblasts are really not looking at the responses of cells, but rather at the release of components those cells originally contained. At a high enough level of oxidative stress (e.g., over 10 mM H_2O_2) all mammalian cell cultures will turn into a necrotic "mess" [15]. Oxidation-induced necrosis may play a significant role in ischemia-reperfussion injures such as heart attacks, strokes, ischemic bowel disease, and macular degeneration. Unfortunately, necrotic cells cause inflammatory responses in surrounding tissues. Such secondary inflammation (also an oxidant stress) may be particularly important in many auto-immune diseases such as rheumatoid arthritis and lupus.

9 Summary

Oxidative stress causes a very wide spectrum of genetic, metabolic, and cellular responses. Only necrosis, which is the most extreme outcome, involves direct cell destruction. Most oxidative stress conditions that cells might actually encounter will modulate gene expression, may stimulate cell growth, or may cause a protective temporary growth-arrest and transient adaptive response. Even the apoptotic response seen at high oxidant exposures appears to protect surrounding cells and issues.

Cultured cells grown in 20% oxygen are essentially pre-adapted or pre-selected to survive under conditions of oxidative stress. If cells are instead grown in 3% oxygen, much closer to physiological cellular levels, they are more sensitive to an oxidative challenge but exhibit far less accumulated oxidant damage.

Cellular redox state is now well-known as a mediator of various signal-transduction pathways [61]. Finally, it now appears that antioxidant compounds, such as vitamin E, may also control the expression of oxidative stress-responsive genes, and regulate key, post-translational, phosphorylation and dephosphorylation steps that regulate signal transduction pathways [60]. This raises the exciting possibility of a very fine tuning of oxidant stress responses, by both positive and negative feedback loops.

Naturally, the full spectrum of responses described in this review can only be seen in proliferating cells such as those from the liver, intestinal lining, skin, etc. Nevertheless, even non-proliferating cells from such organs as brain, heart, and skeletal muscle will exhibit altered gene expression and metabolism at tolerated levels of oxidative stress, and are subject to apoptosis or necrosis at higher stress levels. The full spectrum of cellular responses to oxidative stress must, therefore, be considered when applying the knowledge gained in cell-culture studies to human diseases, and aging.

References

1. Sies H (1985) Oxidative stress: introductory remarks. In: Sies H (ed) Oxidative stress. Academic Press, London, pp 1–8
2. Cadenas E (1989) Biochemistry of oxygen toxicity. Annu Rev Biochem 58:79–110
3. Cadenas E (1995) Mechanisms of oxygen activation and reactive oxygen species detoxification. In: Ahmad S (ed) Oxidative stress and antioxidant defenses in biology. Chapman & Hall, New York, pp 1–61
4. Halliwell B, Gutteridge JMC (1989) Free radicals in biology and medicine. Oxford University Press, Oxford
5. Fridovich I (1995) Superoxide radical and superoxide dismutases. Annu Rev Biochem 64:97–112
6. Davies KJA (1995) Oxidative Stress: the paradox of aerobic life. Biochem Soc Symp 61:1–31
7. Harman D (1956) Aging: a theory based on free radical and radiation chemistry. J Gerontol 11:298–300
8. Pacifici RE, Davies KJA (1991) Protein, lipid, and DNA repair systems in oxidative stress: the free radical theory of aging revisited. Gerontology 37:166–180
9. Grune T, Reinheckel T, Davies KJA (1997) Degradation of oxidized proteins in mammalian cells. FASEB J 11:526–534

10. Van Kuijk FJGM, Sevanian A, Handelman GJ, Dratz EA (1987) A new role for phospholipase A_2: protection of membranes from lipid peroxidation damage. Trends Biochem Sci 12:31–34
11. Wang D, Kreutzer DA, Essigmann JM (1998) Mutagenicity and repair of oxidative DNA damage: insights from studies using defined lesions. Mutation Research 400:99–115
12. Chance B, Sies H, Boveris H (1979) Hydroperoxide metabolism in mammalian organs. Physiol Rev 59:527–605
13. Burdon RH, Rice-Evans C (1989) Free radicals and the regulation of mammalian cell proliferation. Free Radic Res Commun 6:345–358
14. Burdon RH (1995) Superoxide and hydrogen peroxide in relation to mammalian cell proliferation. Free Radic Biol Med 18:775–794
15. Wiese AG, Pacifici RE, Davies KJA (1995) Transient adaptation to oxidative stress in mammalian cells. Arch Biochem Biophys 318:231-240
16. Demple B, Halbrook J (1983) Inducible repair of oxidative DNA damage in *Escherichia coli.* Nature 304:466–468
17. Christman MF, Morgan RW, Jacobson FS, Ames BN (1985) Positive control of a regulon for a defense against oxidative stress and heat shock proteins in *Salmonella typhimurium.* Cell 41:753–762
18. Davies JMS, Lowry CV, Davies KJA (1995) Transient adaptation to oxidative stress in yeast. Arch Biochem Biophys 317:1–6
19. Fornace AJ, Jr, Nebert DW, Hollander MC, Luethy JD, Papathanasiou M, Fargnoli J, Holbrook NJ (1989) Mammalian genes coordinately regulated by growth arrest signals and DNA-damaging agents. Mol Cell Biol 9:4196–4203
20. Bartlett JD, Luethy JD, Carlson SG, Sollott SJ, Holbrook NJ (1992) Calcium ionophore A23187 induces expression of the growth arrest and DNA damage inducible CCAAT/ enhancer-binding protein (C/EBP)-related gene, gadd153. Ca2+ increases transcriptional activity and mRNA stability. J Biol Chem 267:20465–20470
21. Fornace AJ, Jr, Jackman J, Hollander MC, Hoffman-Liebermann B, Liebermann DA (1992) Genotoxic-stress-response genes and growth-arrest genes. gadd, MyD, and other genes induced by treatments eliciting growth arrest. Annals NY Acad Sci 663:139–153
22. Crawford DR, Schools GP, Salmon SL, Davies KJA (1996) Hydrogen peroxide induces the expression of *adapt15*, a novel RNA associated with polysomes in hamster HA-1 cells. Arch Biochem Biophys 325:256–264
23. Crawford DR, Schools GP, Davies KJA (1996) Oxidant-inducible *adapt15* is associated with growth arrest and DNA damage-inducible *gadd153* and *gadd45.* Arch Biochem Biophys 329:137–144
24. Spitz DR, Dewey WC, Li GC (1987) Hydrogen peroxide or heat shock induces resistance to hydrogen peroxide in Chinese hamster fibroblasts. J Cell Physiol 131:364–373
25. Laval F (1988) Pretreatment with oxygen species increases the resistance to hydrogen peroxide in Chinese hamster fibroblasts. J Cell Physiol 201:73–79
26. Crawford DR, Edbauer-Nechamen C, Lowry CV, Salmon SL, Kim YK, Davies JMS, Davies KJA (1994) Assessing gene expression during oxidative stress. Methods in enzymology: oxygen radicals in biological systems, part D 234:175–217
27. Wang Y, Crawford DR, Davies KJA (1996) *Adapt33*, a novel oxidant-inducible RNA from Hamster HA-1 cells. Arch Biochem Biophys 332:255–260
28. Crawford DR, Leahy KP, Wang Y, Schools GP, Kochheiser JC, Davies KJA (1996) Oxidative stress induces the levels of a *mafG* homolog in Hamster HA-1 cells. Free Radical Biol Med 21:521–525
29. Crawford DR, Davies KJA (1997) Modulation of a cardiogenic shock inducible RNA by chemical stress: *adapt73*/PigHep3. Surgery 121:581–587

30. Crawford DR, Leahy KP, Abramova N, Lan L, WangY, Davies KJA (1997) Hamster *adapt78*, mRNA is a Down Syndrome critical region homologue that is inducible by oxidative stress. Arch Biochem Biophys 342:6–12
31. Leahy KP, Davies KJA, Dull M, Kort JJ, Lawrence KW, Crawford DA (1999) *Adapt78*, a stress-inducible mRNA, is related to the glucose-regulated family of genes. Arch Biochem Biophys 368:67–74
32. Storz G, Tartaglia LA (1992) OxyR: a regulator of antioxidant genes. J Nutr 122:627–630
33. Greenberg JT, Monach P, Chou JH, Josephy PD, Demple B (1990) Positive control of a global antioxidant defense regulon activated by superoxide-generating agents in Escherichia coli. Proc Natl Acad Sci USA 87:6181–6185
34. Ermak G, Morgan TE, Davies KJA (2001) Chronic overexpression of the calcineurin inhibitory gene D*SCR1 (Adapt78*) is associated with Alzheimer's disease. J Biol Chem 276:38787–38794
35. Ermak G, Harris C, Davies KJA (2002) The *DSCR1 (Adapt78)* Isoform 1 protein, "Calcipressin 1" inhibits calcineurin and protects against acute calcium-mediated stress damage, including transient oxidative stress. FASEB J 16:814–824
36. Crawford D, Zbinden I, Amstad P, Cerutti PA (1988) Oxidant stress induces the proto-oncogenes c-fos and c-myc in mouse epidermal cells. Oncogene 3:27–32
37. Shibanuma M, Kuroki T, Nose K (1988) Induction of DNA replication and expression of proto-oncogene c-myc and c-fos in quiescent Balb/3T3 cells by xanthine/xanthine oxidase. Oncogene 3:17-21
38. Muehlematter D, Ochi T, Cerutti P (1989) Effects of tert-butyl hydroperoxide on promotable and non-promotable JB6 mouse epidermal cells. Chem Biol Interact 71:339–352
39. Nose K, Shibanuma M, Kikuchi K, Kageyama H, Sakiyama S, Kuroki T (1991) Transcriptional activation of early-response genes by hydrogen peroxide in a mouse osteoblastic cell line. Eur J Biochem 201:99–106
40. Keyse SM, Tyrrell RM (1989) Heme oxygenase is the major 32-kDa stress protein induced in human skin fibroblasts by UVA radiation, hydrogen peroxide, and sodium arsenite. Proc Natl Acad Sci USA 86:99–103
41. Keyse SM, Emslie EA (1992) Oxidative stress and heat shock induce a human gene encoding a protein-tyrosine phosphatase. Nature 359:644–647
42. DeForge LE, Preston AM, Takeuchi E, KenneyJ, Boxer LA, Remick DG (1993) Regulation of interleukin 8 gene expression by oxidant stress. J Biol Chem 268:25568–25576
43. Shull S, Heintz NH, Periasamy M, Manohar M, Janssen YM, Marsh JP, Mossman BT (1991) Differential regulation of antioxidant enzymes in response to oxidants. J Biol Chem 266:24398–24403
44. Kim AT, Sarafian TA, Shau H (1997) Characterization of antioxidant properties of natural killer-enhancing factor-B and induction of its expressionby hydrogen peroxide. Toxicol Appl Pharmacol 147:135–142
45. Guyton KZ, Liu Y, Gorospe M, Xu Q, Holbrook NJ (1996) Activation of mitogen-activated protein kinase by H2O2. Role in cell survival following oxidant injury. J Biol Chem 271:4138–4142
46. Kugelman A, Choy HA, Liu R, Shi MM, Gozal E, Forman HJ (1994) gamma-Glutamyl transpeptidase is increased by oxidative stress in rat alveolar L2 epithelial cells. Amer J Resp Cell Mol Biol 11:586–592
47. Janssen YM, Matalon S, Mossman BT (1997) Differential induction of c-fos, c-jun, and apoptosis in lung epithelial cells exposed to ROS or RNS. Amer J Phys 273:L789–L796
48. Morris BJ (1995) Stimulation of immediate early gene expression in striatal neurons by nitric oxide. J Biol Chem 270:24740–24744

49. Spitz DR, Elwell JH, Sun Y, Oberley LW, Oberley TD, Sullivan SJ, Roberts RJ (1990) Oxygen toxicity in control and hydrogen H_2O_2-resistant Chinese hamster fibroblast cell lines. Arch Biochem Biophys 279:249–260
50. Chen Q, Ames BN (1994) Senescence-like growth-arrest induced by hydrogen peroxide in human diploid fibroblast F65 cells. Proc Natl Acad Sci USA 91:4130–4134
51. Zamzani NT, Hirsch B, Dallaporta P, Petit X, Kroemer G (1997) Mitochondrial implication in accidental and programmed cell death: apoptosis and necrosis. J Bioenerg Biomembr 29:185–193
52. Reed JC (1997) Cytochrome *c*: can't live with it – can't live without it. Cell 91:559–562
53. Kane DJ, Sarafian TA, Anton R, Hahn H, Gralla EB, Valentine JS, Ord T, Bredesen DE (1993) Bcl-2 inhibition of neural cell death: Decreased generation of reactive oxygen species. Science 262:1274–1277
54. Crawford DR, Wang Y, Schools GP, Kochheiser J, Davies KJA (1997) Down-regulation of mammalian mitochondrial RNAs during oxidative stress. Free Radic Biol Med 22:551–559
55. Crawford DR, Lauzon RJ, Wang Y, Mazurkiewicz JE, Schools GP, Davies KJA (1997) 16S mitochondrial ribosomal RNA degradation is associated with apoptosis. Free Radic Biol Med 22:1295–1300
56. Crawford DA, Abramova NE, Davies KJA (1998) Oxidative stress causes a general, calcium dependent degradation of mitochondrial polynucleotides. Free Radic Biol Med 25:1106–1111
57. Kristal BS, Chen J, Yu BP (1994) Sensitivity of mitochondrial transcription to different free radical species. Free Radic Biol Med 16:323–329
58. Ankarcrona M, Dypbukt JM, Bonfoco E, Zhivotosky B, Orrenius S, Lipton S, Nicotera P (1995) Glutamate-induced neuronal death: a succession of necrosis or apoptosis depending on mitochondrial function. Neuron 15:961–973
59. Polyak K, Xia Y, Zweier JL, Kinzler KW, Vogelstein B (1997) A model for p53-induced apoptosis. Nature 389:300–305
60. Farber JL, Kyle ME, Coleman JB (1990) Mechanisms of cell injury by activated oxygen species. Lab Invest 62:670–679
61. Sen CK, Packer L (1996) Antioxidant and redox regulation of signal transduction. FASEB J 10:709–720
62. Ricciarelli R, Tasinato A, Clement S, Ozer NK, Boscoboinik D, Azzi A (1998) alpha-Tocopherol specifically inactivates cellular protein kinase C alpha by changing its phosphorylation state

Part B
Primary Antioxidative Defense

The Handbook of Environmental Chemistry Vol. 2, Part O (2005): 77–90
DOI 10.1007/b101147

Low Molecular Weight Antioxidants

Tilman Grune[1] (✉) · Peter Schröder[1] · Hans K. Biesalski[2]

[1] Research Institute of Environmental Medicine, Heinrich Heine University Düsseldorf, Auf'm Hennekamp 50, 40225 Düsseldorf, Germany
Tilman.Grune@uni-duesseldorf.de

[2] Biological Chemistry and Nutrition, University of Hohenheim70, 70593 Stuttgart, Germany

Abstract Low molecular weight antioxidants are an important part of the antioxidative defense mechanisms of cells and organisms. This chapter gives a short overview of the actions of the main antioxidants, including uric acid, ubichinones, lipoic acid, vitamins C and E, carotenoids, and phenolic compounds. The antioxidative properties of these endogeous and nutritional compounds are discussed in this chapter. However, it is critically discussed whether the antioxidative properties of some of these compounds are really important in vivo.

Keywords Antioxidants · Nutrition · Vitamins

1 Introduction

Free radicals and oxidants produced are trapped in cells and tissues by numerous compounds. Together with enzymatic antioxidative mechanisms, low molecular weight compounds also interact with oxidizing species and detoxify them. Low molecular weight antioxidants are compounds that interact with free radicals or oxidants and deactivate them. They are very often used up by this process. Antioxidants have different chemical structures and are either water- or lipid-soluble, and are either synthesized within the human body or are taken in with the diet.

Since numerous components of the body interact with free radicals and oxidants, it is very often difficult to differentiate whether the interaction of a given compound with an oxidant is a protecting reaction or is due to damage of a body constituent. For example, it is well known that carbohydrates react with reactive oxygen species [1]. Since glucose is present in the body at up to 5 mM, such interactions are likely to occur. On the other hand, this is surely not considered the main function of glucose. Therefore, a number of biological compounds that are present in high concentrations interact with the formed oxidants and detoxify them. Not all of these components are low molecular weight compounds but also include proteins and melanins [2, 3].

This chapter will concentrate on the important functions of the known antioxidants, including the classical vitamins, in the human or mammalian body.

2 Endogenous Antioxidants

2.1 Uric Acid

Uric acid (Fig. 1) in the human body is the end product of purine metabolism. It is produced by the enzymatic conversion of hypoxanthine to xanthine and then to uric acid. The enzyme involved here is xanthine oxidoreductase. This enzyme exists in two forms: xanthine dehydrogenase and xanthine oxidase. The latter is able to produce oxidizing species during enzymatic catalysis [4]. In most organisms uric acid is enzymatically degraded by an enzyme called urate

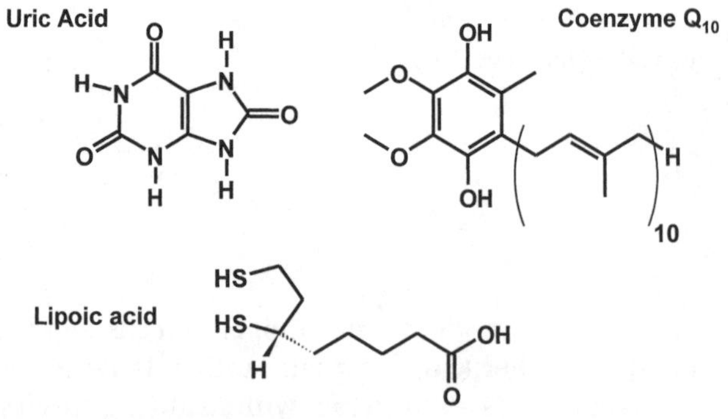

Fig. 1 Structures of endogenous antioxidants

oxidase. The product of this conversion is allantoin. The genetic information of this enzyme is still present in the human genome, but is silenced due to a mutation [5]. Therefore, blood plasma concentrations of uric acid in humans are 200–400 μM. High concentrations of uric acid result in the formation of crystals. This happens in gout patients [6].

Uric acid is a strong scavenger of several oxidants, including the hydroxyl radical, singlet oxygen, ozone, and several organic and nitrogen based oxidants, like the peroxyl radical [7]. Several reaction products of the urate molecule are formed, including the urate radical and allantoin [8]. In several diseases, including Wilson's disease and rheumatoid arthritis, the increase of allantoin in patients was considered to be a measure of enhanced oxidative stress [8]. The question of using allantoin to measure oxidative stress in human subjects has to be addressed carefully, since allantoin might also result from the diet.

Due to the high concentration of urate in vivo, the interaction with reactive oxygen and some nitrogen species is likely to play a role in the antioxidative protection of the hydrophilic environment.

2.2 Coenzyme Q

Coenzyme Q or ubiquinone is synthesized in the human body and also taken up with the diet. The importance of dietary coenzyme Q is still uncertain. The endogenous ubiquinone consists of 10 isoprene units in the hydrophobic side chain, whereas nutritional components might contain various side chain lengths from 6 to 10 isoprene subunits.

Coenzyme Q is an essential part of the mitochondrial electron transport chain and is therefore located in the inner mitochondrial membrane, but it is also present in other cell membranes and in lipoproteins. Coenzyme Q is able to undergo oxidation reactions via a radical intermediate – ubisemiquinone. In vitro, ubiquinol is able to scavenge several free radical intermediates of lipid peroxidation and may also interact with the tocopheryl radical. This recycling of vitamin E might be of potential importance, especially since the tocopherols interact at higher rates with lipid peroxidation products than does coenzyme Q. However, although coenzyme Q is able to interact with several oxidants, the role of this antioxidant in vivo remains obscure. The recycling of oxidized and reduced coenzyme Q in mitochondria enables this compound to have some antioxidative effects there. On the other hand, such recycling has also been suggested as the source of free radicals in vivo.

2.3 Lipoic Acid

Lipoic acid (or thioctic acid) is an essential part of the decarboxylating enzymes of α-ketoacids. It exists in both an oxidized and a reduced form, with both showing antioxidant properties. In vitro lipoic acid is able to interact with numerous radicals and oxidants. Dihydrolipoic acid may reduce oxidized glutathione and can regenerate vitamin E. In vivo the enzyme lipoamide dehydrogenase uses reduced equivalents ($NADPHH^+$ or $NADHH^+$) to produce the reduced form, dihydrolipoic acid. Since the levels of lipoic acid in either the oxidized or the reduced form in vivo are low, it is uncertain how important it is as an antioxidant generally. However, the ability to interact with oxidants and the antioxidative defense machineries at several points makes lipoic acid an important candidate for supplementation strategies. That it penetrates the blood-brain barrier contributes further to that fact.

In the body lipoic acid is metabolized via the β-oxidation pathway. Some of the products also have antioxidative properties.

2.4 Others

Numerous other endogenous compounds are described as having antioxidative properties. One of these, for example, is bilirubin. As a breakdown product of heme, bilirubin is permanently being formed in the body. It is water-insoluble and is transported while bound to proteins. In vitro it is a scavenger of peroxyl radicals and singlet oxygen. However, whether it also fulfils this function in vivo remains obscure. On the other hand, it should be mentioned that bilirubin is also able to form singlet oxygen in the presence of light and that bilirubin in newborn babies is toxic to the nervous system.

Several hormones have strong antioxidant properties. These include steroid hormones, especially estrogens. These are able to inhibit lipid peroxidation as chain-breaking antioxidants. The higher concentration of estrogens in women is considered to be the reason for higher antioxidative properties of female tissue and the decreased aging rate of women before the menopause.

Melatonin, a hormone of the pineal gland which is responsible for the circadian rhythms, is also a strong antioxidant in vitro. However, the concentrations of this hormone in vivo are far below the effective range.

The dipeptide carnosine (β-alanyl-histidine) is able to develop antioxidative properties, at least in vitro [9]. The dipeptide, however, does not enhance the metal-catalyzed free radical production, as does histidine. The high concentrations of this metabolite enable it to act as an antioxidant and as an anti-glycoxidation compound. This makes it an interesting substance in several age-related studies [10].

It is important to notice that most of the compounds mentioned in this part of the chapter might have only limited direct effects in vivo, but that several of them might be involved in the induction of antioxidative enzymes or might be effective in interaction with other antioxidants.

3
Nutritional Antioxidants

3.1
Vitamin C

Vitamin C (ascorbic acid, Fig. 2) is a water-soluble vitamin that dissociates at physiological pH. It is essential as a cofactor of several enzymes, including proline hydroxylase and lysine hydroxylase. Scurvy is known as the result of malnutrition with ascorbic acid. This vitamin deficiency is characterized by instable collagen. This results from insufficient hydroxylation of collagen molecules. Besides this, ascorbic acid has a function as an antioxidant.

Ascorbic acid is a strong reducing agent and is able to reduce Fe^{3+} to Fe^{2+}. This might be important for the role of ascorbic acid in iron uptake in the intestine. Several nitroso-compounds are also reduced by ascorbic acid. This contributes to detoxifying reactions against these substances. These reactions produce the ascorbyl radical, an unstable product. The metal-catalyzed oxidation (in vivo most likely iron- or copper-mediated) of ascorbic acid produces reactive oxygen species [11].

The concentration of ascorbate in the human plasma is 25 μM and above. Cells take up ascorbate by a Na^+-coupled uptake mechanism against a concentration gradient. A marked stereo-selectivity for L-ascorbic acid relative to D-isoascorbic acid in their cellular transport has been shown by Franceschi et al. [12]. The same transport is also important in the intestine. The nutritional supply of ascorbic acid is the only source for this vitamin in humans, primates, and guinea pigs. Other mammals are able to produce ascorbic acid. There exists sufficient evidence for an active role of ascorbate as an antioxidant in vivo. Decreased ascorbic acid will increase lipid peroxidation and decrease vitamin E and is connected with oxidative DNA damage. The supply of ascorbate in some cases will reduce the amount of oxidative damage in diseases that

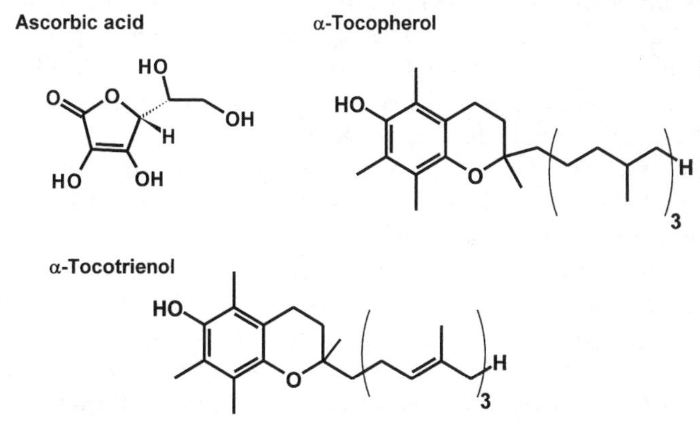

Fig. 2 Structures of vitamins

are connected to increased oxidant production. Oxidized ascorbate can be recycled by NADH- or GSH-mediated enzymatic activities. Whether ascorbate in vivo also acts as a prooxidant is still unknown. The prooxidant action of ascorbate is connected with the availability of free catalytic active metals. However, the availability of free iron and copper is very limited due to an effective sequestering apparatus. Ascorbate on the other hand is able to promote iron sequestering via ferritin [13]. Therefore, even high concentrations of ascorbic acid are non-toxic. The excess concentrations are simply excreted from the body.

The current recommended dietary allowance (RDA) for vitamin C (L-ascorbate) for adult non-smoking individuals is 60 mg day^{-1}, which is based on a mean requirement of 46 mg day^{-1} to prevent the deficiency disease scurvy [14]. It was shown that ascorbate uptake by neutrophils, monocytes, and lymphocytes saturate with a daily supplementation of 100 mg ascorbate and that these cells contain concentrations at least 14-fold higher than those in plasma. Bioavailability was complete for 200 mg of vitamin C as a single dose.

Biomarkers of oxidative damage to DNA bases have given no compelling evidence that ascorbate supplements might decrease the levels of oxidative DNA damage in vivo, except perhaps in subjects with very low vitamin C intakes [15]. Similarly, there is no conclusive evidence from studies of strand breaks, micronuclei, or chromosomal aberrations for a protective effect of vitamin C [16]. Furthermore there is limited evidence that supplements of vitamin C might have beneficial effects in disorders of vascular function, and that diet-derived vitamin C may decrease gastric cancer incidence in certain populations. It is even less clear whether the antioxidant or other properties of ascorbate are responsible for these two actions [15]. It is also possible that the amount of vitamin C required to prevent scurvy is not sufficient to optimally protect against these diseases.

3.2 Vitamin E

Without doubt vitamin E is the most important lipid-soluble chain-breaking antioxidant in the human body. It is able to interact with lipid peroxyl radicals and is also able to react with singlet oxygen. The role of vitamin E as an antioxidant in vivo has been established several times by measuring tissue lipid peroxidation in vitamin E-deficient or in vitamin E-supplemented animals.

The term "vitamin E" does not refer to a single compound, but to a group of substances all able to interfere with lipid peroxidation. All these compounds contain the chromanol ring, although with several substituents in the positions C_5, C_7, or C_8. Either hydrogen atoms or methyl groups can be located there. This results in the so-called α-, β-, γ-, or δ-derivatives. The side chain might be either saturated or contain three unsaturated bonds. This divides the compounds into the tocopherols or the tocotrienols. Since the side chain contains three asymmetric carbon atoms, eight optical isomers are formed. All these compounds taken together form vitamin E. All these compounds have antioxidant properties although to a different degree. In the human body the RRR-α-tocopherol is the most important and active compound.

Malnutrition with vitamin E results in a diffuse array of different symptoms, including muscle degeneration, sterility and hemolysis. These symptoms develop only after long periods of vitamin E deficiency [17]. The only known human disease caused by vitamin E deficiency is ataxia with vitamin E deficiency (AVED), which occurs in somewhat different clinical presentations. This disease shares very similar clinical phenotypes with Friedreich's ataxia [18]. Similar symptoms, referable to diminished absorption of the liposoluble vitamin E, are associated in fat malabsorption situations and in a-beta-lipoproteinemia [19]. Delayed-onset ataxia in mice lacking α-tocopherol transfer protein has been recently demonstrated [20]. Very high concentrations of vitamin E may interfere with platelet aggregation, prostacyclin synthesis, and blood coagulation. Sometimes vitamin E metabolites can also interfere with these processes, like the α-tocopherylquinone with the action of vitamin K [21]. Generally even very high doses of vitamin E are non-toxic. This is also due to the fact that large doses of vitamin E taken up with food or supplements are not absorbed in the intestine. The absorbed vitamin E is transported via chylomicrons and lipoproteins through the body and then distributed to tissues. A number of tocopherol transport and binding proteins are involved in these processes. These proteins have higher or even selective activities to some vitamin E components, most importantly the RRR-α-tocopherol. This seems to be the reason for the high concentration of this vitamin E form in the human plasma. The role of RRR-α-tocopherol as a messenger-like or hormone-like regulator of gene expression has been described by several authors. Plasma and tissue α-tocopherol concentrations are remarkably stable, which suggests that they are regulated. α-Tocopherol transfer protein, tocopherol-associated proteins, and tocopherol-binding proteins all bind α-tocopherol. These proteins might function as tocopherol regulatory proteins, although only the tocopherol transfer protein has been shown to influence vitamin E transport [22]. Tocopherol associated proteins have been described and implicated in the regulation of genes and cell signal transduction [23, 24].

A randomized controlled trial has indicated that vitamin E supplementation enhances certain clinically relevant in vivo indexes of T-cell-mediated function in healthy elderly persons [25]. In addition, epidemiologic studies suggest a number of benefits as a consequence of high intakes of α-tocopherol [26–29].

Although ex vivo studies in α-tocopherol-supplemented humans have generally shown beneficial effects [30], clinical trials with α-tocopherol therapy have produced diverse results [31–36]. If α-tocopherol has no effects in humans with established coronary artery disease then either the role of LDL oxidation in human atherosclerosis or the ability of vitamin E to exert an antioxidant function in vivo may become highly questionable [37].

Thus, although evidence based on epidemiologic data, in vitro studies, animal models, and some clinical trials appears to support a beneficial role for α-tocopherol, supplementation in patients with pre-existing cardiovascular disease gives a very controversial picture. For an extensive discussion see [38].

3.3 Carotenoids

Carotenoids are plant antioxidants with a long system of conjugated double bounds. The basic carbon skeleton of these compounds contains 40 atoms of carbon. This stretched skeleton can be modified by cyclization, hydroxyl groups, or various locations of double bounds (Fig. 3). Most of the compounds can be found in the *trans*-isoform although several *cis*-isomers exist. The yellow, red, or orange color of many plants is derived from the 600 carotenoids known. Approximately 50 of them might be important for human nutrition. Due to their structure, the carotenoids are very lipophilic substances and are distributed in the membranes, lipoproteins, and adipocytes of the human body. Therefore, the highest concentrations of these compounds are in the fat tissue, in liver, in ovary glands, and in the macula of the eye. In plasma, concentrations of carotenoids are as usual below 1 μM. Carotenoids are a possible source for the lipophilic vitamin A. Approximately 40 carotenoids are precursors of this vitamin, which is generated by the 15,15′-dioxygenase. The importance of carotenoids as vitamin A precursors is dependent on the nutritional supply with the various vitamin A forms.

Carotenoids are substances able to interact with reactive oxygen species, especially to quench singlet oxygen. Therefore, these compounds are especially

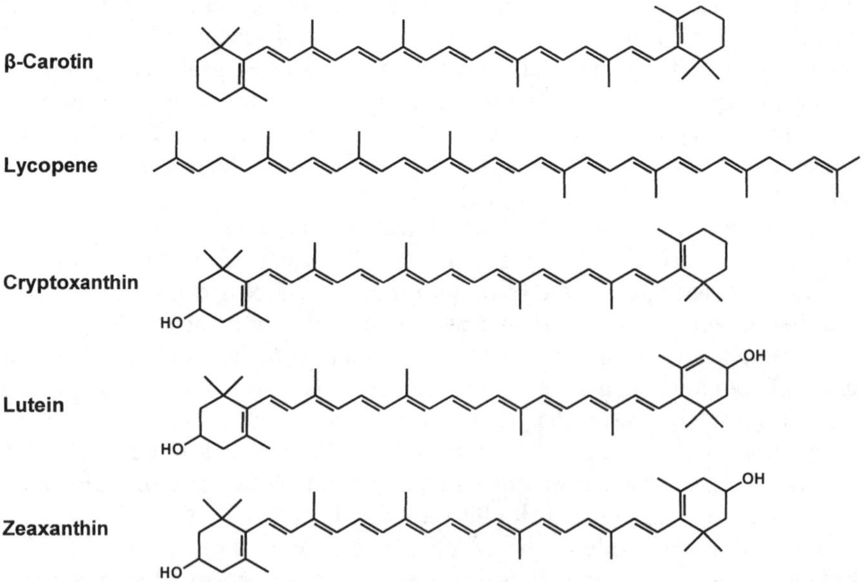

β-Carotin

Lycopene

Cryptoxanthin

HO

Lutein

HO

OH

Zeaxanthin

HO

OH

Fig. 3 Structures of selected carotenoids

important in the UV- and light-exposed areas of the human body, including skin and eye. On the other hand, singlet oxygen can be formed by lipid peroxidation and therefore carotenoids might be important throughout the body. Several oxidants might interact with carotenoids, including carbon-, nitrogen-, and sulfur-centered oxidants, and so carotenoids might therefore also act as free radical scavengers [39, 40]. Since the electron is widely dislocated within the carotenyl radical, this product is relatively stable. It might interact with vitamins C and E and be re-reduced. Radical reactions of carotenoids with oxidants very often lead to bleaching reactions, as a result of the oxidative breakdown of the carbon skeleton of these compounds [41].

Since carotenoids and several of their enzymatic and non-enzymatic breakdown products exert a multitude of biological effects, including the regulation of gene expression, it is very often difficult to discriminate between antioxidant effects and other mechanisms of carotenoid action.

Numerous clinical studies testing for the beneficial effects of carotenoids have been performed. A detailed description of these is beyond the scope of this chapter. However, the results of these studies performed with various carotenoid supplements are remarkably different. Results includes beneficial effects, no effects, and even adverse effects leading to trial interruption. Therefore, no conclusive evidence for the antioxidative or overall beneficial effects of carotenoids in certain situations can be given today.

3.4 Phenolic Compounds

Phenols or phenol groups are a common structure of several secondary plant metabolites. Characterized by at least one hydroxyl function attached to a benzene ring, such molecules can be found in numerous variations in plants, e.g., in flavonoids (Fig. 4) or tocopherols (Fig. 2). The antioxidant capacity of phenolic compounds includes not only direct scavenging of ROS, but also originates from the ability of compounds carrying two or more hydroxyl groups to bind transition metal ions. Such chelation strongly decreases the transition metals' capability to induce environmental oxidative stress.

Flavonoids form a great share of antioxidant-relevant phenolic compounds; they are derived from a tricyclic structure (flavan) whose modifications lead to flavanols (e.g., epicatechin), flavanones (e.g., taxifolin), flavonols (e.g., quercetin), flavones (e.g., apidenin), and anthocyanidins (e.g., malvedin). Phenylpropanoids (e.g., caffeic acid) are also of significant importance in providing antioxidant capacities. A third group of interest consists of gallic acid (a tri-hydroxy benzene) and its derivatives; they may occur solitaire or as gallates, conjugates with other phenolic compounds such as epicatechin and epicatechingallate. A major characteristic of phenolic compounds in general and of flavonoids in particular is that they do not only occur as monomers, but tend to be found as oligomers. Such structures can exceed the size of decamers and can be formed by the same or different polyphenols, e.g., several epicatechin oligomers have been identi-

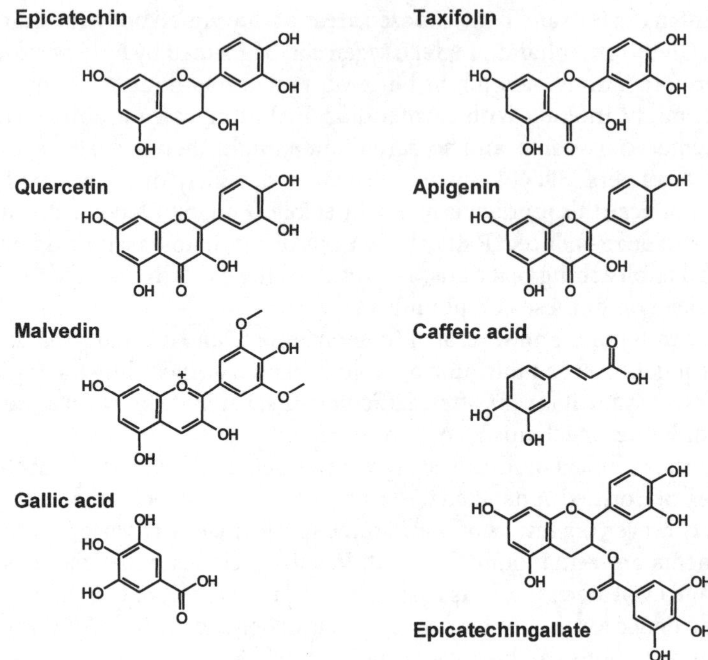

Fig. 4 Structures of phenolic compounds

fied in cocoa [42]. Antioxidative capacities are influenced by oligomer chain length [43]

Direct antioxidant effects of phenolic compounds include scavenging of several ROS. Epicatechin, for example, is suitable for scavenging nitric oxide [44], superoxide [45], peroxynitrite [46], and for preventing UV-A induced oxidative damage in skin cells [47].

Based on in vitro experiments, the mode of antioxidant action of phenolic compounds has been suggested as involving an electron transfer to the phenolic compound; substances with a possible chinon function form a semichinon radical [48, 49]. Further mechanisms/modes of action have been discussed, e.g., the ability of catechin/epicatechin and derivatives to scavenge the tyrosyl radical intermediate involved in ROS-derived nitrotyrosine formation [50].

In vitro experiments provided further proof that epicatechin, for example, is able to protect hydrophilic and hydrophobic target molecules from ROS/RNS effects [50]. Results showed that epicatechin was enriched in living cells during incubation in cell culture media, which led to protection against ROS/RNS even if no epicatechin was actually present in the buffer during ROS/RNS treatment [51].

The basis for investigations of the antioxidant capacity of phenolic compounds was laid by the "French paradox", the low cardiovascular mortality of Mediterranean populations in spite of a high consumption of red wine and saturated fatty acids [52, 53].

Phenolic compounds are available to the human body mainly from the diet. After ingestion of appropriate food or drink (e.g., wine, tea, chocolate) [54], phenolic compounds can be taken up via the gastrointestinal tract with varying efficiency [55]. Mechanisms of uptake have been investigated and results point to active as well as passive uptake, depending on chemical formula and other factors such as food composition [56–8]. Retention periods and the macroscopic and microscopic whereabouts of phenolic compounds inside the body reveal a complex relationship between the physicochemical properties of the compound, the rate and kind of metabolic modifications by xenobiotic pathways, and the rate of excretion [59–61]. Oligomerization of phenolic compounds (see above) can affect uptake, as larger oligomers have been shown to be taken up only in very small amounts. However, it is very likely that acidic cleavage of oligomers in the stomach results in dimers and monomers, which could be taken up [62–64].

4 Interaction of Antioxidants

At several places in this chapter it has been mentioned that several antioxidants may interfere with other antioxidative molecules. This electron transfer is connected with a re-reduction of the initial antioxidant. Several possible interactions between antioxidants exist, some of which are presented in Fig. 5. This

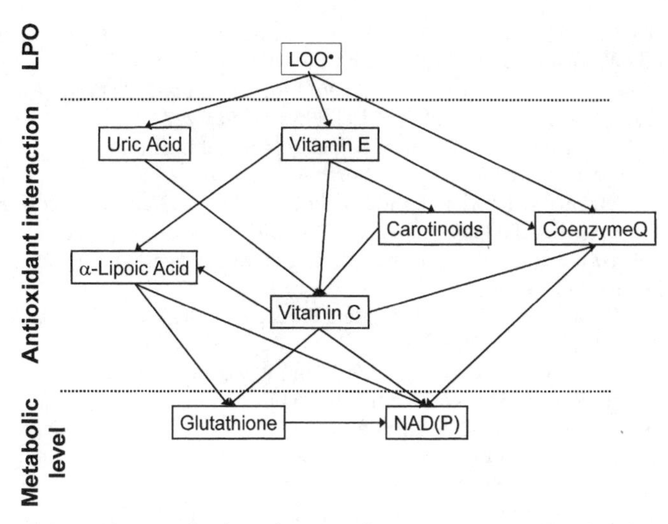

Fig. 5 Interaction scheme of selected antioxidants. Starting with a lipid peroxidation (*LPO*)-driven lipid peroxide radical, detoxification via electron transfer can take different paths. Electrons can primarily be transferred to, e.g., uric acid, vitamin E, or directly to coenzyme Q. Further electron flow can involve carotenoids, other vitamins, or lipoic acid. The final metabolic level involves the oxidation of NAD(P) or glutathione

antioxidantive network is only focused on the detoxification of a single radical species: the lipidhydroperoxide radical. Since several oxidants and free radicals have to be detoxified, a high number of interactions with the antioxidant network becomes more complicated. Furthermore, it is obvious that the electron flux through this network not only depends on the redox potential of the compounds involved, but also on concentrations and distributions, on availability (bound to proteins or not), and on the ability to diffuse in a certain compartment etc. Furthermore, that the demonstrated compounds in Fig. 5 are only the main antioxidants has to be considered. Furthermore, this scheme does not take into account that the carotenoids and vitamin E are a multitude of compounds with different reactivities and availabilities. Taking into consideration that this whole scheme is expanded by hundreds (if not thousands) of phenolic compounds, one should realize that an almost inestimable number of interactions is possible. Interestingly, the overall direction stays the same: from the lipid peroxidation, and perhaps also from protein and DNA oxidation, via a network of electron flux directions to the interference points with normal metabolism. Therefore, the recycling of antioxidants results in the usage of metabolic compounds, production of which requires energy. Therefore, one has to take into consideration that antioxidative defense via low molecular antioxidants requires metabolic activity and energy for production of the compounds or intestinal adsorption and for the detoxification itself.

References

1. Schiller J, Arnhold J, Schwinn J, Spritz H, Brede O, Arnold K (1998) Free Radic Res 28:215
2. Naskalski JW, Bartosz G (2000) Adv Clin Chem 35:161
3. Sarna T, Pilas B, Lund EJ, Truscott TG (1986) Biochem Biophys Acta 883:162
4. Li C, Jackson RM (2002) Am J Physiol Cell Physiol 282:C227
5. Oda M, Satta Y, Takenaka O, Takahata N (2002) Mol Biol Evol 19:640
6. Monu JU Pope TL (2004) Radiol Clin North Am 42:169
7. Ames BN, Cathcart R, Schwiers E, Hochstein P (1981) Proc Natl Acad Sci USA 78:6858
8. Grootveld M, Halliwell B (1987) Biochem J 243:803
9. Babizhayev MA, Seguin MC, Gueyne J, Evstigneeva RP, Ageyeva EA, Zheltukhina GA (1994) Biochem J 304:509
10. Hipkiss AR, Brownson C, Bertani MF, Ruiz E, Ferro A (2002) Ann NY Acad Sci 959:285
11. Porter WL (1995) Tox Ind Health 9:93
12. Franceschi RT, Wilson JX, Dixon SJ (1995) Am J Physiol 268:C1430
13. Toth I, Bridges KR (1995) J Biol Chem 270:19540
14. Carr A, Frei B (1999) Am J Clin Nutr 69:1086
15. Halliwell B (2001) Mutat Res 475:29
16. Carr A, Frei B (1999) FASEB J 13:1007
17. Muller DPR, Goss-Sampson MA (1990) Crit Rev Neurobiol 5:239
18. Ben Hamida C, Doerflinger N, Belal S, Linder C, Reutenauer L, Dib C, Gyapay G, Vignal A, Le Paslier D, Cohen D (1993) Nat Genet 5:195
19. Triantafillidis JK, Kottaras G, Sgourous S, Cheracakis P, Driva G, Konstantellou E, Parasi A, Choremi H, Samouilidou E (1998) J Clin Gastroenterol 26:207

20. Yokota T, Igarashi K, Uchihara T, Jishage K, Tomita H, Inaba A, Li Y, Arita H, Mizusawa H, Arai H (2001) Proc Natl Acad Sci USA 98:15185
21. Dowd P, Zheng ZB (1995) Proc Natl Acad Sci USA 92:8171
22. Blatt DH, Leonard SW, Traber M (2001) Nutrition 17:799
23. Zimmer S, Stocker A, Sarbolouki MN, Spycher SE, Sassoon J, Azzi A (2000) J Biol Chem 275:25672
24. Zingg JM, Ricciarelli R, Azzi A (2000) IUBMB Life 49:397
25. Meydani SN, Meydani M, Blumberg JB, Leka LS, Siber G, Loszewski R, Thompson C, Pedrosa MC, Diamond RD, Stollar BD (1997) JAMA 277:1380
26. Albanes D, Heinonen OP, Huttunen JK, Taylor PR, Virtamo J, Edwards BK, Haapakoski J, Rautalahti M, Hartman AM, Palmgren J (1995) Am J Clin Nutr 62:1427
27. Gey KF (1998) Biofactors 7:113
28. Rimm EB, Stampfer MJ, Ascherio A, Giovannucci E, Colditz GA, Willett WC (1993) N Engl J Med 328:1450
29. Stampfer MJ, Hennekens CH, Manson JE, Colditz GA, Rosner B, Willett WC (1993) N Engl J Med 328:1444
30. Jialal I, Traber M, Devaraj S (2001) Curr Opin Lipidology 12:49
31. Marchioli R, Valagussa F (2000) Eur Heart J 21:949
32. Stephens NG, Parsons A, Schofield PM, Kelly F, Cheeseman K, Mitchinson MJ (1996) Lancet 347:781
33. Marchioli R (1999) Pharmacol Res 40:227
34. Pruthi S, Allison TG, Hensrud DD (2001) Mayo Clin Proc 76:1131
35. Hoogwerf BJ, Young JB (2000) Clin J Med 67:287
36. Salonen JT, Nyyssonen K, Salonen R, Lakka HM, Kaikkonen J, Porkkala-Sarataho E, Voutilainen S, Lakka TA, Rissanen T, Leskinen L (2000) J Intern Med 248:377
37. Heinecke JW (2001) Arterioscler Thromb Vasc Biol 21:1261
38. Ricciarelli R, Zingg JM, Azzi A (2001) FASEB J 15:2314
39. Everett SA, Dennis MF, Patel KB, Maddix S, Kundu SC, Willson RL (1996) J Biol Chem 271:3988
40. Liebler DC, McClure TD (1996) Chem Res Toxicol 9:8
41. Krinski NI, Yeum KJ (2003) Biochem Biophys Res Commun 305:754
42. Hammerstone JF, Lazarus SA, Mitchell AE, Rucker R, and Schmitz HH (1999) Identification of procyanidins in cocoa (Theobroma cacao) and chocolate using high-performance liquid chromatography/mass spectrometry. J Ag Food Chem 47:490
43. Lotito SB, Actis-Goretta L, Renart ML, Caligiuri M, Rein D, Schmitz HH, Steinberg FM, Keen CL, Fraga CG (2000) Biochem Biophys Res Commun 276:945
44. van Acker SA, Tromp MN, Haenen GR, van der Vijgh WJ, Bast A (1995) Biochem Biophys Res Commun 214:755
45. Robak J, Gryglewski RJ (1988) Biochem Pharmacol 37:837
46. Arteel GE, Sies H (1999) Protection against peroxynitrite by cocoa polyphenol oligomers. FEBS Lett 462:167
47. Basu-Modak S, Gordon MJ, Dobson LH, Spencer JP, Rice-Evans C, Tyrrell RM (2003) Free Radic Biol Med 35:910
48. Bors W, Michel C, Stettmaier K (2000) Arch Biochem Biophys 374:347
49. Bors W, Michel C (1999) Free Radic Biol Med 27:1413
50. Schroeder P, Zhang H, Klotz LO, Kalyanaraman B, Sies H (2001) Biochem Biophys Res Commun 289:1334
51. Schroeder P, Klotz LO, Sies H (2003) Biochem Biophys Res Commun 307:69
52. Renaud S, Gueguen R (2000) Novartis Found Symp 216:208
53. Formica JV, Regelson W (1995) Food Chem Toxicol 33:1061

54. Hammerstone JF, Lazarus SA, Schmitz HH (2000) J Nutr 2000 130:2086S
55. Scalbert A, Williamson G (2000) J Nutr 130:2073S
56. Frei B, Higdon JV (2003) J Nutr 133:3275S
57. Schramm DD, Karim M, Schrader HR, Holt RR, Kirkpatrick NJ, Polagruto JA, Ensunsa JL, Schmitz HH, Keen CL (2003) Life Sci 73:857
58. Baba S, Osakabe N, Natsume M, Muto Y, Takizawa T, Terao J (2001) J Agric Food Chem 49:6050
59. Carbonaro M, Grant G, Pusztai A (2001) Eur J Nutr 40:84
60. Vaidyanathan JB, Walle T (2003) J Pharmacol Exp Ther 307:745
61. Meng X, Sang S, Zhu N, Lu H, Sheng S, Lee MJ, Ho CT, Yang CS (2002) Chem Res Toxicol 15:1042
62. Zhu QY, Holt RR, Lazarus SA, Ensunsa JL, Hammerstone JF, Schmitz HH, Keen CL (2002) J Agric Food Chem 50:1700
63. Spencer JP, Chaudry F, Pannala AS, Srai SK, Debnam E, Rice-Evans C (2000) Biochem Biophys Res Commun 272:236
64. Spencer JP, Schroeter H, Rechner AR, Rice-Evans C (2001) Antioxid Redox Signal 3:1023

The Handbook of Environmental Chemistry Vol. 2, Part O (2005): 91–108
DOI 10.1007/b101148

Glutathione

Juan Sastre · Federico V. Pallardo · Jose Viña (✉)

Departamento de Fisiologia, Facultad de Medicina, Avenida Blasco Ibáñez 17,
46010 Valencia, Spain
juan.sastre@uv.es, federico.v.pallardo@uv.es, jose.vina@uv.es

Abstract Glutathione is the most abundant non-protein thiol in cells. It is a tripeptide with two important structural features: the thiol group and the gamma-glutamyl peptide bond between glutamate and cysteine. It is a major antioxidant, able to reduce peroxides (due to its action as substrate of glutathione peroxidases). As a result it is oxidised to the disulphide form (GSSG), which in turn is reduced back to GSH by glutathione reductase. Thus glutathione is a major player in maintaining physiological redox status in cells.

Some of the functions of glutathione depend on the presence of the gamma-glutamyl bond, for instance its role in the regulation of amino acid transport. But the majority of the functions of glutathione are related to its role in redox regulation in cells and in detoxification of xenobiotics.

Some areas of research of special interest on this molecule are glutathionylation of proteins, the cellular compartmentation and the role of this interesting molecule in disease.

Keywords Glutathione · Redox · Cysteine · Drug detoxication · Metabolism · Gamma-glutamyl · Thiol

1 Glutathione: Introduction

Glutathione is the most abundant non-protein thiol in cells [1]. Its intracellular concentration is around 5 μmol per gram of tissue, i.e. similar to that of glucose in cells [2]. Glutathione was discovered in the late 19th century by deRey-Pailhade [3] in Montpellier, France. Its structure was studied by Hopkins who thought that it was a dipeptide composed of glutamate and cysteine. Harrington and Mead, in 1935, finally described the correct structure of the tripeptide: gamma-glutamyl cysteinyl glycine. The name glutathione was introduced by Hopkins in 1921. An excellent review on historical aspects of the discovery of glutathione has been published by Meister [4].

Work on glutathione was greatly boosted after a critical paper in Nature by E. M. Kosower and N. S. Kosower [5]. The title of that paper "Lest I forget thee glutathione" indicates the relatively scarce number of papers published on glutathione in those days. Research on this molecule boosted thereafter and over 40,000 papers have now been published on this molecule [6]. Books on glutathione have also been published [7].

Meister, who contributed significantly to the understanding of the enzymology and physiological functions of glutathione published a comprehensive review on this topic in 1983 [8].

Glutathione, as it will be apparent in this review, has many physiological functions. They are derived from its peculiar chemical characteristics: the γ-glutamyl bond between the glutamate and cysteine residues and the presence of the free thiol group. This latter characteristic makes glutathione one of the best redox buffers in cells, given the high concentration of the reduced form of glutathione (GSH) and the presence of enzymes catalyzing the oxidation of glutathione (glutathione peroxidases) and its reduction (glutathione redutases).

The level of GSH in cells is very high (i.e. in the millimolar range). The level of the oxidised form, GSSG, is significantly lower. The GSH/GSSG ratio is around 10–100. If one assumes that the glutathione reductase is at equilibrium in vivo, and given the well-known NADP/NADPH ratio in the cytosol and the equilibrium constant for the glutathione reductase, the GSH/GSSG ratio should be even higher, closer to 10,000 [2]. However, after considerable experimental care, ratios of this order are not found. Thus, we may conclude that glutathione reductase may not be at equilibrium in vivo [2].

2 Glutathione Redox Ratio: A Sensor of Oxidative Stress

2.1 Methodological Aspects

We have already stated that the GSH/GSSG ratio in many cells is 10–100. This means that there is about 100 times more GSH than GSSG in living cells. Given the relatively easy oxidation of GSH to form GSSG, the determination of these two compounds, when present in the same biological fluid, poses the problem that a 2% oxidation of GSH yields a 100% increase in GSSG. Thus, sample preparation and handling is extremely important to minimise spontaneous (or metal-catalysed) oxidation of glutathione in these samples. If oxidation of glutathione is not brought down to less than 0.1%, very significant methodological errors occur and the levels of GSSG obtained may be seriously mistaken. Thus, determination of the GSH/GSSG ratio becomes meaningless. In general terms, ratios lower than 10 become suspect. In our group, we developed [9] a method to accurately determine GSSG in the presence of GSH. This has been applied to many physiological and pathological situations. We have been able to correlate changes in the redox status of cells with physiological changes in several situations. For instance, we were able to find a direct relationship between oxidation of glutathione and increase in blood lactate levels during exhaustive exercise [10]. We have also studied the relationship between oxidation of glutathione in mitochondria of aged animals with the oxidation of mitochondrial DNA [11]. By using this methodology, the relationship between oxidation of glutathione and the occurrence of apoptosis also became apparent [12]. The problem of the accurate determination of glutathione has also been tackled by other groups. Recently a new chromatographic method to determine glutathione in blood has been published [13].

3 Glutathione Biosynthesis and Its Regulation

Reduced glutathione (GSH) is synthesised from its constituent amino acids in the cytosol of all mammalian cells [4]. It takes place following two enzymatic steps:

$$\text{L-glutamate} + \text{L-cysteine} + \text{ATP} \rightarrow \gamma\text{-glutamyl-L-cysteine} + \text{ADP} + \text{Pi} \quad (1)$$

The first step is catalysed by γ-glutamylcysteine synthetase (GCS) and is rate-limiting for GSH biosynthesis [14]. The second step is catalysed by glutathione synthetase.

GCS is composed of a heavy catalytic subunit (73 kDa) and a light regulatory subunit (30 kDa), which are encoded by different genes both in rats and in humans [15, 16]. The catalytic subunit exhibits the feedback inhibition by

GSH [17]. The physiological regulation of GCS activity relies on the availability of L-cysteine [18] and on the feedback inhibition by GSH [4, 14, 19]. The intracellular cysteine concentration approximates the apparent K_m value of GCS for cysteine (0.1–0.3 mM in rats and humans), whereas intracellular glutamate levels are several fold higher than the K_m value of GCS for glutamate (1.8 mM) [20, 21]. Consequently, the availability of intracellular cysteine determines the rate of GSH synthesis under physiological conditions.

Both GCS subunits are regulated at the transcriptional level by oxidative stress [14]. Oxidative stress induced by numerous agents – such as hydrogen peroxide, quinones, ionising radiation, TNF-α or 4-hydroxy-2-nonenal – enhances GCS activity as well as the transcription of the heavy GCS subunit [14, 22, 23]. Transcription of the light GCS subunit is also increased in response to *tert*-butylhydroquinone or 4-hydroxy-2-nonenal [14, 22, 24, 25]. Furthermore, 4-hydroxy-2-nonenal stabilises the mRNA of both GCS subunits [24].

The up-regulation of GCS subunits by oxidative stress is mediated by different transcription factors such as nuclear factor kappa B (NF-kB), Sp-1, activator protein-1 (AP-1), activator protein-2 (AP-2) and antioxidant response elements (ARE) or electrophile responsive elements (EpRE).

The hormonal regulation of GCS expression has special physiological relevance. Phenylephrine, glucagon or dibutyryl cAMP inhibit GCS activity, which decreases glutathione synthesis and leads to glutathione depletion in rat hepatocytes [26, 27]. The loss of GCS activity induced by these stress hormones is mediated by phosphorylation of the catalytic GCS subunit due to activation of protein kinase A, protein kinase C or Ca^{2+}-calmodulin kinase [28]. Consequently, the stress response diminishes GSH synthesis, which may increase the availability of cysteine for the synthesis of stress proteins [14].

In contrast to this acute response, hydrocortisone or insulin need several hours to increase GCS activity and glutathione levels in rat hepatocytes [21]. In these cases, the increase in GCS activity is mediated by an up-regulation in the gene expression of the catalytic GCS subunit [29]. Glutathione depletion and low GCS activity were observed in streptozotocin-treated or adenalectomised rats, whereas glucocorticoid administration had no effect on GCS activity or GSH levels in sham-operated rats [21]. Consequently, these hormones are required for normal expression of GCS and GSH levels. Furthermore, the regulation of glutathione levels during stress seems to occur in two phases: firstly some stress hormones, such as glucagon and phenylephrine, decrease glutathione levels in the short term, and secondly other stress hormones, such as cortisol, replace glutathione levels in the medium or long term.

Glutathione synthetase is composed of two identical subunits and is not subject to feedback inhibition by GSH [30]. The rat kidney enzyme is glycosylated, but the role of glycosylation in the regulation of this enzyme activity remains poorly understood [14, 30]. There is a coordinate induction of GCS subunits and glutathione synthetase by glutathione-depleting agents, such as diethyl maleate, buthionine sulphoximine or *tert*-butylhydroquinone [31]. Furthermore, AP-1 is required for this coordinate induction as well as for basal expression of GCS and glutathione synthetase [32]. In hepatocellular carcinoma, an increase in the mRNA levels of both the catalytic GCS subunit and glutathione synthetase has been reported. This leads to high GSH levels and facilitates cell proliferation [33].

4
Glutathione Turnover: The γ-Glutamyl Cycle

The γ-glutamyl cycle was postulated by Meister in the mid-seventies [34]. It accounts for the biosynthesis and degradation of glutathione. The biosynthesis has been described in an earlier paragraph. In summary, glutathione is synthesised by two intracellular enzymes γ-glutamyl cysteinyl synthetase and glutathione synthetase. The degradation of glutathione, i.e. the catabolic part of the cycle, takes place partially extracellularly and partially inside cells. Indeed, glutathione is released from the cell through a specific carrier. It is then accessible to the active site of γ-glutamyl transpeptidase, which catalyses the breakdown of the γ-glutamyl bond of glutathione and the synthesis of a γ-glutamyl amino acid. Cysteinyl glycine is also formed. The γ-glutamyl amino acid is taken up by cells through a specific transport mechanism. Cysteinyl glycine is also taken up by cells. Inside the cell, the γ-glutamyl amino acid is hydrolysed by γ-glutamyl cyclo transferase and converted into oxoproline and a free amino acid. Oxoproline is a cyclic form of glutamate and is converted into glutamate via oxoprolinase. For a complete picture of the cycle, see Fig. 1.

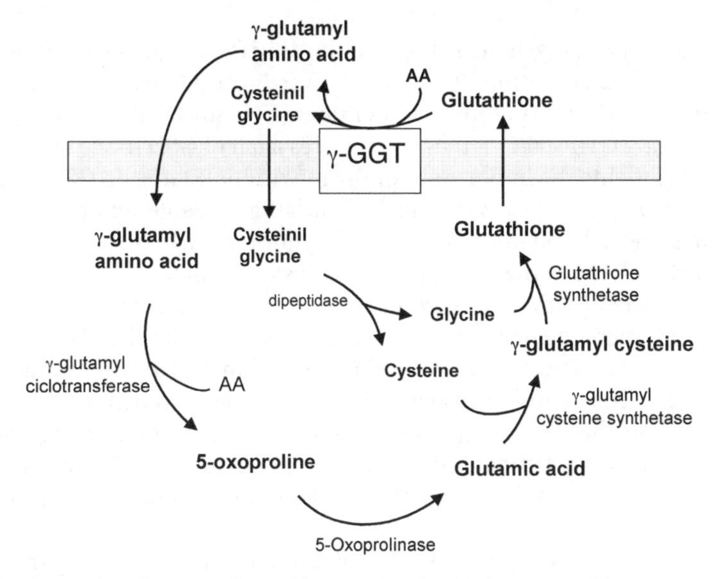

Fig. 1 γ-Glutamyl cycle. *AA* amino acid, *γ-GGT* γ-glutamyl transpeptidase

The γ-glutamyl cycle was postulated by Meister [34] as a mechanism for amino acid transport. However, this presents major problems. The most important is the energetic one. The γ-glutamyl cycle requires the use of three ATP molecules per turn of the cycle. Thus, the uptake of an amino acid would require the use of three high-energy phosphate bonds. In favour of the cycle was the fact that addition of γ-glutamyl transpeptidase inhibitors in vivo caused a decrease in amino acid transfer into cells. We have proved this in the mammary gland of lactating rats and also in placenta. We then postulated an alternative hypothesis, i.e. that γ-glutamyl amino acids or oxoproline could be signalling molecules to activate the transport of amino acids through membranes. We checked this hypothesis and found that oxoproline indeed catalytically activates the uptake of amino acids through the placental barrier. We also found that the transfer of amino acids through the blood–brain barrier is activated by oxoproline [35]. Thus, the γ-glutamyl cycle, apart from explaining the synthesis and degradation of glutathione, may serve as a generator of signals to activate amino acid transport into cells [35].

Glutathione turnover may be considered as a multi-organ process. In fact, in liver, an organ in which glutathione synthesis is most active, the degradation is very slow due to the very low activity of γ-glutamyl transpeptidase. However, in kidney, γ-glutamyl transpeptidase is very high. Thus, the γ-glutamyl cycle may be considered as a multi-organ cycle in which the synthetic part occurs in liver and the catabolic part occurs in kidney amongst other tissues.

5 Glutathione Precursors

The availability of L-cysteine is the rate-limiting factor for GSH synthesis under physiological conditions. Cysteine derives from the diet, from proteolysis or alternatively is synthesised from methionine via the transulphuration pathway [36] (Fig. 2). The fact that fasting induces a fall of GSH levels to 50% of the controls highlights the importance of the nutritional status on GSH synthesis. Indeed, fasting for 48 h causes a marked glutathione depletion in the liver and GSH levels are restored upon refeeding [37]. Moreover, hepatic GSH levels are particularly related to the cysteine and/or cystine content of the diet.

Cysteine gets into the cells through specific transporters for neutral amino acids [38]. In hepatocytes, it is transported primarily by the Na^+-dependent ASC system and hence is subject to *cis*-inhibition and *trans*-stimulation by the other substrates of this system [38, 39]. Cystine is transported in the anionic form by the Na^+-independent Xc-carrier, glutamate being the only other substrate [38]. Although this system is not expressed in normal hepatocytes, it is up-regulated by glutathione depletion induced by electrophilic agents [40].

The transulphuration pathway takes place in several tissues but primarily in the liver. This pathway involves five steps:

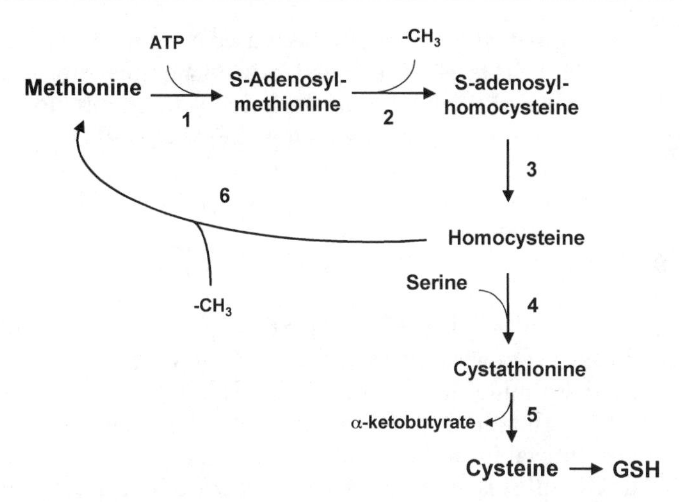

Fig. 2 Transulphuration pathway. Enzymes involved: *1* methionine S-adenosyltransferase, *2*methyltransferase, *3* S-adenosylhomocysteine hydrolase, *4* cystathionine β-synthase, *5* β-cystathionase, *6* methionine synthase

1. Activation of methionine to S-adenosylmethionine
2. Demethylation
3. Removal of the S-adenosyl moiety to yield homocysteine
4. Formation of cystathionine from homocysteine and serine
5. Cleavage of cystathionine to give cysteine and α-ketobutyrate

The key steps in this pathway are the first one catalysed by methionine S-adenosyltransferase and the last one, which is irreversible, catalysed by β-cystathionase.

The transulphuration pathway is impaired or even absent in several physiological and pathological conditions leading to glutathione depletion. It is critically hindered in liver cirrhosis, in which GSH depletion is associated with reduced expression and activities of methionine adenosyltransferase, cystathionine β-synthase and γ-cystathionase [41–43]. Of special clinical relevance is hyperhomocysteinemia, which is due to cystathionine β-synthase deficiency and leads to mental retardation and thrombovascular dysfunction [44]. α-Cystathionase activity is markedly reduced and becomes rate limiting for GSH synthesis in foetal life [45], premature infants [46], surgical stress [47], aging [48] and AIDS [49]. *N*-Acetyl cysteine might be used as a glutathione precursor to restore GSH levels in these situations as the transulphuration pathway does not provide enough cysteine for GSH synthesis [50]. Nonetheless, the daily dose of NAC in chronic treatments must be carefully chosen to avoid accumulation of intracellular cysteine. Free cysteine may suffer auto-oxidation to generate damaging free radicals [51]. This might occur if cysteine is provided in excess since GSH synthesis in inhibited by physiological levels of this tripeptide.

Glutathione mono-ethyl ester has also been used often in research to rapidly restore intracellular GSH levels. The drawback to using GSH ester as a therapy for diseases associated with glutathione depletion is that it provides intracellularly equimolecular amounts of ethanol together with GSH.

6 Glutathione Transport

Membrane transport plays a key role in the inter-organ turnover of glutathione. This serves to take into account the short half-life of glutathione in blood plasma – just a few minutes in rats and humans [52, 53]. The liver is the major organ for releasing glutathione to plasma, whereas the kidney is the main one for glutathione removal from the circulation [54–58].

Sinusoidal GSH efflux is the major determinant of GSH, cysteine and cystine levels and thiol-disulphide status in blood plasma [59]. Indeed, the highest rate of GSH efflux occurs in hepatocytes across the sinusoidal membrane (15 nmol $min^{-1}g^{-1}$ rat liver), the rate of GSH biliary excretion being much lower (1–3 nmol $min^{-1}g^{-1}$ rat liver) [60, 61]. Glutathione transport through the sinusoidal membrane is Na^+-independent and unidirectional under physiologic conditions, i.e. no re-uptake of GSH or GSSG by intact hepatocytes occurs [54, 62]. The efflux of GSH may also occur in exchange for uptake of γ-glutamyl compounds and organic anions [63]. Adrenaline, phenylephrine or glucagon increase GSH efflux from the liver across the sinusoidal membrane [64, 65]. Consequently, the release of glutathione from the liver would be induced during stress to fulfil increased demands from other tissues [54, 66].

Biliary GSH excretion occurs through an electrogenic Na^+-independent ATP-independent canalicular transport system which is *cis*-inhibited and *trans*-stimulated by certain organic anions and induced by phenobarbital [67].

Mitochondria cannot synthesise GSH (see below). Thus, the mitochondrial GSH pool comes from cytosol GSH through a mitochondrial carrier. This GSH transporter identified in rat liver mitochondria is different from the canalicular and the sinusoidal carriers [68]. Chronic ethanol feeding impairs the mitochondrial GSH transport leading to mitochondrial GSH depletion [69]. An increase in microviscosity of mitochondrial membranes due to high cholesterol content appears to account for this impairment in GSH transport [70].

Under physiological conditions there is a very low biliary GSSG excretion (0.4 nmol $min^{-1}g^{-1}$ rat liver) [61]. GSH and GSSG are also released from extrahepatic tissues, such as the heart, although at a much lower rate [71]. Erythrocytes exhibit two transport systems to export GSSG – both ATP-dependent – but they lack GSH efflux capacity [72, 73].

Oxidative stress increases the release of GSSG from cells and tissues, including liver, lung, heart and erythrocytes [74]. Pro-oxidant agents raise the cellular efflux of GSSG into the bile, which parallels the increase in intracellular GSSG concentration [74]. Biliary GSSG excretion appears to be mediated by an active transport, since it occurs against a concentration gradient [61]. The low capacity of GSSG export from other tissues, such as the heart, would render these tissues more susceptible to oxidative stress.

GSSG efflux is inhibited by glutathione-S conjugates in liver, heart and erythrocytes [71, 75, 76]. Accordingly, a common translocator is involved in the transport of both GSSG and GS-conjugates. Moreover, the formation of glutathione-S conjugates might contribute to the accumulation of intracellular GSSG during oxidative stress.

The activity of the sinusoidal GSH carrier is also influenced by the thiol-disulphide status [77]. Thiol oxidation favours inward transport by inhibiting efflux and consequently changes the direction of GSH transport through this carrier.

Uptake of GSH has been observed in the kidney, lung and intestinal cells [57, 78, 79]. About 80% of plasma GSH present in the renal artery is extracted after a single pass through the kidney [57]. This capacity for removal is in part due to the high renal γ-glutamyl transpeptidase activity, located mainly in the luminal side of the brush border membrane [80]. GSH present in the glomerular filtrate is degraded to its constituent amino acids, which are reabsorbed across the luminal membrane. Glomerular filtration and subsequent GSH breakdown by γ-glutamyl transpeptidase account for 20–30% of the renal extraction [55]. Renal uptake of GSH also occurs through basolateral membranes by a Na^+-dependent GSH transport system [81].

Na^+-independent and also Na^+-dependent GSH transport systems have been found in lens epithelium, retinal Muller cells, brain endothelial cells and astrocytes [82–84]. The Na^+-dependent transport mediates GSH uptake, whereas the Na^+-independent carrier appears to be mainly involved in GSH efflux. It is worth noting that these transport systems allow GSH transport across the blood–brain barrier in vivo [85].

7 Glutathione in Cell Compartments: Mitochondrial Nuclear and Endoplasmic Glutathione

The many cellular functions of glutathione (see below), which occur in several cell compartments, indicate that the level and the redox state of glutathione might be different in different cell compartments.

Using cell separation techniques (reviewed in [86]), the rich variety of glutathione levels in cell compartments has become more and more apparent. Of particular interest is the distribution of glutathione in mitochondria, endoplasmic reticulum and in the nucleus.

The importance of mitochondrial glutathione was first noticed by Jocelyn and Kamminga [87]. The actual distribution and the precise levels of glutathione in mitochondria as related to the cytosol were first described by Sies and colleagues [88]. Glutathione has a different rate of turnover in mitochondria and in the cytosol [89]. Glutathione is not synthesised within the mitochondria, but rather it is transported in these organelles by a specific transport system from cytosol (see above). Mitochondrial glutathione has important cellular functions both in physiology and in disease [90]. Of special importance is its function in ageing, which was first postulated by Miquel et al. [91]. We found that total cellular glutathione is slightly decreased with age. This decrease accounts for approximately 30% of all cellular glutathione. However, when we measured changes of mitochondrial glutathione with ageing we found that it decreases very significantly and that this fall may account for all the missing glutathione in the ageing cell [92]. Furthermore, we found that there is a correlation between mitochondrial glutathione depletion and an increase in oxidative damage to mitochondrial DNA [93]. Mitochondrial glutathione has a role in several other disease states such as AIDS (mainly due to the toxicity of antiretroviral agents), and also in relevant physiological states. For instance, the mitochondrial pathway of cellular apoptosis is triggered by an oxidation of mitochondrial glutathione [12].

Endoplasmic reticulum is another cell compartment in which glutathione plays an important role, particularly due to its relevance in the folding of secreted proteins. Early observations by Trimm et al. [94] showed that oxidation of endoplasmic glutathione leads to calcium release by the reticulum. A critical paper by Hwang and colleagues appeared in Science [95]. These investigators used an ingenious method to determine the glutathione redox state in cell compartments and, rightly, reported that the cytosolic glutathione redox ratio (GSSG/GSH) was around 60. However, glutathione is very significantly oxidised in the endoplasmic reticulum and the GSSG/GSH ratio in the reticulum was around 1. These researchers attributed the high oxidation level of glutathione to a significant transport of oxidised glutathione from the cytosol into the endoplasmic reticulum. The importance of this highly oxidative state in endoplasmic reticulum for the correct folding of proteins has been suggested.

Nuclear glutathione also deserves attention. The major problem when studying nuclear glutathione is that measurements of its redox status are very difficult. The structural characteristics of a nuclear membrane, particularly its pore complex which allows diffusion of low molecular weight molecules, make it very difficult to obtain a reasonable GSH/GSSG ratio. Studies using lipid fractionation techniques on focal microscopy [96] suggest that GSH concentration in nuclear GSH is similar to that in cytoplasm. On the other hand, Bellomo et al. [97], using fluorescent probes, showed that glutathione concentration in the nucleus is higher in that in the cytosol of isolated hepatocytes. These authors concluded that an active transport of glutathione from cytosol into the nucleus must exist. Furthermore, the nuclear pool of glutathione is more resistant to depletion than the cytosolic one, suggesting that the tripeptide plays an essential role in the physiology of the nucleus. Indeed, glutathione is involved in the regulation of nuclear matrix organisation, in the regulation of chromatin structure [98] and in DNA synthesis. More recently a number of studies have emphasised the importance of oxidative stress in the regulation of nuclear transcription factors like NF-KB and AP1 (for a review see [99]).

Studies aiming to solve the importance of the cellular compartmentation of glutathione are a very interesting field of research, as pointed out by Sies [6].

8
Physiological Functions of Glutathione

The many physiological functions of this tripeptide are a direct consequence of its two most important structural features: the γ-glutamyl bond and the presence of a free thiol group. The γ-glutamyl bond of glutathione makes it inaccessible to the usual cellular peptidases. It is almost exclusively degraded by γ-glutamyl transpeptidase (see above). The importance of glutathione in the regulation of amino acid transport into cells has been discussed above.

Glutathione does not show signs of toxicity in cells. This contrast with cysteine, which must be kept at low concentrations because of its cellular toxicity – particularly due to its very fast auto-oxidation rate which generates free radicals [51, 100]. Thus, glutathione serves as a reservoir of cysteine and is useful for maintaining a high concentration of low molecular weight thiols in the cells. Glutathione serves to detoxify peroxides (mediated by glutathione peroxidase). It also serves to degrade xenobiotics (catalysed by glutathione-S-transferases) and to maintain the correct three-dimensional structure of many proteins (which are frequently S-glutathionylated). Furthermore, glutathione serves as a co-enzyme of an interesting enzyme system, the glyoxalase system whose function is as yet not well known.

Protein glutathionylation is a key mechanism by which glutathione redox status regulates enzyme activity [6]. Mixed disulphides formed between GSH and the cysteines of proteins by reversible S-glutathionylation affect the activity of many enzymes. Most of them, such as glyceraldehyde-3-phosphate dehydrogenase, creatine kinase, enolase, 6-phosphoglucunolactone, protein tyrosin phosphatase 1 B, protein phosphatase 2A, cAMP-dependent protein kinase, thioredoxin, tyrosine hydroxylase and α-ketoglutarate dehydrogenase are inhibited by S-glutathionylation [101–109]. Under physiological conditions, two key reducing systems related to glutaredoxin and thioredoxin keep protein cysteines in the reduced state. The glutaredoxin system (which includes NADPH, glutathione reductase, GSH and the small protein glutaredoxin) transfers reducing equivalents from NADPH to thiol groups of enzymes via GSH [110]. Thioredoxin is a small dithiol protein that acts as a hydrogen donor for enzymes reducing sulphate and methionine sulphoxides and for a protein disulphide reductase [111]. Oxidised thioredoxin, which contains a disulphide, is in turn reduced by NADPH and thioredoxin reductase [111]. Oxidative stress induces glutathione oxidation and S-glutathionylation of proteins. Furthermore, oxidation of mitochondrial GSH to GSSG leads to glutathionylation of complex I, which increases mitochondrial superoxide formation and may further enhance oxidative stress [112].

S-Glutathionylation may also regulate gene expression by modulating the DNA-binding ability of transcription factors, such as c-Jun or NF-kB [113, 114]. Thus, the DNA binding activity of NF-kB can be inhibited by oxidative and nitrosative agents through glutathionylation of the p50 subunit [114]. On the other hand, nitric oxide and glutathione may form S-nitrosoglutathione, which inhibits c-Jun DNA binding by S-glutathionylation [113]. These mechanisms might be involved in the effects of oxidative and nitrosative stress on gene expression.

The many functions of glutathione have been reviewed in the past by us [7] and also by other authors. An excellent review, with an emphasis on the future foreseeable trends has recently been written by Helmut Sies [6].

9 Concluding Remarks

Glutathione, whose existence has been known for just over a century, is a very interesting molecule with critical functions in cells. Research into the function of this molecule, particularly the glutathionylation of proteins, the cellular compartmentation, and its function in pathophysiological states, promise exciting future developments.

References

1. Jocelyn PC (1973) Biochemistry of the SH group. Academic Press
2. Viña J, Hems R, Krebs HA (1978) Maintenance of glutathione content in isolated hepatocytes. Biochem J 170 (3):627–630
3. DeRey-Pailhade J (1888) Sur un corps d'origine organique hydrogenant le soufre a froid. CR Acad Sci 106:1683–1684
4. Meister A (1988) On the discovery of glutathione. TIBS 13:185–189
5. Kosower EM, Kosower NS (1969) Lest I forget thee, glutathione. Nature 224:117–120
6. Sies H (1999) Glutathione and its role in cellular functions. Free Rad Biol Med 27: 916–921
7. Viña J (1990) Glutathione: metabolism and physiological functions. CRC, Boca Raton, USA
8. Meister A (1983) Glutathione. In: Arias IM, Jakoby WB, Popper H, Schachter D, Shafritz DA (eds) The liver: biology and pathobiology, 2nd edn. Raven, New York, pp 401–417
9. Asensi M, Sastre J, Pallardó FV, García de la Asunción J, Estrela JM, Viña J (1994) A high-performance liquid chromatography method for measurement of oxidized glutathione in biological samples. Anal Biochem 217:323–328
10. Sastre J, Asensi M, Gascó E, Ferrero JA, Furukawa T, Viña J (1992) Exhaustive physical exercise causes and oxidation of glutathione status in blood. Prevention by antioxidant administration. Am J Physiol 263:R992–R995
11. De la Asunción JG, Millán A, Plá R, Bruseghini L, Esteras A, Pallardó FV, Sastre J, Viña J (1996) Mitochondrial glutathione oxidation correlates with age-associated oxidative damage to mitochondrial DNA. FASEB J 10:333–338

12. Esteve JM, Mompó J, De la Asunción JG, Sastre J, Boix J, Viña JR, Viña J, Pallardó FV (1999) Oxidative damage to mitochondrial DNA and glutathione oxidation in apoptosis studies in vivo and in vitro. FASEB J 13:1055–1064
13. Giustarini D, Dalle-Donne I, Colombo R, Milzani A, Rossi R (2003) An improved HPLC measurement for GSH and GSSG in human blood. Free Radic Biol Med 35:1365–7230.
14. Lu SC (1999) Regulation of hepatic glutathione synthesis: current concepts and controversies. FASEB J 13:1169–1183
15. Huang C, Anderson ME, Meister A (1993) Amino acid sequence and function of the light subunit of rat kidneyγ-glutamylcysteine synthetase. J Biol Chem 268:20578–2058330
16. Gipp JJ, Chang C, Mulcahy RT (1992) Cloning and nucleotide sequence of a full length cDNA for human liver γ-glutamylcysteine synthetase. Biochem Biophys Res Commun 185:29–3530.
17. Seelig GF, Simondsen RP, Meister A (1984) Reversible dissociation of γ-glutamylcysteine synthetase into two subunits. J Biol Chem 259:9345–9347
18. Tateishi N, Higashi T, Shinya S, Naruse A, Sakamoto Y (1974) Studies on the regulation of glutathione level in rat liver. J Biochem (Tokyo) 75(1):93–103
19. Huang CS, Moore WR, Meister A (1988) On the active site thiol of γ-glutamylcysteine synthetase: relationship to catalysis, inhibition, and regulation. Proc Natl Acad Sci USA 85:2464–246830
20. Fernández-Checa J, Lu SC, Ookhtens M, LeLeve L, Runnegar M, Yoshida H, Saiki H, Kannan R, García-Ruiz C, Kuhlenkamp JF, Kaplowitz N (1992) The regulation of hepatic glutathione. In: Tavoloni N, Berk PD (eds) Hepatic anion transport and bile secretion: Physiology and pathophysiology. Dekker, New York, pp 363–395
21. Lu SC, Ge J, Kuhlenkamp J, Kaplowitz N (1992) Insulin and glucocorticoid dependence of hepatic γ-glutamylcysteine synthetase and GSH synthesis in the rat: studies in cultured hepatocytes and in vivo. J Clin Inv 90:524–532
22. Cai J, Huang Z, Lu SC (1997) Differential regulation of γ-glutamylcysteine synthetase heavy and light subunit gene expression. Biochem J 326:167–172
23. Shi MM, Kugelman A, Iwamoto T, Tian L, Forman HJ (1994) Quinone-induced oxidative stress elevates glutathione and induces γ-glutamylcysteine synthetase activity in rat lung epithelial L2 cells. J Biol Chem 269:26512–317
24. Liu RM, Gao L, Choi J, Forman HJ (1998)γ-Glutamylcysteine synthetase: mRNA stabilization and independent subunit transcription by 4-hydroxy-2-nonenal. Am J Physiol 275:L861–L869
25. Tian L, Shi MM, Forman HJ (1997) Increased transcription of the regulatory subunit of γ-glutamylcysteine synthetase in rat lung epithelial L2 cells exposed to oxidative stress or glutathione depletion. Arch Biochem Biophys 342:126–133
26. Estrela JM, Gil F, Vila JM, Viña J (1988) Alpha-adrenergic modulation of glutathione metabolism in isolated rat hepatocytes. Am J Physiol 255:E801–5
27. Lu SC, Kuhlenkamp J, García-Ruiz C, Kaplowitz N (1991) Hormone-mediated down-regulation of hepatic glutathione synthesis in the rat. J Clin Inv 88:260–9
28. Sun W, Huang Z, Lu SC (1996) Regulation of γ-glutamylcysteine synthetase by protein phosphorylation. Biochem J 320:321–328
29. Cai J, Sun WM, Lu SC (1995) Hormonal and cell density regulation of hepatic gamma-glutamylcysteine synthetase gene expression. Mol Pharmacol 48:212–8
30. Oppenheimer L, Wellmer VP, Griffith OW, Meister A (1979) Glutathione synthetase. Purification from rat kidney and mapping of the substrate binding sites. J Biol Chem 254:5184–90

31. Huang ZA, Yang H, Chen C, Zeng Z, Lu SC (2000) Inducers of gamma-glutamylcysteine synthetase and their effects on glutathione synthetase expression. Biochim Biophys Acta 1493:48–5530
32. Yang H, Zeng Y, Lee TD, Yang Y, Ou X, Chen L, Haque M, Rippe R, Lu SC (2002) Role of AP-1 in the coordinate induction of rat glutamate-cysteine ligase and glutathione synthetase by tert-butylhydroquinone. J Biol Chem 277:35232–9
33. Huang ZZ, Chen C, Zeng Z, Yang H, Oh J, Chen L, Lu SC (2001) Mechanism and significance of increased glutathione level in human hepatocellular carcinoma and liver proliferation. FASEB J 15:19–2130
34. Meister A (1973) The enzymology of amino acid transport. Science 180:33–39
35. Viña JR, Palacín M, Puertes IR, Hernández R, Viña J (1989) Role of the gamma glutamyl cycle in the regulation of amino acid translocation. Am J Physiol 257:E916–E922
36. Viña JR, Pallardó FV, García C, Triguero A, Martín, JA, Boix J, Pellín A, Viña J, Sastre J (1996) Metabolic implications of β-cystathionase activity in mammals. In: Mato JM, Caballero A (eds) Methionine metabolism. molecular mechanism and clinical implications. CSIC, Jarpyo Editores, Madrid, pp 111–120
37. Tateishi N, Sakamoto Y (1983) Nutritional significance of glutathione in rat liver. In: Sakamoto Y et al. (eds) Glutathione storage, transport and turnover in mammals. Scientific Socoetu Press, Tokyo, pp. 13–38
38. Bannai S (1984) Transport of cystine and cysteine in mammalian cells. Biochim Biophys Acta 779:289–306
39. Kilberg MS, Christensen HN, Handlogten ME (1979) Cysteine as a system-specific substrate for transport system ASC in rat hepatocytes. Biochem Biophys Res Commun 88:744–751
40. Bannai S, Ishii T, Takada A, Noriko T (1989) Regulation of glutathione level by amino acid transport. In: Taniguchi N, Higashi T, Sakamoto Y, Meister A (eds) Glutathione centennial. Academic Press, San Diego, pp 407–421
41. Horowitz JH, Rypins EB, Henderson JM, Heymsfield SB, Moffitt SD, Bain RP, Chawla RK (1981) Evidence for impairment of transsulfuration pathway in cirrhosis. Gastroenterology 81:668–67530
42. Ávila MA, Berasain C, Torres L, Martín-Duce A, Corrales FJ, Yang H, Prieto J (2000) Reduced mRNA abundance of the main enzymes involved in methionine metabolism in human liver cirrhosis and hepatocellular carcinoma. J Hepatol 33:907–914
43. Serviddio G, Pereda J, Pallardó FV, Carretero J, Borrás C, Cutrin J, Vendemiale G, Poli G, Viña J, Sastre J (2004) Ursodeoxycholic acid protects against secondary biliary cirrhosis via up-regulation of γ-glutamyl cysteine synthetase and prevention of mitochondrial oxidative stress. Hepatology (in press)
44. Finkelstein JD (2000) Homocysteine: A history in progress. Nutr Rev 58:193–204
45. Pallardó FV, Sastre J, Asensi M, Rodrigo F, Viña J (1991) Physiological changes in glutathione metabolism in foetal and newborn rat liver. Biochem J 274:891–893
46. Viña J, Vento M, García-Sala F, Puertes IR, Gascó E, Sastre J, Asensi M, Pallardó FV (1995) L-cysteine and glutathione metabolism is impaired in premature infants due to cystathionase deficiency. Am J Clin Nutr 61:1067–1069
47. Viña J, Giménez A, Puertes IR, Gascó E, Viña JR (1992) Impairment of cysteine síntesis from methionine in rats exponed to surgical stress. British J Nutr 68:421–429
48. Ferrer JV, Gascó E, Sastre J, Pallardó V, Asensi M, Viña J (1990) Age-related changes in glutathione synthesis in the eye lens. Biochem J 269:531–534
49. Martín JA, Sastre J, García de la Asunción J, Pallardó FV, Viña J (2001) Hepatic β-cystathionase deficiency in patients with AIDS. JAMA 285:1444–5

50. Sastre J, Asensi M, Rodrigo F, Pallardó FV, Vento M, Viña J (1994) Antioxidant administration to the mother prevents oxidative stress associated with birth in the neonatal rat. Life Sci 54:2055–2059
51. Viña J, Sáez GT, Wiggins D, Roberts AF, Hems R, Krebs HA (1983) The effect of cysteine oxidation on isolated hepatocytes. Biochem J 212(1):39–44
52. Ammon HPT, Meliem MCM, Verspohl EJ (1986) Pharmacokinetics of intravenously administered glutathione in the rat. J Pharm Pharmacol 38:721
53. Wendel A, Cikryt P (1980) The level and half-life of lutahtione in human plasma. FEBS Lett 120:209
54. Akerboom T, Sies H (1990) Glutathione transport and its significance in oxidative stress. In: Viña J (ed) Glutathione: metabolism and physiological functions. CRC, Boca Raton, Florida, pp 45–55
55. Griffith OW, Meister A (1979) Glutathione: interorgan translocation, turnover and metabolism. Proc Natl Acad Sci USA 76:560630.
56. Häberle D, Wahlländer A, Sies H (1979) Assessment of the kidney function in maintenance of plasma glutathione concentration and redox state in anaesthetized rats. FEBS Lett 108:33530
57. Inoue M (1985) Interorgan metabolism and membrane transport of glutathione and related compounds. In: Kinne RKH (ed) Renal biochemistry. Elsevier, Amsterdam, p 225
58. McIntyre TM, Curthoys NP (1980) The interorgan metabolism of glutathione. Int J Biochem 12:545
59. Ookhtens M, Kaplowitz N (1998) Role of the liver in interorgan homeostasis of glutathione and cyst(e)ine. Semin Liver Dis 18:313–29
60. Akerboom TPM, Bilzer M, Sies H (1982) The relationship of biliary glutathione disulfide efflux and intracellular glutathione disulfide content in perfused rat liver. J Biol Chem 257:4248
61. Bartoli GM, Sies H (1978) Reduced and oxidized glutathione efflux from liver. FEBS Lett 86(1):89-91
62. Hahn R, Wendel A, Flohé L (1978) The fate of extracellular glutathione in the rat. Biochim Biophys Acta 539:32430
63. García-Ruiz C, Fernández-Checa JC, Kaplowitz N (1992) Bidirectional mechanism of plasma transport of reduced glutathione in intact rat hepatocytes and membrane vesicles. J Biol Chem 267:22256–64
64. Lu SC, García-Ruiz C, Kuhlenkamp J, Ookhtens M, Salas-Prato M, Kaplowitz N (1990) Hormonal regulation of glutathione efflux. J Biol Chem 265:16088–95
65. Sies H, Graf P (1985) Hepatic thiol and glutathione efflux under the influence of vasopressin, phenylephrine and adrenaline. Biochem J 226:545
66. Pyke S, Lew H, Quintanilha A (1986) Severe depletion in liver glutathione during physical exercise. Biochem Biophys Res Commun 139:926
67. Fernández-Checa JC, Ookhtens M, Kaplowitz N (1993) Selective induction by phenobarbital of the electrogenic transport of glutathione and organic anions in rat liver canalicular membrane vesicles. J Biol Chem 268:10836–41
68. García-Ruiz C, Morales A, Colell A, Rodes J, Yi JR, Kaplowitz N, Fernández-Checa JC (1995) Evidence that the rat hepatic mitochondrial carrier is distinct from the sinusoidal and canalicular transporters for reduced glutathione. Expression studies in Xenopus laevis oocytes. J Biol Chem 270:15946–9
69. Fernández-Checa JC, García-Ruiz C, Ookhtens M, Kaplowitz N (1991) Impaired uptake of glutathione by hepatic mitochondria from chronic ethanol-fed rats. J Biol Chem 87:397

70. Lluis JM, Colell A, García-Ruiz C, Kaplowitz N, Fernández-Checa JC (2003) Acetaldehyde impairs mitochondrial glutathione transport in HepG2 cells through endoplasmic reticulum stress. Gastroenterology 124:708–24
71. Ishikawa T, Sies H (1984) Cardiac transport of glutathione disulfide and S-conjugate. Studies with isolated perfused rat heart during hydroperoxide metabolism. J Biol Chem 259:3838
72. Kondo T, Dale GL, Beutler E (1981) Studies on glutathione transport utilizing inside-out vesicles prepared from human erythrocytes. Biochim Biophys Acta 645:132
73. Srivastava SK, Beutler E (1969) The transport of oxidized glutathione from human erythrocytes. J Biol Chem 244:9
74. Sies H, Akerboom TPM (1984) Glutathione disulfide efflux from cells and tissues. Methods Enzymol 105:445
75. Akerboom TPM, Bilzer M, Sies H (1982b) Competition between transport of glutathione disulfide (GSSG) and glutathione-S-conjugates from perfused rat liver into bile. FEBS Lett 140:73
76. Kondo T, Murao M, Taniguchi N (1982) Glutathione S-conjugate transport using inside-out vesicles from human erythrocytes. Eur J Biochem 125:551
77. Lu SC, Kuhlenkamp J, Ge JL, Sun WM, Kaplowitz N (1994) Specificity and directionality of thiol effects on sinusoidal glutathione transport in rat liver. Mol Pharmacol 46:578–85
78. Hagen TM, Brown LA, Jones DP (1986) Protection against paraquat-induced injury by exogenous GSH in pulmonary alveolar type II cells. Biochem. Pharmacol 35:453730
79. Lash LH, Hagen TM, Jones DP (1986) Exogenous glutathione protects intestinal epithelial cells from oxidative injury. Proc Natl Acad Sci USA 83:4641
80. Horiuchi S, Inoue M, Morino Y (1978) Gamma-glutamyl transpeptidase: sidedness of its active site on renal brush-border membrane. Eur J Biochem 87:42930
81. Lash LH, Jones DP (1984) Renal glutathione transport. Characteristics of the sodium-dependent system in the basal-lateral membrane. J Biol Chem 259:14508
82. Kannan R, Yi JR, Tang D, Zlokovic BV, Kaplowitz N (1996) Identification of a novel, sodium-dependent, reduced glutathione transporter in the rat lens epithelium. Invest Ophthalmol Vis Sci 37:2269–75
83. Kannan R, Bao Y, Wang Y, Sarthy VP, Kaplowitz N (1999) Protection from oxidant injury by sodium-dependent GSH uptake in retinal Muller cells. Exp Eye Res 68:609–16
84. Kannan R, Chakrabarti R, Tang D, Kim KJ, Kaplowitz N (2000) GSH transport in human cerebrovascular endothelial cells and human astrocytes: evidence for luminal localization of Na^+-dependent GSH transporter in HCEC. Brain Res 852:374–82
85. Kannan R, Kuhlenkamp JF, Ookhtens M, Kaplowitz N (1992) Transport of glutathione at blood-brain barrier of the rat: inhibition by glutathione analogs and age-dependence. J Pharmacol Exp Ther 263:964–70
86. Jones DP (2002) Redox potential of GSH/GSSG couple: assay and biological significance. Methods Enzymol 348:93–112
87. Jocelyn PC, Kamminga A (1974) The non-protein thiol of rat liver mitochondria. Biochem Biophys Acta 343:356–362
88. Wahllander A, Soboll S, Sies H, Linke I, Muller M (1979) Hepatic mitochondrial and cytosolic glutathione content and the subcellular distribution of GSH S-transferases. FEBS Lett 97(1):138–140
89. Griffith OW, Meister A (1985) Origin and turnover of mitochondrial glutathione. Proc Natl Acad Sci USA 82:4668–467230.
90. Linnane A, Marzuki S, Ozawa T, Tanaka M (1989) Mitochondrial DNA mutations as an important contributor to ageing and degenerative diseases. Lancet 8639:642–645

91. Miquel J, Economos AC, Fleming J, Johnson Jr JE (1980) Mitochondrial role in cell aging. Exp Gerontol 15:575–591
92. Sastre J, Pallardo FV, Pla R, Pellin A, Juan G, O'Connor JE, Estrela JM, Miquel J, Viña J (1996) Aging of the liver: age-associated mitochondrial damage in intact hepatocytes. Hepatology 24:1199–205
93. De la Asunción JG, del Olmo ML, Sastre J, Millan A, Pellin A, Pallardo FV, Viña J (1998) AZT treatment induces molecular and ultrastructural oxidative damage to muscle mitochondria. Prevention by antioxidant vitamins. J Clin Invest 102(1):4–9
94. Trimm JL, Salama G, Abramson JJ (1986) Sulfhydryl oxidation induces rapid calcium release from sarcoplasmic reticulum vesicles. J Biol Chem 261:16092–16098
95. Hwang C, Sinskey AJ, Lodish HF (1992) Oxidized redox-state of glutathione in the endoplasmic reticulum. Science 257:1496–150230
96. Voehringer DW, McConkey DJ, McDonnell TJ, Brisbay S, Meyn RE (1998) Bcl-2 expression causes redistribution of glutathione to the nucleus. Proc Natl Acad Sci USA 95: 2956–2960
97. Bellomo G, Vairetti M, Stivala L, Mirabelli F, Richelmi P, Orrenius S (1992) Demonstration of nuclear compartmentalization of glutathione in hepatocytes. Proc Natl Acad Sci USA 89:4412–4416
98. De Capoa A, Ferraro M, Lavia P, Pelliccia F, Finazzi-Agro A (1982) Silver staining of the nucleolus organizer regions (NOR) requires clusters of sulfhydryl groups. J Histochem Cytochem 30:908–11
99. Droge W (2002) Free radicals in the physiological control of cell function. Physiol Rev 82:47–95
100. Sáez G, Thornalley PJ, Hill HAO, Hems R, Bannister JV (1982) The production of free radicals during the autoxidation of cysteine and their effect on isolated rat hepatocytes. Biochim Biophys Acta 719:24–31
101. Barrett WC, DeGnore JP, Konig S, Fales HM, Keng YF, Zhang ZY, Yim MB, Chock PB (1999) Regulation of PTP1B via glutathionylation of the active site cysteine 215. Biochemistry 38:6699–705
102. Casagrande S, Bonetto V, Fratelli M, Gianazza E, Eberini I, Massignan T, Saltona M, Chang G, Holmgren A, Ghezzi P (2002) Glutathionylation of human thioredoxin: a possible crosstalk between the glutathione and thioredoxin systems. Proc Natl Acad Sci USA 99:9745–9749
103. Borges CR, Geddes T, Watson JT, Kuhn DM (2002) Dopamine biosynthesis is regulated by S-glutathionylation. Potential mechanism of tyrosine hydroxylase inhibition during oxidative stress. J Biol Chem 277:48295–302
104. Fratelli M, Demol H, Puype M, Casagrande S, Eberini I, Salmona M, Bonetto V, Mengozzi M, Duffieux F, Miclet E, Bachi A, Vandekerckhove J, Gianazza E, Ghezzi P (2002) Identification by redox proteomics of glutathionylated proteins in oxidatively stressed human T lymphocytes. Proc Natl Acad Sci USA 99:3505–10
105. Humphries KM, Juliano C, Taylor SS (2002) Regulation of cAMP-dependent protein kinase activity by glutathionylation. J Biol Chem 277:43505–11
106. Mohr S, Hallak H, de Boitte A, Lapetina EG, Brune B (1999) Nitric oxide-induced S-glutathionylation and inactivation of glyceraldehydes-3-phosphate dehydrogenase. J Biol Chem 274:9427–30
107. Nulton-Persson AC, Starke DW, Mieyal JJ, Szweda LI (2003) Reversible inactivation of alpha-ketoglutarate dehydrogenase in response to alterations in the mitochondrial glutathione status. Biochemistry 42:4235–4242
108. Rao RK, Clayton LW (2002) Regulation of protein phosphatase 2A by hydrogen peroxide and glutathionylation. Biochem Biophys Res Commun 293:610–6

109. Reddy S, Jones AD, Cross CE, Wong PS, Van Der Vliet A (2000) Inactivation of creatine kinase by S-glutathionylation of the active-site cysteine residue. Biochem J 347:821–7
110. Holmgren A (1990) Glutaredoxin: structure and function. In: Viña J (ed) Glutathione: metabolism and physiological functions. CRC, Boca Raton, Florida, pp 145–15430
111. Holmgren A (1984) Enzymatic reduction-oxidation of protein disulfides by thioredoxin. Methods Enzymol 107:295–300
112. Taylor ER, Hurrell F, Shannon RJ, Lin TK, Hirst J, Murphy MP (2003) Reversible glutathionylation of complex I increases mitochondrial superoxide formation. J Biol Chem 278:19603–10
113. Klatt P, Molina EP, Lamas S (1999) Nitric oxide inhibits c-Jun DNA binding by specifically targeted S-glutathionylation. J Biol Chem 274:15857–64
114. Pineda-Molina E, Klatt P, Vazquez J, Marina A, García de Lacoba M, Pérez-Sala D, Lamas S (2001) Glutathionylation of the p50 subunit of NF-kappaB: a mechanism for redox-induced inhibition of DNA binding. Biochemistry 40:14134–42

The Handbook of Environmental Chemistry Vol. 2, Part O (2005): 109–149
DOI 10.1007/b101149

Superoxide Dismutases and Catalase

Grzegorz Bartosz (✉)

Department of Molecular Biophysics, University of Lodz, Banacha 12/16, 90–237 Lodz, Poland and Department of Biochemistry and Cell Biology, University of Rzeszów, Rejtana 16 C, 35-959 Rzeszov, Poland
gbartosz@biol.uni.lodz.pl

Abstract Superoxide dismutases (SODs) and catalase represent the primary enzymatic defense against reactive oxygen species. Both enzymes are present in virtually all types of aerobic cells. Both are metalloproteins, employing efficient dismutation reactions to dispose of the two most common reactive oxygen species formed, the superoxide radical anion and hydrogen peroxide. These very fast reactions do not require reducing equivalents and thus energy input. Induction of expression of these enzymes in cell culture and in whole organisms provides protection against deleterious effects of oxidative stress in various situations. Transgenic organisms provided many clues concerning the physiological significance of superoxide dismutases. In the human, mutations of CuZn-superoxide dismutase underly cases of familial amyotrophic sclerosis; pathological aspects of polymorphism of superoxide dismutases have also been extensively studied. Application of the enzymes in therapy and for analysis of the level of reactive oxygen species has been suggested.

Keywords Superoxide dismutase · Catalase · Metalloproteins · Reactive oxygen species · Antioxidant defense

Abbreviations

ALS	Amyotrophic lateral sclerosis
AP	Activator protein
CAT	Catalase
EC-SOD	Extracellular superoxide dismutase
IL	Interleukin
INF	Interferon
NF	Nuclear factor
SA	Salicylic acid
SOD	Superoxide dismutase
SOR	Superoxide reductase
Sp	Specificity protein
TNF	Tumor necrosis factor

1 Superoxide Dismutases

Superoxide dismutases (SODs; EC 1.15.1.1) are a family of metalloenzymes that catalyze the dismutation of superoxide radical anion to hydrogen peroxide and oxygen:

$$O_2^{\bullet -} + O_2^{\bullet -} + 2H^+ \rightarrow H_2O_2 + O_2 \quad (1)$$

1.1 Diversity of Superoxide Dismutases

In animals and fungi, there are three types of SOD, a dimeric cytoplasmic CuZnSOD (SOD1), a tetrameric mitochondrial MnSOD (SOD2) and a tetrameric extracellular CuZnSOD (EC-SOD, SOD3, Fig. 1). Although all these proteins catalyze the same reaction, they are encoded by various genes and differ in their structure and localization.

CuZnSOD is present mainly in the cytoplasm of animal cells but was also detected in lysosomes, peroxisomes, nucleus, and intermembrane space of mitochondria [1,2]. MnSOD is located in the matrix space of mitochondria [3]. However, in decapod crustacea, which contain no CuZnSOD, MnSOD is present both in the mitochondria and in the cytosol [4]. EC-SOD is a secretory form of SOD present on the surface of many cells and outside the cells in blood plasma, lymph, ascites, and cerebrospinal fluid. However, the primary location of EC-SOD in tissues is the extracellular matrix and on cell surfaces where it is found at 20 times the concentration present in blood plasma. Tissue EC-SOD is thought to account for 90–99% of the EC-SOD of the body [5,6]. The content of EC-SOD is the highest in blood vessels, epididymis, heart, lung, kidney, testis, Sertoli and germ cells, and uterus [5, 6]. EC-SOD is also synthesized and secreted by a

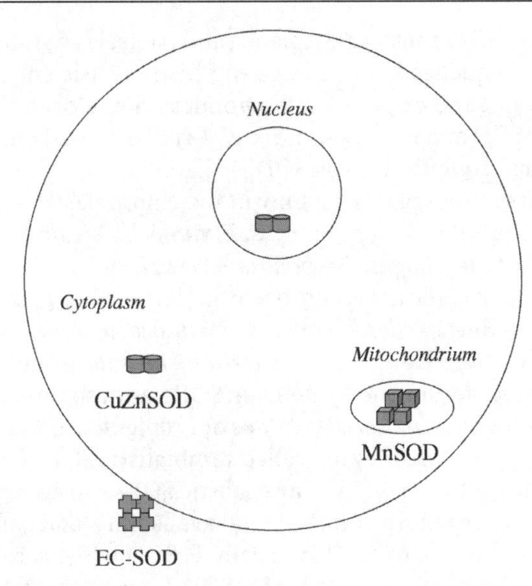

Fig. 1 Three types of SODs present in animal cells

variety of fibroblast, glial, and endothelial cell lines [6, 7]. EC-SOD present in the vasculature is due mainly to secretion by vascular smooth muscle cells [8]. In the brain the expression of EC-SOD is much lower than in other tissues; in the mouse brain the enzyme is localized mainly in the hippocampus, striatum, suprachiasmatic nuclei, and the habenula [9]. In vitro, expression of SOD3 is restricted to only a few cell types. High levels of SOD3 expression were found in alveolar type II cells, proximal renal tubular cells, vascular smooth muscle cells, lung macrophages and a few cultured fibroblast cell lines [10]. Among mammals, there are species which are high-expressing EC-SOD (man, pig, sheep, and cow) and those which are low-expressing (dog, cat, and rat) [11]. In mammalian lungs, the level of EC-SOD ranges from less than 1% (rat, cat, and dog) to about 10% (mouse and human) of total SOD activity.

The presence of extracellular-type SOD, though of different structure (dimers of 18 kDa), has been reported in plants [12].

CuZnSOD constitutes 85–90% of total SOD activity of mammalian tissues while MnSOD usually accounts for 10–15%. The MnSOD/CuZnSOD activity ratio of tissue homogenates depends on the cellular content of mitochondria. Total SOD activity of mammalian erythrocytes, having no mitochondria, corresponds to CuZnSOD; in rat liver, MnSOD is responsible for about 10% of total SOD activity [13].

The cytoplasm of *Escherichia coli* contains dimeric MnSOD, dimeric FeSOD, and a hybrid enzyme composed of FeSOD and MnSOD subunits [3]. The amino acid sequences of MnSOD and FeSOD are similar; however, in most cases they are active only with the correct metal ion at the active site. Though the eukaryotic MnSODs are tetrameric, and bacterial mostly dimeric, amino acid sequences of MnSOD from animals, plants, and bacteria show homologies and are unrelated to sequences of CuZnSODs. This is consistent with the endosymbiotic theory of the origin of mitochondria. Some bacteria (like *E. coli*) contain both MnSOD and FeSOD, some (e.g., *Bacillus cereus*) only FeSOD, other (e.g., *Streptococcus sanguis*) only FeSOD [13].

FeSODs occurs also in plants (mainly in the chloroplasts) and are common in protozoa, e.g., *Tetrahymena pyriformis, Entamoeba histolytica, Plasmodium falciparum, Leishmania chagasi, Trypanosoma cruzi* and *Trichomonas vaginalis.*

SODs are found in some anaerobic bacteria (*Bacteroides fragilis, Bacteroides thetaiotaomicron, Bacteroides gingivalis, Streptococus mutans, Propionibacterium shermanii*) but also in *Mycobacterium smegmatis* and in the aerobic methylolotrophic bacterium *Methylomonas.* SODs from some bacteria, e.g., *Porphyromonas gingivalis, Strptococcus mutans* or *Propionibacterium shermanii* are active with either Fe or Mn; they are called cambialistic SODs (capable of making a cofactor substitution) [14, 15]. Cambialistic SODs can accommodate either iron or manganese, depending on the metal availability, and conserve activity. In the anaerobes, the anaero-SOD is usually FeSOD while aero-SOD (formed during transient exposure to oxygen) is MnSOD. It seems that under conditions of oxidative stress when iron becomes unavailable (as Fe^{3+} precipitates), the ability to substitute Fe for Mn enables the bacteria to survive.

Initially CuZnSOD was believed to occur only in eukaryotes but subsequently it was detected in some bacteria (firstly in the luminescent *Photobacterium leiognathi*, living in symbiosis with the ponyfish). It has been suspected that the bacterium acquired the gene coding for CuZnSOD from its host by gene transfer. However, CuZnSOD was subsequently detected in free-living (non-symbiotic bacteria): *E. coli, Caulobacter crescentis, Pseudomonas diminuta, Pseudomonas maltophila, Brucella abortus, Paracoccus denitrificans, Salmonella typhimurium, Legionella pneumophila*, several species of *Haemophilus, Actinobacillus* and *Pasteurella* (HAP) group, and in *Neisseria meningitidis* [3, 16]. In *E. coli*, small amounts of CuZnSOD are present in the periplasmic space (i.e., the space between cell wall and bacterial cell membrane). They are related to eukaryotic CuZnSODs but differ from them by the presence of a 22–23-residue leader sequence.

Novel types of SOD, a Ni-containing enzyme, and an iron- and zinc-containing enzyme, have been found in the filamentous Gram-positive bacterium *Streptomyces* ssp. [17–19].

1.2 Structure

Eukaryotic CuZnSODs are dimers composed of two identical 16 kDa subunits, each containing one catalytically essential Cu^{2+} ion bridged by a His residue to solvent-inaccessible Zn^{2+} ion. However, the CuZnSOD from *E. coli* is completely stable as a monomer [20].

Each 151-residue subunit of CuZnSOD has an eight-stranded Greek β-barrel key scaffold plus three external loops of non-repetitive structure. The active-site Cu ion is at the bottom of a deep channel on the outside of the β-barrel between two large loops. The Cu^{2+} is partly exposed while the Zn^{2+} is completely buried. The metal ligands form the floor of the narrow part of the channel (0.3×1.5 nm), which lies between Lys 134 and Arg 141. Eighteen solvent-exposed residues form the molecular surface of the active-site channel, which has an area of ca 6.6 nm^2 (11% of the total exposed surface). The surface shows two pits forming specific binding sites in the narrowest part of the active-site channel floor. One pit (the "Cu site") is formed by the exposed surface of the Cu^{2+} and parts of His 61, His 118, Thr 135, and Arg 141. The adjacent pit (the "water site") is formed by parts of the exposed surface of Thr 135, Gly 136, Ala 138, Gly 139, and Cu ligands His 44, and His 118.

MnSODs and FeSODs from most procaryotic organisms are dimeric while MnSODs from mitochondria and some thermophilic bacteria are tetrameric [21]. However, mitochondrial MnSOD from *Caenorhabditis elegans* was found to be dimeric [22]. Eukaryotic MnSOD is a tetrameric protein encoded in the nucleus, synthesized in the cystosol, and imported post-translationally into the mitochondrial matrix. The 25 kDa precursor protein has a mitochondrial transit peptide that is cleaved to produce the mature 22 kDa subunit. The mature protein exists as a tetramer, each subunit containing one Mn ion. The mitochondrial MnSOD primary sequences are highly homologous to the prokaryotic Mn- and FeSOD, but has no resemblance to the CuZnSODs [23]. In MnSODs and FeSODs, the metal ions are coordinated by three His N atoms and one Asp O atom [24].

EC-SODs are usually a tetrameric (dimeric in some species) glycoproteins composed of four identical subunits, each containing one copper and one zinc atom. EC-SOD exists in mammalian cells as a homodimer of molecular weight 135 kDa with high affinity for heparin [25]. The molecular weight of a non-glycosylated subunit of human EC-SOD is 24 kDa. Each tetramer is composed of two dimers linked through disulfide bridges formed betweeen C-terminal Cys residues located in each subunit. In addition to dimers and tetramers, active larger oligomers can also be formed [26]. The subunits are synthesized as a 240-amino acid propeptide, containing an 18-signal peptide that is cleaved to yield a 222-residue mature protein [27]. The structure of the active site is similar to that of CuZnSOD but the amino acid sequences of both types of SOD differ.

Overall, the cytoplasmic CuZnSOD and the active-site containing domain of EC-SOD have roughly 50% amino acid sequence identity. Enzymes of both classes have conserved residues that are required for activity and metal binding: the four His residues involved in copper binding, the three His and one Asp residue involved in zinc binding, the two Cys involved in the formation of the disulfide bridge, and Arg in the entrance to the active site [28].

Each subunit of EC-SOD is composed of four domains: an N-terminal signal peptide (removed during maturation of the protein), a glycosylation domain (amino acids 1–95 in human EC-SOD) containing Asn 89 as the site of *N*-glycosylation, the domain containing the active site (residues 96–193), and a hydrophilic C-terminal heparin-binding domain (residues 194–222) containing a cluster of nine positively charged residues (three Lys and six Arg), responsible for binding to heparin and to sulfated proteoglycans on cell surfaces [29]. Glycosylation contributes to the solubility of the enzyme [27, 30]. Interestingly, it has been demonstrated that the heparin-binding domain of EC-SOD may act as a nuclear localization signal in some cell types [31].

When subjected to heparin-sepharose chromatography, blood plasma EC-SOD divides into three fractions: one with no affinity for heparin (fraction A), one with weak affinity for heparin (fraction B), and one with strong affinity (fraction C). The A and B forms of EC-SOD are derived from the C form by post-translational proteolytic cleavage at the carboxyl terminus. Most tissue EC-SOD exists as type C [32] although rat EC-SOD exists only in the A and B forms [33]. Fraction C consists of tetramers containing subunits with intact C-terminal heparin-binding domains. Fraction B corresponds to heterogenous EC-SOD tetramers, some subunits containing intact heparin-binding domains and some with the third domain removed. Fraction A consists of homotetramers in which all subunits have the heparin-binding domain removed by proteolysis [29]. EC-SOD C exists in equilibrium between the endothelial cell surfaces and blood plasma; injection of heparin leads to an increase in the plasma EC-SOD content, mostly of type C [34]. EC-SOD-C has a half-life of about 85 h, while truncated EC-SOD-A of only 7 h [35].

Two major kinds of SODs, CuZnSODs and Fe/MnSODs do not seem to have a common ancestor protein. Extracellular SOD diverged from the cytosolic form at early stages of evolution, before the differentiation of fungi, plants, and metazoa [36]. Interestingly, comparison of sequences demonstrated that during the last billion years, MnSOD evolved at a relatively constant rate while CuZnSOD evolved initially very slowly but unusually quickly in the last 100 million years [37].

NiSOD from *Streptomyces* contains about 1 atom of nickel per subunit and is present as homo-tetramers of 13 and 22 kDa subunits [17–19].

1.3 Properties

CuZnSOD is an unusually stable protein, showing high resistance to heating, proteolysis, and denaturing agents such as sodium dodecyl sulfate, guanidinium chloride, and urea, and being active in a broad range of pH. It survives the Tsuchihashi procedure (treatment with chloroform and ethanol); in fact, this procedure is employed for isolation of the enzyme and estimation of its activity in erythrocytes where hemoglobin must be removed prior to the assay [3]. Unlike mammalian CuZnSOD, the *E. coli* enzyme is extremely sensitive to proteases [38].

CuZnSOD is inhibited by azide and cyanide and (like other copper enzymes) inactivated by diethyldithiocarbamate, which binds to the copper and removes it from the active site of the enzyme, and by H_2O_2. EC-SOD is also very stable

Table 1 Inhibitors of various superoxide dismutases

Inhibitor	Diethyldithiocarbamate	H_2O_2	KCN
CuZnSOD	Inhibits	Inhibits	Inhibits
EC-SOD	Inhibits	Inhibits	Inhibits
MnSOD	–	–	–
FeSOD	–	Inhibits	–

and shows considerable resistance to high temperatures, pH extremes, and high urea concentrations [39], and is inhibited like CuZnSOD [5]. Stability of MnSODs and FeSODs to physical agents is generally lower. MnSOD is more susceptible than CuZnSOD to heat and detergents but is not inhibited by cyanide, diethyldithiocarbamate, or H_2O_2. FeSOD, unlike MnSOD, is inhibited by H_2O_2 (Table 1). Azide inhibits the enzymes in the following order: FeSOD >MnSOD>CuZnSOD. Treatment with 2% sodium dodecyl sulfate at 37 °C for 30 min inhibits MnSOD activity but not CuZnSOD activity [40, 41]. Enzymes from extremophiles may be exceptions to the rules of stability to physical agents; Fe-SOD of the hyperthermophilic archaeon *Acidianus ambivalens* has an extraordinarily high melting temperature of 128 °C, which is probably the highest melting temperature of a native protein. The purification procedure of this protein included a 2-h heat denaturation of contaminating proteins at 95 °C [42].

In order to identify the type of SOD after separation by native polyacrylamide gel electrophoresis, 2 mM KCN and 4 mM H_2O_2 can be added to the staining mixture to study the inhibition pattern.

1.4 Mechanism of Action

SODs accelerate, by four orders of magnitude, the spontaneous reaction of dismutation of superoxide radical anion by a cyclic oxidation-reduction mechanism of an active site metal ion. In the case of CuZnSOD, the Cu ion is involved in the catalysis. In the catalytic cycle, the Cu is reversibly oxidized and reduced by successive encounters with $O_2^{\bullet-}$ to yield O_2 and H_2O_2:

$$E\text{-}Cu^{2+} + O_2^{\bullet-} + H^+ \xrightarrow{k_1} E'\text{-}Cu^+ + O_2 \tag{2}$$

$$E'\text{-}Cu^+ + O_2^{\bullet-} + H^+ \xrightarrow{k_2} E\text{-}Cu^{2+} + H_2O_2 \tag{3}$$

Summarizing:

$$2O_2^{\bullet-} + 2H^+ \rightarrow O_2 + H_2O_2$$

The reaction is very fast and virtually diffusion-limited. The reactions of $O_2^{\bullet-}$ with either the reduced or oxidized form of the enzyme have extremely high and equal rate constants $k_{+1}=k_{+2}=k=2.3\times10^9\ M^{-1}\ s^{-1}$ [43].

In the first catalytic step, the transition is facilitated by relaxation (mediated through the His 61 bridging the Cu and Zn ions, and the pair of H-bonds to Asp 122) and distortions in Zn-ligand geometry found for the Cu^{2+} enzyme (Fig. 2). In the second step, the roles of Zn are to ensure protonation of the nitrogen atom of His 61 and to position the proton correctly for H-bonding and subsequent transfer to the second incoming $O_2^{\bullet-}$, so completing the catalytic cycle. Thus, the zinc atom is not involved in the redox cycle but maintains the configuration of the active site and facilitates the oxidation step.

The narrow width of the channel (<0.4 nm) above the Cu means that the Cu ion is accessible to Cl^- (0.362 nm) and Br^- (0.390 nm) but not I^- (0.432 nm). The catalytic rate of native SOD is independent of pH in the range 5.0–9.5 [44] indicating that the protons required by the reaction stoichiometry do not come from bulk water [45].

At physiological pH, mammalian CuZnSOD has a net negative charge that varies with species; pI values range from 4.6 to 6.8 [46]. The electrostatic barrier to incoming superoxide resulting from the net negative charge of the enzyme molecule should diminish with increasing ionic strength thus accelerating the reaction. Instead, raising the ionic strength slows catalysis in the native enzyme implying that local electrostatic forces facilitate the reaction. Indeed, the distribution of electrostatic potential due to the presence of ionized amino acid residues shows no organized pattern over most of the surface of the enzyme molecule but reveals a striking positive region that extends over the long, deep channel above the catalytic Cu ion. The positive potential near the Cu ion is very high, forming a complementary binding site for the negatively charged $O_2^{\bullet-}$ substrate. The arrangement of electrostatic charges near the active site provides an electrostatic guidance mechanism directing the negatively charged substrate to the highly positive catalytic binding site at the bottom of the active-site channel. For bovine enzyme, the positive charges of ε-amino groups of Lys 120 and Lys 134 and the negative charge γ-carboxylic group of Glu 131 seem to have important roles in directing the long-range approach of $O_2^{\bullet-}$ while the guanidinium group of Arg 141 has local orienting effects [47, 48]. The rate constant for SOD dismutation of $O_2^{\bullet-}$ ($2.3\times10^9\ M^{-1}\ s^{-1}$) is about 10% of the maximum diffusion-controlled limit calculated from collision theory, neglecting steric and orientation factors, while the catalytic Cu ion forms only about 0.1% of the enzyme's molecular surface. However, the entire active-site channel forms about 10% of the surface, so the observed rate constant could be rationalized by assuming that all collisions with the active-site channel are productive. It is possible due to pre-collision orientation and/or enhancement of diffusion provided by the electrostatic guidance mechanism [47].

In the human enzyme, Glu 132, Glu 133, and Lys 136 (which are involved in the electrostatic guidance) form a hydrogen-bonding network at the rim of the active-site channel, with the positively charged nitrogen of Lys 136 tethered between the negatively charged carboxylates of Glu 132 and Glu 133. Site-specific mutants of the enzyme, in which Glu 132 or Glu 133 residues were replaced

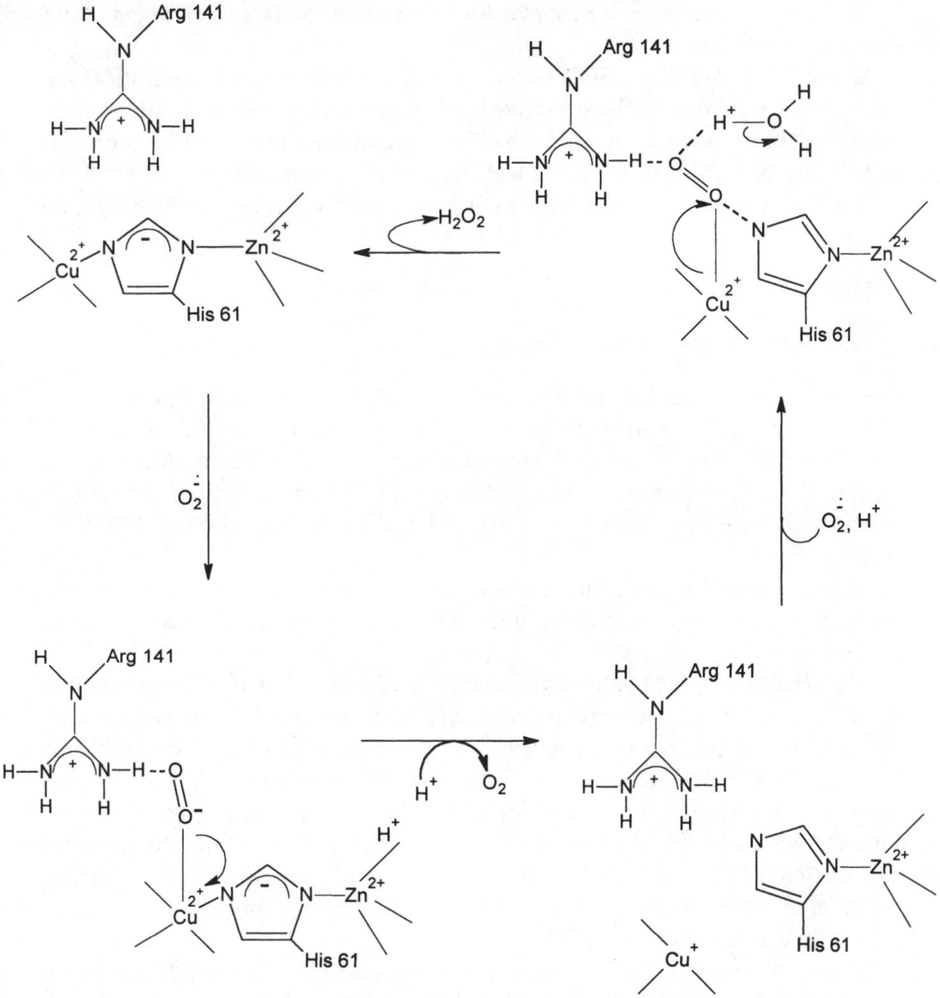

Fig. 2 Mechanism of catalytic action of CuZnSOD. After [45], modified (copyright agreement pending). Superoxide radical anion binds with one oxygen atom to the Cu^{2+} while its second oxygen forms a hydrogen bond to a guanidinium nitrogen of Arg 141. Bound superoxide reduces Cu^{2+} to Cu^{+} simultaneously breaking the bond between His 61 and Cu^{+}. An oxygen molecule is released. A second superoxide binds with one oxygen to Cu^{+} and its second oxygen forms a hydrogen bond with Arg 141. Zn^{2+} bound to His 61 raises the pK_a of its nitrogen to about 13 to make it always protonated at physiological pH. Electron transfer from Cu^{+} coupled with proton transfer from His 61 forms Cu^{2+}-hydroperoxide. Addition of a second proton from an active-site water yields the neutral hydrogen peroxide molecule, which leaves the active site

by Gln, have higher activity than wild-type human SOD and stronger ionic strength dependence [49].

Bacterial (*E. coli*) CuZnSOD lacks most of the charged residues critical for electrostatic guidance of the substrate but has a compensatory and functionally equivalent mechanism contributed by Lys 60 providing a reaction rate constant, $(1.4\pm0.1)\times10^9\ M^{-1}\ s^{-1}$, not much lower than mammalian enzymes [50].

MnSOD dismutates superoxide owing to reactions of the active site manganese:

$$Mn^{3+} + O_2^{\bullet-} \rightarrow Mn^{2+} + O_2 \quad (4)$$

$$Mn^{2+} + O_2^{\bullet-} + 2H^+ \rightarrow Mn^{3+} + H_2O_2 \quad (5)$$

FeSOD acts in an analogous manner. MnSOD is slightly less efficient than FeSOD, with k_{cat}/K_M values of $5.6\times10^7\ M^{-1}\ s^{-1}$ and $3\times10^8\ M^{-1}\ s^{-1}$, respectively [24]. However, while at pH 7.0 the reaction rates of CuZnSOD and MnSOD are similar, the reaction rates of MnSOD and FeSOD decrease at higher pH (so assays of tissue extracts at alkaline pH underestimate the activity of MnSOD) [13].

As in CuZnSOD, electrostatic surface potentials at the active site funnel steer the negatively charged substrate into the active site of MnSOD and FeSOD [24, 51].

There are conflicting data on the redox potential of the Cu^{2+}/Cu^+ transition in CuZnSOD. It has been argued that in oxidized SOD, the presence of Zn and the tetrahedral distortion of the Cu^{2+} site increase the redox potential of the Cu^{2+} to 0.42 V [52], which is much higher than the value of 0.17 V for an aqueous Cu^{2+}/Cu^+ pair or the 0.01 V for Cu^{2+}-His/Cu^+-His complexes. However, a much lower value (0.12 V) has also been reported [53]. The reduction potentials of FeSOD of *E. coli* and human MnSOD are lower; the apparent redox potential of the Fe^{3+}/Fe^{2+} and Mn^{3+}/Mn^{2+} transition were reported to be –0.07 V and –0.29 V, respectively [53, 54].

It has been suggested that spontaneous dismutation of superoxide generates singlet oxygen and that SOD-catalyzed dismutation liberates ground-state (triplet oxygen); however, further studies dismissed this view [13].

1.5 Other Functions of Superoxide Dismutase

A key feature of CuZnSOD is the maintenance of high dismutase activity while shielding the active site copper from other redox transformations [55]. However, it has been repeatedly suggested that apart from its main activity, CuZnSOD shows other enzymatic activities. One reason for these suggestions involved the role of CuZnSOD mutations in the pathogenesis of amyotrophic lateral sclerosis (see Sect. 1.9).

It has been claimed that CuZnSOD could catalytically scavenge singlet oxygen [56]. Although SODs, as other proteins, may react with singlet oxygen, this reaction is due to the oxidation of His, Try, and Met residues present in the apoenzyme; no enzyme is known to enzymatically scavenge singlet oxygen [13].

CuZnSOD has been demonstrated to act as a peroxidase, oxidizing various substrates, among them nitrite (to NO_2) [57] and relatively bulky molecules such as 5,5-dimethyl-1-pyrroline *N*-oxide (DMPO; to DMPO-OH), tyrosine (to dityrosine) or 2,2′-azino-bis(3-ethylbenzothiazoline-6-sulfonic acid) (ABTS; to a cation radical). However, the enzyme becomes inactivated in the presence of H_2O_2. Apparently, hydrogen peroxide reduces Cu^{2+} at the active site of the enzyme. Subsequent reaction of H_2O_2 with Cu^+ generates a potent oxidant (Cu^+-O or Cu^{2+}-OH) that can attack the adjacent histidine residue and thus inactivate the enzyme, or oxidize an alternative substrate [3]. Most authors assume that the mechanism of the peroxidative action of CuZnSOD, significantly accelerated in the presence of bicarbonate, consists in oxidation of bicarbonate to a carbonate radical anion able to oxidize other substrates [58, 59]:

$$\text{E-Cu}^+ + H_2O_2 \rightarrow \text{E-Cu}^{2+} - OH + HO^- \tag{6}$$

$$\text{E-Cu}^{2+} - OH + HCO_3^- \rightarrow \text{E-Cu}^{2+} + H_2O + CO_3^{\bullet-} \tag{7}$$

CuZnSOD catalyzes nitration of free tyrozine and tyrosyl residues in proteins by peroxynitrite [60, 61]. Interestingly, MnSOD, which is not an efficient catalyst of this reaction, is inactivated by nitration of Tyr 34 [62, 63]. CuZnSOD has been reported to catalyze breakdown of nitrosothiols, releasing NO [64].

Some anaerobes that lack SOD contain superoxide reductase (SOR; EC 1.15.1.2) which catalyzes only the reduction of superoxide to hydrogen peroxide:

$$SOR_{red} + O_2^{\bullet-} + 2H^+ \rightarrow SOR_{ox} + H_2O_2 \tag{8}$$

The oxidized form of superoxide reductase formed in this reaction is reduced back by rubredoxin, dependent ultimately on reduced pyridine nucleotides via intermediate electron carriers [65]. The reaction of SOR with superoxide is also very fast, the reaction rate constant being of an order of 10^9 M^{-1} s^{-1}. It has been demonstrated that CuZnSOD can also function as superoxide reductase reducing superoxide at the expense of oxidation of ferrocyanide, or as superoxide oxidase, oxidizing superoxide at the expense of reducing ferricyanide [66]. Both ferri- and ferrocyanide are unphysiological substrates but the enzyme can also act as superoxide reductase with nitroxyl anion oxidizing it to nitric oxide [67].

$$\text{SOD-Cu}^+ + O_2^{\bullet-} \rightarrow \text{SOD-Cu}^{2+} + H_2O_2 \tag{3}$$

$$\text{SOD-Cu}^{2+} + NO^- \rightarrow \text{SOD-Cu}^+ + NO^\bullet \tag{9}$$

CuZnSOD (but not MnSOD) has an activity of thiol oxidase:

$$\text{2R-SH} + O_2 \xrightarrow{\text{SOD}} \text{R-SS-R} + H_2O_2 \tag{10}$$

Cysteamine and then cysteine are the best substrates for this reaction; K_m of bovine CuZnSOD for cysteine is 1.4 mM and V_{max} is 0.49 μM s^{-1} with 1 μM enzyme. Other thiols, including glutathione, are much less reactive [55].

These "side reactions" of superoxide dismutase are generally of no biological significance (except for a possibility in pathology, see Sect. 1.9). Nevertheless, they may be important under peculiar conditions; e.g., due to its peroxidase activity, SOD may increase rather than decrease oxidation of a detector of reactive oxygen species [68]. The oxidation rate of a popular probe used for detection of reactive oxygen species, 2′,7′-dichlorofluorescin, is considerably augmented by CuZnSOD, especially in the presence of bicarbonate [67].

1.6 Molecular Biology and Genetics

In procaryotes, short upstream regulatory sequences (generally less than 100 bp) known as antioxidant response elements (AREs) control the expression of antioxidant enzymes. In *E. coli*, up-regulation of SODs involves *soxR* and *soxS* gene products [69].

The human CuZnSOD (SOD1) gene has been localized to chromosome 21 (region 21q22) [10]. Patients with Down syndrome (trisomy 21) show a 50% increase in SOD1 activity due to elevation of the level of CuZnSOD protein but the role of SOD1 in the pathology associated with this disease remains questionable [70].

The SOD1 gene has 5 exons and 4 introns. The promoter region of the gene contains TATA and CCAAT boxes and several putative binding sites for nuclear factor-1 (NF-1), specificity protein-1 (Sp-1), activator protein-1 and -2 (AP-1, AP-2), glucocorticoid response element (GRE), heat shock transcription factor (HSF), and nuclear factor κB (NF-κB) and the metal responsive element [10, 71, 72]. The synthesis of CuZnSOD has been shown to be increased by heat shock in rat lungs [73] and by shear stress in aortic endothelial cells [74]. Biosynthesis of the enzyme is generally not induced by cytokines or oxidant stress in vivo and in vitro [75] although in keratinocytes, CuZnSOD expression is increased by endogenous or exogenous nitric oxide [76].

MnSOD (SOD2) gene is localized in human chromosome 6 (region 6q25) [10, 77]. The SOD2 gene has 5 exons and 4 introns. Its promoter region contains no upstream TATA or CAAT box elements and CG-rich regions which is typical of "housekeeping" genes, but has putative sites for NF-κB, Sp-1 and AP-2 [10]. Sp-1 element positively promotes transcription while AP-2 proteins repress the promoter activity [78].

In the human, MnSOD is unique among SODs since it is up-regulated in response to oxidative stress, probably as a key compensatory response. MnSOD has been demonstrated to be induced in various cell types by ionizing radiation, cigarette smoke, asbestos fibers, ozone, hyperoxia [79], and paraquat and following treatment with endotoxin [80], TNF-α, INF-γ, interleukin-1 (IL-1), IL-6, thiol-reducing agents and some drugs [81–83]. Cytokine-induced activation of MnSOD is rapid (seen already after 2 h, with maximum between 24 and 48 h) and usually requires low concentrations of cytokines (0.1 to 1 ng mL^{-1}) [84]. In A549 lung epithelial cells treated with TNF-α, the mRNA level for MnSOD was observed to increase 20-fold after a 48-h incubation, specific activity increasing by 520% [85]. MnSOD is induced in cultured rat colonic smooth muscle cells by IL-1 (over 20-fold), *E. coli* lipopolysaccharide, and TNF-α and in epithelial, neuronal, and smooth muscle cells after acetic acid-induced colitis [86]. In the cerebral vasculature, increase MnSOD expression is increased by lipopolysaccharide and in bacterial meningitis [87, 88].

Induction of MnSOD is different in various phases of cell growth, the highest expression being observed in confluent cells [84]. Expression of MnSOD in transformed fibroblasts is up-regulated in the presence of mutant forms of the tumor suppressor protein p53, which normally acts as a negative regulator of the enzyme [89]. Circadian rhythms in the expression of genes and activity of both CuZnSOD and MnSOD activities were found in some organs of experimental animals, e.g., in the rat brain cortex and cerebellum, and intestines, activity peaks being usually observed during the night although not for all organs and species examined [90].

EC-SOD (SOD3) gene has been localized to chromosome 4 (region 4p-q21) in the human [91]. The EC-SOD gene is ca 60% homologous to that of CuZn-SOD and shows minimal homology with that of MnSOD [10]. The human gene is approximately 5,900 bp long and consists of three exons and two introns. The 720 bp coding region is located entirely within exon 3. The promoter region of the gene lacks classical TATA or CCAAT boxes and contains various regulatory elements including a metal regulatory element, glucucorticoid response element, xenobiotic response elements, two potential antioxidant response elements (RE), AP-1 binding sites, and NF-κB motifs [10, 92, 93].

The EC-SOD is less responsive to oxidant stress than MnSOD. However, in human fibroblasts, the level of SOD3 is elevated by interferon (IFN)-γ and IL-1 [94]. In rat alveolar type II pneumocytes, TNF-α and INF-γ elevate SOD3 expression via activation of NF-κB [95]. In vascular smooth muscle cells, expression of EC-SOD is induced by INF-γ and IL-4 [96]. Angiotensin II up-regulates EC-SOD synthesis in vascular smooth muscle cells [30].

EC-SOD is down-regulated by IL-1 and transforming growth factor-β in cultured fibroblasts [11] and by TNF-α, epidermal growth factor, platelet-derived growth factor, and granulocyte monocyte-colony stimulating factor in vascular smooth muscle cells [96]. It suggests a possibility of impaired antioxidant defense in vivo under pathological conditions when the levels these cytokines and growth factors are elevated [97].

In eukaryotes, copper chaperone proteins, CCS and LYS7 have been identified that specifically deliver copper for the synthesis of human and *S. cerevisiae* CuZnSOD, respectively [98]. LYS57 knockout yeasts produce copper-free SOD and display oxygen sensitivity similar to that found in CuZnSOD knockout strains [99]. The human CCS metallochaperone has a polypeptide region bearing high resemblance to SOD1; homology of this part of CSS and SOD1 approaches 50% identity and 60% similarity [100]. On the basis of this conservation of sequence, CCS was originally postulated to be the fourth mammalian SOD (SOD4; Gen-Bank gene accession number 1608528). The sequence similarity seems to enable formation of metallochaperone-SOD heterodimers, which may initiate the metal transfer [101].

1.7 Genetic Polymorphism

Polymorphism of CuZnSOD has been found in human and animal populations. A variant of CuZnSOD (SOD-2; an unfortunate designation because "SOD-2" is used nowadays for MnSOD, or SOD-A^2 as opposed to the prevalent SOD-A^1) has been shown to occur in a small fraction of the Scandinavian population and to be present in the Mormons (Utah, USA) of Scandinavian origin, but it is also present in Central Africa, Italy, and Iraq. The variant protein has a different electrophoretic behavior and slightly lower specific activity [102, 103].

Several genetic variations have been described for human SOD2 gene. The substitution of Ile58 to Thr leads to a two- to threefold decrease in the enzymatic activity and increase of heat sensitivity of the protein [23, 104]. The Ile58 variant was found to have higher 'tumor-suppressing activity" (see Sect. 1.9), apparently due to higher specific activity [104].

At least 3 mutations in the proximal promoter region of human SOD2 have been identified and linked to reduced transcriptional activity in transient transfection experiments [105]. The main polymorphism of MnSOD concerning the mitochondrial targeting sequence (MTS) consists of one base pair transition (T→C), which results in a Val→Ala amino acid change at –9 position of MTS. It has been suggested (though not proven) that this mutation may lead to inefficient targeting to mitochondria and/or mistargeting of MnSOD. Structural studies point to a disturbance of the α-helix conformation by the mutation. Secondary structure seems to be of importance for targeting of the enzyme to mitochondrial matrix [106], which makes the postulated mistargeting probable. The frequency of this substitution has been found to differ among various ethnic groups, the frequency of the normal (Ala) gene being lower in Chinese population than in European ones [107]. The MnSOD AA genotype has been associated with increased (1.5-fold) breast cancer risk, especially among premenopausal women with a low consumption of antioxidants [108], and in postmenopausal women on estrogen replacement therapy or smoking cigarettes [109]. Individuals homozygous for Ala (especially females) also have a higher risk of being affected by motor neuron disease (2.9 and 5 times, respectively) [110] and age-related macular degeneration (10 times) [111]. Individuals with VV and VA genotypes had a higher risk for lung cancer than the AA genotype (odds ratio of 1.67 and 1.4, respectively) [112]. A similar tendency was observed for mesothelioma (odds ratio of 2.0 and 1.4) [113]. Pregnant women with AA or VA genotypes were found to excrete more 8-hydroxydeoxyguanosine in urine than those with the VV genotype, suggesting higher oxidative DNA damage [114]. No difference was observed in the genotypic distribution of V and A between controls and schizophrenics but, within schizophrenia patients, decreased A mutant was found among patients with tardive dyskinesia (odds ratio 0.29). This suggests that the A allele may protect against susceptibility to tardive dyskinesia in schizophrenics [115].

A polymorphism (Arg213Gly) in SOD3 has been found in 3–5% of the population. Approximately 94–98% of the population exhibit the normal (low) level of serum EC-SOD activity (93.6% of the Japanese population, 96.8% of white Australians, and 97.8% of Swedes) [116–118]. The arginine-to-glycine replacement in the hydrophobic C-terminal part of the enzyme that is responsible for the binding of EC-SOD to negatively charged endothelial surfaces and macromolecules weakens the binding of the enzyme to polyanionic surfaces and results in a tenfold increase in its activity circulating in the plasma. The frequency of this mutation is about two times higher in patients undergoing renal dialysis, which suggests that diminished adherence of EC-SOD to cell surfaces may be a risk factor for renal failure [119].

The majority of subjects carrying this mutation had no clinical abnormalities, but the mutation was found to be associated with increased occurrence of familial amyloidotic polyneuropathy type 1 [120] and increased levels of cholesterol and serum triglycerides [121].

Two additional mutations of EC-SOD have been described in humans, neither of them linked with any particular disease: a Thr 40 to Ala 40 substitution and a silent substitution mutation at amino acid 280 [116].

1.8 Biological Significance

The biological importance of SODs has been the subject of heated debate since their discovery. Initially they were postulated to be the main primary defense against oxygen toxicity. Subsequently, it has been argued that superoxide is not too reactive and its deleterious effects can be lower when compared with hydrogen peroxide, which can be a direct source of hydroxyl radical in the Fenton reaction [52]. Superoxide has been postulated to be the reductant of iron oxidized in the Fenton reaction and in this way to drive the metal-catalyzed Haber–Weiss reaction. However, it appears that other cellular reductants, present at much higher concentrations than superoxide, can be at least as effective. On the ground of these arguments, the role of SODs might appear obsolete; however, their ubiquitous presence in aerobic cells speaks against such a view [122].

Strict anaerobes were initially found to have no SOD activity and, generally, no catalase activity. Aerobes were found to contain both SOD and catalase activity, while aerotolerant aerobes had SOD and no catalase activity [123]. However, aerobes lacking SOD have been found subsequently, including *Neisseria gonorrhoeae*, *Leptospira* strains and *Mycoplasma pneumoniae* [13].

In *E. coli* FeSOD is expressed constitutively, even under anaerobic conditions, while MnSOD is induced by aerobiosis and iron limitation; FeSOD seems to be useful upon unexpected exposure to oxygen until MnSOD is expressed. Moreover, FeSOD appears more effective in protecting cytoplasmic enzymes while MnSOD, binding non-specifically to DNA and colocalizing with nucleoids, is more effective in preventing mutagenic lesions [3, 124]. The peripherally located CuZnSOD plays a major role in defense against extracellular superoxide from the environment or host [16] although bacteria themselves release superoxide to the periplasmic space, most probably from the respiratory chain [122]. The CuZnSOD activity is low in the logarithmic phase and increases at least 100-fold upon entry into stationary phase [38]. This SOD may be important in the bacterial defense against superoxide produced by phagocytic cells of the host [20].

In pathogenic bacteria, SOD may be a virulence factor. Strains of *Mycobacterium tuberculosis* and *Nocardia asteroides* that secrete FeSOD during logarithmic phase growth are resistant to phagocytic attack, while non-secreting strains are less virulent.

The growth of *Escherichia coli* null mutants unable to produce MnSOD (*SodA*) and FeSOD (*SodB*), and believed to be devoid of SOD activity (the presence of CuZnSOD in the periplasmic space of *E. coli* was unknown by then) was found to be only slightly impaired in a rich medium but completely inhibited in minimal medium, unless all amino acids were provided. This multiple amino acid auxotrophy results from damage to different superoxide targets, one of which is dihydroxy-dehydratase, an enzyme catalyzing the penultimate step in the biosynthesis of branched amino acids containing a 4Fe-4S cluster. Important cellular targets of superoxide include iron-sulfur clusters of dehydratases of the aconitase class; the second-order rate constants for the inactivation of these enzymes by superoxide are of the order of 10^6–10^7 M^{-1} s^{-1} [125]. The mutants showed enhanced sensitivity to oxygen and hydrogen peroxide and increased (41-fold) spontaneous mutation rates when grown on rich medium under aerobic (but not anaerobic) conditions. Additionally, membrane permeability appeared to be modified, leading to loss of metabolites. The mutants were more sensitive than wild-type bacteria to a moderate heat shock. The growth defects could be eliminated when the SOD-deficient bacteria were transfected with a gene coding for any SOD (including mammalian CuZnSOD) that showed neither structural homology nor prosthetic metal common with the missing protein. This suggests that the superoxide dismutase activity is important [3, 126, 127].

Yeast *Saccharomyces cerevisiae* lacking CuZnSOD are sensitive to oxygen and superoxide-generating agents such as paraquat or menadione, and require methionine and lysine when grown under aerobic conditions [128]. They have disturbed iron metabolism and higher levels of weakly bound iron [129]. Increased iron demand in the Δ*sod1* mutant may be a reflection of the cells' efforts to reconstitute iron-sulfur cluster-containing enzymes that are continuously inactivated under conditions of excess superoxide [130]. Yeast lacking CuZnSOD have decreased tolerance to hyperosmotic shock [131] and to freezing and thawing (which suggests that superoxide is generated under these conditions)

[132]. Yeast lacking MnSOD are also more sensitive to increased concentrations of oxygen [133]. MnSOD is essential for ethanol tolerance of diauxic-shift and post-diauxic-phase cells, as could be expected from the increased generation of superoxide in mitochondria after turning on oxidative metabolism of ethanol [134].

Transgenic animals with overexpression or knockdown of SOD genes have provided interesting information concerning the biological significance of these enzymes in mammals. Somewhat surprisingly, homozygous $Sod1^{-/-}$ mice (with complete absence of CuZnSOD) and heterozygous $Sod1^{-/+}$ mice (with about 50% of CuZnSOD activity of the wild-type $Sod1^{+/+}$ animals) develop normally and show no evidence of overt oxidative damage. The indices of lipid peroxidation and protein oxidation (carbonyl groups) in their brain tissue are normal. There is no compensatory increase in the MnSOD activity [135]. The animals show no increased mortality after exposure to 100% oxygen but are more sensitive to paraquat and myocardial ischemia-reperfusion injury [136, 137] and develop more severe hippocamp injury after global ischemia [138]. Females have reduced fertility, due to a marked increase in embryonic lethality [139]. CuZnSOD-knockout mice also show augmented hearing loss, loss of spiral ganglion cells and hair cell loss associated with aging [140], and show augmented deletions in mitochondrial DNA [141].

Transgenic mice overexpressing CuZnSOD are more resistant to lung oxygen caused by hyperoxia [142, 143] or allergens [144]. They are more resistant to heart postischemic injury, showing a twofold higher increase in the contractile function and a 2.2-fold decrease in infarct size [145]. Cells from these animals are more resistant to superoxide-generating agents like paraquat [146]. Pregnant mice with experimentally induced diabetes show lower fetal damage by elevated glucose but the animals may display some neurological defects reminiscent of those found in Down syndrome: abnormal neuromuscular junctions in the tongue, impaired muscle function, and decreased serotonin uptake by the platelets [143].

Interestingly, various authors demonstrated that excessive expression of CuZnSOD in bacterial cells or mammalian cell lines, without a simultaneous augmentation of hydrogen peroxide-scavenging mechanisms, can be harmful in some cases. In *E. coli*, overexpression of SODA was reported to paradoxically increase paraquat sensitivity of the cells. The most probable explanation of this phenomenon involves interference by SOD with the induction of the *soxR* regulon and prevention of a balanced adaptation to oxidative stress [147]. In mammalian cells, overexpression of CuZnSOD can enhance cell senescence and apoptosis [148]. Mammalian cells transfected with SOD1 were found to have increased rates of lipid peroxidation [149], and in many cases bell-shaped dose-response curves were obtained, especially when superoxide dismutase was used to protect against lipid peroxidation. Various explanations have been offered to explain these phenomena [150], the most common being that overexpression of SOD increases intracellular levels of hydrogen peroxide. However, experimental evidence for the increase in H_2O_2 levels in such cells is controversial and it has been argued that, depending on the ratio between different types of H_2O_2-producing and -consuming reactions, SOD overexpression may increase, decrease, or have no effect on the cellular level of hydrogen peroxide [151]. A recent proposal to explain a prooxidative action of high doses of superoxide dismutase points to the fact that the enzyme is unique in being present in cells at a concentration (of an order of 1 μM) higher than its substrate. It may lead to a situation where the active site Cu^+ is oxidized by a reaction with the substrate and, before encountering another superoxide anion which would reduce Cu^{2+} back to Cu^+, the oxidized form of the enzyme oxidizes another intracellular substrate (i.e., acts as a superoxide reductase), imposing oxidative damage [152].

Apparently independently of its enzymatic activity, CuZnSOD also plays a role in the regulation of 80S-ribosome activity in *S. cerevisiae* and was identified as a new inhibitor regulating activity of PK60S kinase, forming an inactive complex with the kinase [153, 154].

Mitochondria consume more than 90% of oxygen utilized by the cell and MnSOD protects the cell against the main source of superoxide. In *S. cerevisiae* MnSOD transcription is activated about tenfold by growth on a non-fermentable carbon source that forces the yeast to turn on mitochondria for respiration [155].

Transgenic mice lacking MnSOD (produced be deletion of exon 3) die within the fist 10 days of life showing mitochondrial damage (especially in the heart), decreased aconitase and succinate dehydrogenase activities, cardiac abnormalities, metabolic acidosis, and fat accumulation in the liver and skeletal muscle. No compensative increase in CuZnSOD activity is found in these animals [156]. Mice lacking MnSOD due to deletion of exons 1 and 2 survive up to 18 days and show severe anemia, degeneration of neurons in the basal ganglia and brain stem, and progressive motor disturbances (weakness, rapid fatigue, circling behavior) [157]. Mice surviving for more than 7 days show extensive mitochondrial injury within degenerative neurons and cardiac myocytes. $MnSOD^{-/-}$ mice show severe defects in the electron transport chain in skeletal muscle and heart, inactivation of mitochondrial aconitase in these organs, and increased excretion of markers of oxidative DNA damage.

Heterozygous ($Sod2^{-/+}$) mice have a MnSOD level of about 50% of the normal and show no oxygen toxicity under normal air atmosphere, or increased mortality when exposed to 100% oxygen. They show no ultrastructural abnormality in the heart; thus, 50% of MnSOD activity seems sufficient for normal resistance to air and 100% oxygen toxicity in mice [158].

Transfection with MnSOD protects various mammalian cells against hyperoxia [159], ionizing radiation [160, 161], injury by asbestos fibers [162], cigarette smoke [163], paraquat [164], drugs such as doxorubicin and mitomycin C, IL-1, and TNF-α [160]. Transgenic mice overexpressing human MnSOD under a protein C promoter have higher survival in 95% oxygen [165].

Nitric oxide, the main vasodilating factor and inhibitor of proliferation of vascular smooth muscle cells and platelet aggregation, reacts very rapidly with superoxide, the rate constant for this reaction ($k=6.7\times10^9\ M^{-1}\ s^{-1}$) being even higher than that for reaction of superoxide with superoxide dismutases. The reaction between $O_2^{\bullet-}$ and NO leads not only to the loss of NO bioactivity but also to the formation of a strong oxidant, peroxynitrite $ONOO^-$, involved in many pathological processes. The reaction can take place in the extracellular space and EC-SOD is there an important factor preventing this reaction [166, 167].

SOD3 knockout mice are apparently healthy but are more sensitive to hyperoxic lung injury [168], to ischemia-reperfusion injury [169], and to the diabetogenic action of alloxan [170].

Transgenic mice overexpressing EC-SOD have better neurological outcome and cognitive performance after severe cranial impact [171]. Such animals were, however, more susceptible to hyperbaric oxygen than control animals, an effect ascribed to sparing excessive levels of nitric oxide and thus blocking the normal vasoconstrictive response of the cerebral vasculature to hyperoxia [172].

It has been suggested that neural EC-SOD plays a role in the cognitive function. Transgenic mice with both elevated and decreased levels of EC-SOD have impaired learning and memory, a finding suggesting that brain extracellular superoxide may contribute to adequate learning function, especially under low motivational states. These effects may invoke a signaling role of superoxide in normal neuronal function or reflect abnormal level of nitric oxide due to altered extracellular superoxide levels in the transgenic animals [173–175].

There is a plethora of examples demonstrating alterations in SOD activity under various situations involving endogenous or exogenous oxidative stress. On the contrary, changes in the level of SOD may be an indication of oxidative stress. To mention only a few examples: Expression of MnSOD and CuZnSOD closely correlates with steroidogenesis in the human ovary [176]. Endurance training was found to affect EC-SOD more than other SOD enzymes, decreasing EC-SOD activity in blood plasma while acute exercise after training increased both the plasma Mn-SOD and extracellular SOD. This suggests that endurance training increases the reserve of EC-SOD in tissues. The results also suggest the possibility of plasma EC-SOD assay as a new index of endurance training [177]. Cold acclimation of plants was reported to increase SOD activity in concert with improved freezing tolerance [178].

1.9 Pathologies

More than 90 mutations in the SOD1 gene have been found in a dominantly-inherited form of amyotrophic lateral sclerosis (ALS) or the Lou Gehrig's disease), a fatal disorder causing degeneration of motor neurons in the corticospinal tracts and brain stem. The primary characteristics of the disease is the selective degeneration of upper and lower motor neurons, initiating in mid-adult life and almost invariably progressing to paralysis and death. The age of disease onset is about 55 years, death occurring usually between 3 to 5 years after onset. Familial inheritance has been shown in 10–20% of ALS cases; among them about 20 have mutations in the SOD1 gene. Over 90 different SOD1 mutations have been described; they are found in all exons but mostly in exon 4 [179, 180]. Expression of CuZnSOD mutants in mice results in selective killing of motor neurons.

The mechanism of the pathological sequelae of mutations of SOD1 gene is unclear. Originally increased superoxide level has been invoked, due to decreased enzymatic activity of the mutant proteins. However, some mutants did not show decreased enzymatic activity and complete inactivation of SOD1 in knockout mice did not cause any motor neuron abnormalities [135]. Two hypotheses have been put forward to explain why CuZnSOD mutations lead to selective killing of motor neurons in spite of unchanged or elevated CuZnSOD activity. A "gain-of-function" theory has been proposed implying that mutant SODs have increased their side functions, such as peroxidase activity or catalysis of nitration of tyrosyl residues by peroxynitrite [62, 63]. An "aggregation hypothesis" postulates that misfolding of the mutant and formation of intracellular aggregates leads to the neurodegenerative disease [181, 182].

Cancer cells are usually low in MnSOD and catalase activity, and often low in CuZnSOD activity, due to decreased expression of these enzymes caused by methylation of particular sequences in the intronic region [183] and elevated levels of AP-2 [78]. Reduced levels of MnSOD have been reported in cancers from a variety of cell types and etiologies [184]. Transfection of *MnSOD* cDNA into cultured human cancer cell lines caused suppression of the malignant phenotype in several tumor cell lines [185, 186]. Elevated MnSOD activity was correlated with a loss of metastatic capacities of tumor cells [184]. These findings were the basis for the view of "tumor-suppressing activity" of MnSOD [78] and ascribed mainly to an increased level of H_2O_2 over a cytotoxic threshold [186]. However, other authors also found augmented expression of MnSOD in some tumors including brain tumors, cervical carcinoma, and colon and thyroid carcinoma, associated with accumulation of mutated p53. Transformed cells lacking p53 showed deregulated expression of MnSOD while transient transfection with wild-type p53 led to a significant reduction in MnSOD level. It has been suggested therefore that MnSOD can play an important role in the maintenance of tumor cell viability and promote rather than decrease their survival, especially under stressful conditions [89]. Recent discoveries concerning the central role of mitochondria in the initiation of apoptosis have supported the notion that MnSOD, through the surveillance of mitochondrial integrity, can directly influence the cell fate of cells subjected to various damaging factors [187].

EC-SOD expression is substantially reduced in patients with coronary artery disease and appears to contribute to endothelial dysfunction in patients with this disease [188]. Overexpression of EC-SOD in vascular endothelial cells can protect against the oxidation of LDL, a major factor contributing to the development of atherosclerosis [189]. However, a positive correlation was found between the plasma level of EC-SOD and plasma homocysteine level (a risk factor for increased incidence of coronary vascular disease) [190]; other risk factors for coronary artery disease such as male gender and smoking, also correlated with lower EC-SOD [191].

Glycation is known to inhibit the activity of cytoplasmic CuZnSOD which may be responsible for the lowering of SOD activity in diabetes [192]. Glycation of EC-SOD at Lys residues located in the heparin-binding domain does not affect the enzyme activity but results in the loss of heparin-binding activity [193].

There are numerous reports on changes in SOD and catalase activities in various diseases [30, 41, 97, 150], e.g., reduced SOD activity in lung cells (by about 50% in bronchial epithelial cells) was found in asthma and suggested to be a marker of the inflammation characterizing asthma [194]. Loss of SOD occurs within minutes of an acute asthmatic response [195].

2 Catalase

Catalase (H_2O_2: H_2O_2 oxidoreductase, EC 1.11.1.6) dismutates hydrogen peroxide to water and oxygen:

$$2H_2O_2 \rightarrow 2H_2O + O_2 \tag{11}$$

Catalase (CAT) is an intracellular enzyme present in most aerobic cells. In animal tissues it is mainly localized in peroxisomes and in the cytoplasm. With the exception of rat myocardial cells, CAT is not detected in mitochondria [196].

2.1 Diversity of Catalases

This large group of oxidoreductases is not homogenous. Catalases are divided into three groups, basing on their physical and biochemical properties. One group with enhanced peroxidative activity is called catalase-peroxidases. Another subgroup with manganese in the active center are non-heme catalases. The third group are “true” catalases, i.e., monofunctional heme catalases. Catalase-peroxidases are widely distributed in procaryotes and are also found in lower eukaryotes, Mn-catalases were reported only in procaryotes while true catalases are found ubiquitously in eukaryotes, but also in many procaryotes.

Catalase-peroxidases were detected in all three living kingdoms, although in eukaryotes only in fungi. Probably these hydroperoxidases are successors of the first ancestral hydrogen-peroxide degrading enzymes. Their molecular weight ranges from 120 to 340 kDa. They are generally homodimers although homotetramers were also reported. They show much higher sequence homology with heme proxidases than with typical catalases. A characteristic feature of catalase-peroxidases is their bifunctional behavior. The maximal catalatic turnover of catalase-peroxidases is two or three orders of magnitude lower than that of typical catalases. However, their K_m for hydrogen peroxide is in the millimolar range, one or two orders of magnitude lower than the apparent K_m values of typical catalases. The k_{cat}/K_m ratio of catalase-peroxidases at low substrate concentrations is only about one order of magnitude below that of typical catalases, i.e., they are fairly equivalent under physiological conditions. As peroxidases, they accept a broad range of organic substrates. However, the maximal reaction rates are usually much below that of horseradish peroxidase. In contrast to typical catalases, they show a sharp pH optimum around pH 6.5. Catalase-peroxidases are also more vulnerable to inactivation by high temperatures, pH, and H_2O_2 than typical catalases. Unlike typical catalases, they are not inhibited by aminotriazole but their heme iron is readily reduced by sodium dithionite.

Manganese catalases, sometimes referred to as pseudocatalases, are found in lactic acid bacteria and in thermophilic bacteria. The molecular weight of these enzymes ranges from 170 to 210 kDa. They may form unusual oligomeric structures like homopentamers and homohexamers. Unlike heme catalases, they are not inhibited by CN^- or N_3^- [197].

Monofunctional heme catalases are usually homotetramers, of molecular weight 200–340 kDa. They have been identified in organisms from bacteria to human. Their typical features include rather strong absorption at the Soret band and irreversible inhibition by a suicide inhibitor, 3-amino-1,2,4-triazole. They are very efficient as catalases (k_{cat}=4×10^7 M^{-1} s^{-1} for human erythrocyte catalase). They can also catalyze 2-electron oxidation of short-chain aliphatic alcohols at reasonable rates [198].

2.2 Structure

Typical CATs are tetrameric molecules. Human CAT is composed of four identical subunits. It seems that for human and other CATs tetramerization is essential for function; perhaps it ensures that the active site is sequestered and the enzyme can complete the reaction cycle rather than allow generation of hydroxyl radicals from exposed heme.

Each subunit contains four distinct structural regions: an N-terminal arm comprising the first 70 amino acids, the β-barrel domain (positions 71–379, the domain connection (wrapping domain; residues 380–438) and the α-helical domain (positions 440–503). A small subgroup of CATs also contains a 150-residue long C-terminal region (a flavodoxin-like domain). The β-barrel domain is well conserved in all CATs from lower procaryotes to eukaryotes. It is an 8-stranded antiparallel β-barrel with 6 α-helical insertions in the turns between the strands. Internal parts of the barrel harbor several essential amino acid residues (His 70, Ser 109, and Asn 143). The imidazole ring of His 70 is oriented parallel to the plane of the prosthetic heme group. Ser 109, the nearest neighbor of His 70, is essential for maintaining the proper orientation and the nucleophilic character of His 70 and thus stabilizes the distal heme pocket structure. The coessential asparagine (Asn 143) orients the peroxide substrate and influences the redox potential of heme iron. This domain contains the main substrate channel connecting the heme group to the molecular surface. The heme is at the bottom of a 2.5 nm-long channel extending from the enzyme surface [199].

Three channels have been implicated as potentially having a role in the access to the active site of CAT. The perpendicular (relative to the plane of the heme) or main channel has been considered as an access channel. The lateral or minor channel has been shown to be important, possibly as an inlet channel, in HPII and in small-subunit enzymes. A third channel has been identified, leading from the heme to the central cavity of the tetramer [200]. All three channels are quite narrow, particularly as they approach the heme-containing active-site cavity, and this restricts accessibility to relatively small molecules, generally not much larger than H_2O_2. Subtle differences in the size and shape of the channels may influence substrate entry or product exit, contributing to the differences in the reaction rates between different CATs [201].

The active site of CAT contains the porphyrin ring and iron, which is attached to nitrogen atoms of the porphyrin by four coordination bonds. The fifth valency of iron is bound to tyrosine and the sixth position remains free [202]. In bovine liver CAT, up to 50% of its heme group is degraded to biliverdin and bilirubin [197]. *E. coli* hydroperoxidase II and *Penicillium vitale* CAT contain a chlorin, i.e., a partially saturated porphyrin macrocycle, rather than protoporphyrin IX (heme B) as the prosthetic group. This modified porphyrin (heme d) incorporates a *cis*-hydroxy gamma-spirolactone at the saturated pyrrole ring III, yielding a characteristic spectrum.

Human CAT (as well as CATs from numerous other species studied) has been demonstrated to bind NADPH (one molecule per subunit) [203]. Each tetrameric molecule of human or bovine CAT binds four molecules of NADPH, with a dissociation constant lower than 10 nM. The order of affinity for binding of nicotinamide-adenine dinucleotides is NADPH>NADH≫$NADP^+$>NAD^+. NADPH is bound at a cleft between the helical domain and the β-barrel on the surface of the molecule, with the redox active C4 atom of nicotinamide distant about 1.9 nm from the nearest heme iron.

2.3 Properties

Heme CATs are fairly heat-stable, with T_m of 56 °C for bovine liver CAT and 82 °C for *E. coli* HPII [201]. They are inhibited by a number of compounds that interact with the active-site heme, including cyanide, azide, hydroxylamine, 3-amino-1,2,4-triazole, and mercaptoethanol (Table 2). A specific inhibitor is 3-amino-1,2,4-triazole, which covalently modifies a heme-binding histidine (His 75 in the human enzyme) when CAT is in the form of Compound I (see Sect. 2.4) [204].

Table 2 Some characteristics of human erythrocyte catalase. Adapted from [201]

K_m^{app} (calculated)	80 mM
V_m^{app} (calculated)	587,000 µmol H_2O_2 (µmol heme)$^{-1}$ s^{-1}
K_{cat}/K_m	7.34×10^6 s^{-1} M^{-1}
Concentrations of inhibitors needed for 50% inhibition	
NaCN	20 µM
NaN3	1.5 µM
Aminotriazole	30 mM
NH_2OH	2.0 µM
Time to 50% inactivation	
At 65 °C	0.2 min
In 1 mM 2-mercaptoethanol	3 min

2.4 Mechanism of Action

Mn CATs from *Thermus thermophilus* and *Lactobacillus plantarum* contain a binuclear manganese cluster. The mechanism of catalysis involves two-electron redox cycling of the binuclear manganese cluster between the divalent and trivalent states Mn^{2+}-$Mn^{2+} \leftrightarrow Mn^{3+}$-$Mn^{3+}$ [205]:

$$\text{CAT-Mn}^{2+}\text{-Mn}^{2+} + H_2O_2 + 2H^+ \rightarrow \text{CAT-Mn}^{3+}\text{-Mn}^{3+} + 2H_2O_2 \quad (12)$$

$$\text{CAT-Mn}^{3+}\text{-Mn}^{3+} + H_2O_2 + 2H^+ \rightarrow \text{CAT-Mn}^{2+}\text{-Mn}^{2+} + 2H^+ + O_2 \quad (13)$$

The catalytic mechanism of heme CATs is a two-step reaction. In the first step, interaction between ferriCAT (Fe^{3+}) and hydrogen peroxide leads to the heterolytic cleavage of the O–O bond in the substrate molecule and oxygen is bound to the 6th valency of the porphyrin iron. In this reaction Compound I (in which formal oxidation state of Fe is +5), a spectroscopically distinct and enzymatically active form of CAT, is formed. Compound I is an oxyferryl species in which one oxidation equivalent is removed from the iron and one from the porphyrin ring to generate a porphyrin π-cation radical [206]:

$$\text{CAT (Porphyrin-Fe}^{3+}) + H_2O_2 \xrightarrow{k_1} \text{Compound I (Porphyrin}^{\bullet+}\text{-Fe}^{4+}{=}O) + H_2O \quad (14)$$

In the second step, Compound I oxidizes a second peroxide molecule to molecular oxygen and releases the ferryl oxygen species as water, restoring ferriCAT:

$$\text{Compound I (Porphyrin}^{\bullet+}\text{-Fe}^{4+}{=}O) + H_2O \xrightarrow{k_2} \text{CAT (Porphyrin-Fe}^{3+}) + H_2O_2 + O_2 \quad (15)$$

Alternative substrates (AH_2) for the second part of the cycle are short-chain alcohols (which react with Compound I to yield ferriCAT, a corresponding aldehyde and water) and nitrite (which is oxidized to nitrate).

$$\text{Compound I (Porphyrin}^{\bullet+}\text{-Fe}^{4+}\text{=O)} + \text{AH}_2 \xrightarrow{k_3} \text{CAT (Porphyrin-Fe}^{3+}) + \text{H}_2\text{O}_2 + \text{A} \quad (16)$$

The rate constant of reaction of Compound I with H_2O_2 is much higher than that with alcohols [207]. No free radicals are formed in the typical cycle of CAT. Organic peroxides, such as peroxyacetic acid, can substitute for hydrogen peroxide by slowly forming Compound I; however, they do not reduce CAT back to the resting state.

Whenever Compound I is present, the enzyme undergoes slow inactivation by a gradual one-electron reduction of Compound I to Compound II (formal oxidation state of Fe is +4) by external electron donors or, in their absence, even by elements of the enzyme molecule itself; in this case an amino acid free radical is formed on the tyrosine residue distant from the active site (Tyr 370 in the human enzyme). Compound II is an inactive form of CAT and cannot be recycled by hydrogen peroxide. NADPH prevents formation of Compound II by serving as an electron donor to reduce Compound I. When NADPH is unavailable, the one-electron reduction by an external donor or of the porphyrin π-cation by an electron from Tyr 370 and formation of Compound II appears to be the next alternative [199].

$$\text{Compound I (Porphyrin}^{\bullet+}\text{-Fe}^{4+}\text{=O)} + \text{AH} \rightarrow \text{Compound II (Porphyrin}^{+}\text{-Fe}^{4+}\text{=O)} + \text{H}^+ + \text{A} \quad (17)$$

The rates of these reactions are slow and some CATs virtually do not form Compound II [208].

In principle, Compound II can be reduced by some substrates

$$\text{Compound II} + \text{AH} \rightarrow \text{CAT-Fe}^{3+} + \text{H}^+ + \text{A} \quad (18)$$

but this reaction is very slow (Fig. 3).

CAT may react with superoxide forming Compound III which is a resonance hybrid between the forms $CAT\text{-}Fe^{3+}\text{-}O_2^-$ (predominating) and $CAT\text{-}Fe^{2+}\text{-}O_2$ [209].

The specificity of mammalian CATs for peroxides is high; only hydrogen, and methyl and ethyl hydroperoxides give appreciable activity while *t*-butyl hydroperoxide (a glutathone peroxidase substrate) does not react with CAT. Similarly, lower alcohols can be donors for the peroxidative activity of CAT, with high activities for methyl and ethyl, but also some unusual alcohols (Table 3).

CAT reacts with hydrogen peroxide with a rate constant of an order of 10^7 M^{-1} s^{-1} in a broad range of pH [210]. Reactions 14 and 15 are very fast, the values of k_1 and k_2 being 1.7×10^7 M^{-1} s^{-1} and 2.6×10^7 M^{-1} s^{-1}, respectively, for rat liver CAT [207]. The turnover number (number of peroxide molecules decomposed per CAT molecule per second) is in the case of CAT and hydrogen peroxide 3.5×10^6 [211].

Solving the structure of CAT enabled elucidation of its reaction mechanism. The hydrophobic narrow lower part of the channel allows only small substrates

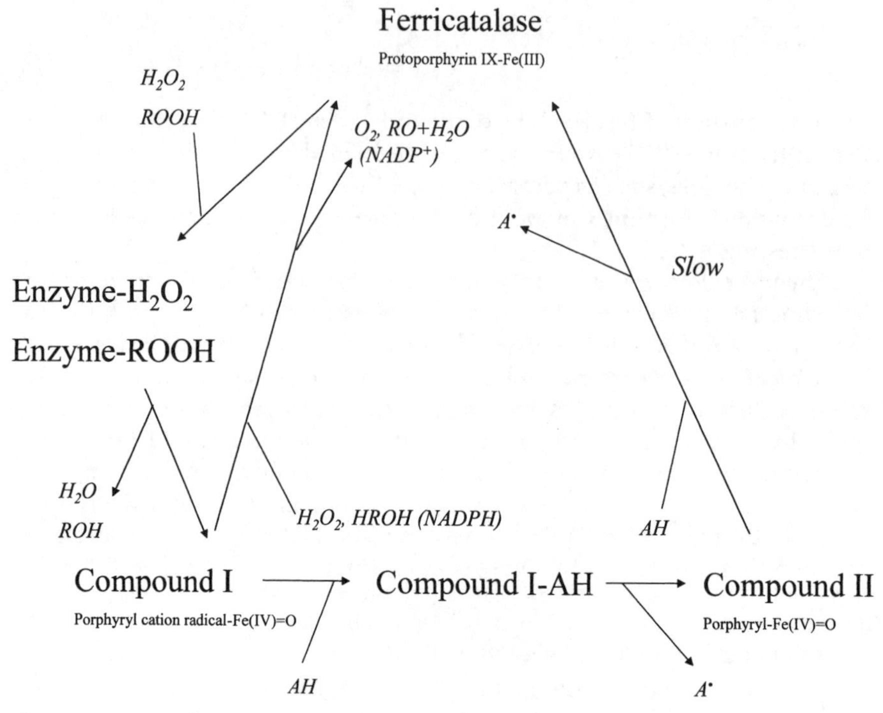

Fig. 3 Reactions of catalase. After [197], modified

Table 3 Values of the rate constants for the reactions of Compound I of horse catalase with various alcohols [207]

Alcohol	k_3^{app} (M^{-1} s^{-1})
Ethyl	1,020
Propyl	6.5
Butyl	0.4
Allyl	330
Propargyl	2,500

to reach the buried heme groups blocking the passage of large molecules. The discrimination between hydrogen peroxide and rather similar water molecules is accomplished by a "molecular ruler" mechanism. Four water molecules are bound immediately above the active site at a hydrophobic constriction of the channel. In human CAT, this site is lined by hydrophobic side chains of Val 74, Val 116, Pro 129, Phe 153, Phe 154 and Trp 168 constricting the distal channel to 0.2–0.3 nm, allowing only water, hydrogen peroxide and other small molecules to reach the heme. At either end of the hydrophobic constriction, water molecules form hydrogen bonds to the channel walls and are well ordered. One water molecule forms hydrogen bonds to His 75 and Asn 148, another to Gln 168 and Asp 128. The remaining two water molecules form no hydrogen bonds to the protein, are more mobile and cannot bridge the channel with water–water hydrogen bonds, leaving a gap of about 0.37 nm. This distance is too short to fit an additional water molecule into the channel. However, a hydrogen peroxide molecule, with an O–O bond of about 0.145 nm would enable a better filling of this volume and formation of a full network of hydrogen bonds, which would be more stable due to a lower dipole moment of hydrogen peroxide as compared to water.

Peroxide forms an iron coordination complex at the active site, with one oxygen atom ligating the heme iron and the other forming hydrogen bonds to His 75 and Asn 148. Ligation to the metal opens the channel and probably promotes binding of second peroxide to the ruler, which helps to avoid the decay of unreacted Compound I via the tyrosine radical. The enzyme uses two asymmetric interactions with the substrate to make a heterolytic cleavage of the symmetric peroxide bond. Coordination to heme iron with simultaneous hydrogen bond formation by the second oxygen atom requires stretching of the peroxide bond, facilitating its cleavage (Fig. 4).

There is an electrical potential between the negatively charged Asp (or Glu) at position 181 and the positively charged heme iron 1.2 nm distant. The electric field acts upon the substrate orienting it optimally for interaction with the active site [212].

The catalytic reaction in which H_2O_2 reacts with both ferriCAT and Compound I does not obey Michaelis–Menten kinetics; the rate of substrate decomposition increases linearly with hydrogen peroxide concentration over wide concentration range [207]. However, at substrate concentrations below 200 mM small-subunit CATs do exhibit a Michaelis–Menten-like dependence of velocity on H_2O_2 concentration and on this basis apparent V_m and K_m values are often determined. These values do not correspond to reality since at high substrate concentrations CATs are inactivated by the substrate. Large-subunit CATs from *E. coli* and *Aspergillus niger* are more stable at high (molar) concentrations of the substrate. The apparent K_m values of CATs from various sources range from 70 mM to 1.63 M, and turnover rates from 54,000 to 833,000 reactions s^{-1} [201]. The catalytic activity of the enzyme is independent of pH over a broad range of pH (5–9) [213].

Removal of NADPH does not abolish the catalytic function of the enzyme; however, during prolonged exposure to H_2O_2 NADPH becomes oxidized to $NADP^+$ and the activity of CAT drops. Apparently, the function of NADPH is to protect the enzyme against inactivation by its own substrate by preventing and reversing the accumulation of Compound II (by reacting with the cation radical on the polypeptide chain of Compound I) [214]. When NADPH is oxidized, $NADP^+$ dissociates due to the much lower affinity of the enzyme for the oxidized form of the dinucleotide and can be reduced back to NADPH [215]. In addition, binding of NADPH stabilizes the structure of the enzyme [197]. The exact mechanism of NADPH action is unclear; it seems that it does not di-

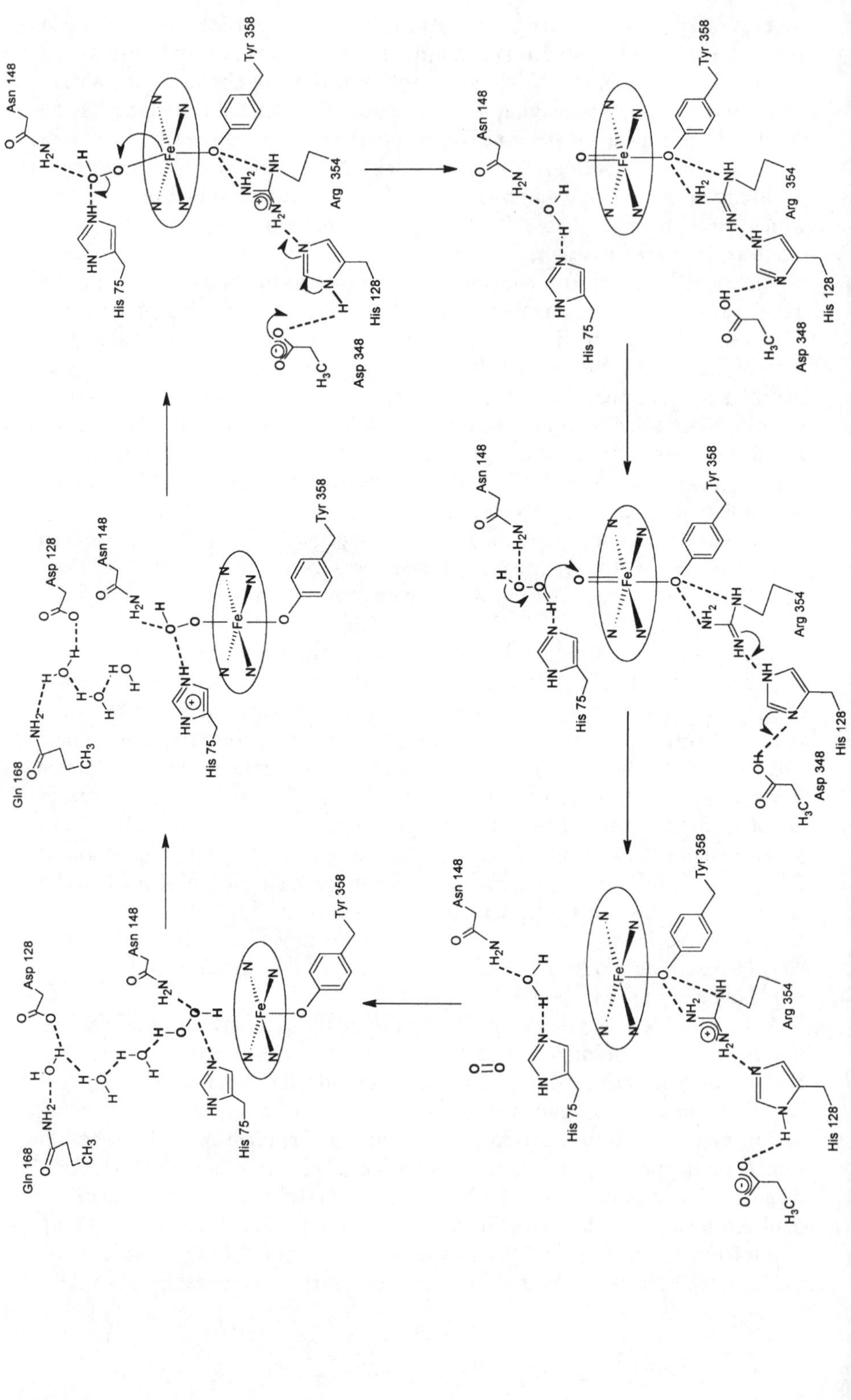
Asn 148
Tyr 358
His 75
Arg 354
His 128
Asp 348
Asp 128
Gln 168
Fe

rectly reduce Compound II but rather prevents it from occurring by being a better reductant of Compound I [214].

2.5 Molecular Biology and Genetics

In *E. coli* catalases are up-regulated by the redox-sensitive *oxyR* protein, which in its oxidized state binds to AREs, while up-regulation of SODs involves *soxR* and *soxS* gene products [69].

In *S. cerevisiae*, STRE sequences mediate the repression of catalase T gene when high levels of cAMP are present, through a Ras-dependent pathway and its derepression during nutrient starvation and oxidative stress [216]. There is no indication of a participation of any molecular chaperone in the catalase assembly in yeast [197].

In plants, multiple catalases, encoded by a small family of genes consisting of several classes have been found [217, 218].

The gene coding for catalase in the human consists of 13 exons and is located on chromosome 11p13 The human catalase promoter is GC-rich, lacks a TATA box, and has several putative Sp-1 binding sites [219]. The peroxisomal targeting signal is located approximately eight amino acids from the C-terminus in mammalian catalases and are SHI, SHM, and SHL for rat and mouse liver catalases and human kidney catalase, respectively [220].

Lipid peroxides induce the expression of catalase in vascular cells [221].

High intersex differences and ultradian (period of 6–12 h) oscillations in the activity of catalase were found in the Harderian gland of Syrian hamster, with peak activities in females exceeding those in males by several 100-fold [222]. In oxidative stress response, CAT is phosphorylated at Tyr 231 and Tyr 386 by tyrosine kinases c-Abl and Arg [223].

2.6 Genetic Polymorphism

Several rare mutations/polymorphisms have been reported in the CAT gene in humans, most of them being associated with acatalasemia, characterized by erythrocyte CAT levels of 0.2–4% of normal (see Sect. 2.8). A common C-T

◄

Fig. 4 Mechanism of catalytic action of human catalase. After [199], modified (copyright agreement pending). Hydrogen peroxide is fixed by hydrogen bonding to His 75 and Arg 148, and a fixed water molecule. Its binding to the heme opens the channel for the next peroxide molecule. Heterolytic cleavage of the substrate is driven through interaction with the electron-rich ferric ion and electron-withdrawing His 75 and Arg 148 residues. A charge relay system of Asp 348 and His 218 increases electron density at the active site and removes charge–charge repulsion between Arg 354 and the porphyrin π-cation radical. Next, peroxide is fixed by His 75 and Arg 148 with release of the oxygen molecule, and reversal of the charge relay brings the active site to the initial state

polymorphism has been found in the 5′-flanking region of the promoter of the CAT gene, located –262 base pairs from the transcription start site [224].

2.7 Biological Significance

Most aerobic organisms contain CAT activity although there are exceptions including *Bacillus popilliae*, *Mycoplasma pneumoniae*, algae *Euglena* and *Gleocapsa*, and some parasitic helminths, among them the liver fluke. Anaerobic bacteria generally do not contain CAT, but there are also some exceptions including *Propionibacterium shermanii* [13].

Escherichia coli contain two CATs: hydroperoxidase I which is a CAT-peroxidase and hydroperoxidase II which is a hexameric heme CAT (not binding NADPH) [225]. Hydroperoxidase I (HPI) is a bifunctional CAT-peroxidase active as a homotetramer of 80,049-Da subunits, with sequence similarity to plant peroxidases, containing protoheme IX encoded by *katG*. Its induction by hydrogen peroxide is regulated by *OxyR*. Alkyl hydroperoxidase I is the primary scavenger of endogenous H_2O_2.

Hydroperoxidase II (HPII) is a homotetramer of 84,118-Da subunits encoded by katE. HPII contains heme d with two *cis* -OH groups in ring III, which can arise from heme b (protoheme IX) by epoxide formation on ring III followed by hydrolysis to the *cis* diol. Hydroperoxidase II is not induced by H_2O_2 but its level increases six- to tenfold in the stationary phase as compared with exponential phase cultures [225]. It is thought to be important for long-term survival of the bacteria [226].

The majority of eukaryotic CATs were found in the peroxisomes, sometimes CAT being the main peroxisomal protein. There are two CATs in the yeast *Saccharomyces cerevisiae*: CAT T (typical), localized in the cytoplasm and CAT (A) atypical, localized in peroxisomes. CAT T is more important since CAT T is expressed only when proliferation of peroxisomes is induced, e.g., by fatty acids. Generally, disruption of CAT genes in bacteria or yeast has less dramatic effects on normal growth as compared with disruption to SOD genes [13]. However, several examples seem to illustrate various biological roles played by CATs. In *S. cerevisiae*, expression of both CAT and CuZnSOD is increased after hyperosmotic stress [131]. In *Neurospora crassa*, conidia have 60-fold higher CAT activity than exponentially growing hyphae. The isoenzyme pattern changes during the development; CAT-1 is predominant in conidia, during germination and early exponential growth while CAT-3 dominates during late exponential growth and at the start of the conidiation process (this species has four CAT isoenzymes) [227].

In the human tissues, the highest CAT activity has been found in the liver and then in erythrocytes [13]. In the erythrocytes, CAT can perhaps not only remove hydrogen peroxide produced during hemoglobin autoxidation, but also protect other tissues from this compound, which easily penetrates cellular membranes.

In plant leaves, the major part of CAT activity is localized in the peroxisomes; CAT detoxified hydrogen peroxide produced by photorespiration. Reduction of peroxisomal CAT1 activity in *Nicotiana plumbaginifolia* leaves by use of an antisense construct down to 10% of normal activity leads to a normal phenotype at low light intensity but the plants develop white necrotic lesions upon exposure to light [218]. Similar symptoms have been observed in other CAT-deficient plants [228].

CATs are believed to play a role in plant defense in combination with salicylic acid (SA). Two effects of SA on plant CATs have been suggested. At low concentrations of H_2O_2, in healthy tissues, CAT is inhibited by SA, which allows for an increase in the concentration of H_2O_2, acting as a second messenger to activate expression of defense-related genes. In tissues adjacent to necrotizing cells, at high H_2O_2 concentrations, higher concentrations of SA protect CAT by preventing accumulation of Compounds II and III, serving as an electron donor, and thus supporting or substituting for the protective function of CAT-bound NADPH [229]. Contradictory data have been published on whether salicylic acid can also inhibit mammalian CATs [229, 230].

In mammalian cells, CAT is located mainly in the peroxisomal matrix and can leak out when peroxisomes are damaged. Some CAT is present in the matrix of rat heart mitochondria but their contribution to the removal of hydrogen peroxide in the mitochondria is not significant [231].

CATs are thought to protect SODs against higher levels of H_2O_2 (and vice versa; SOD protects catalase from inactivation by superoxide) [3]. The relative role of CAT with respect to glutathione peroxidase in the disposal of hydrogen peroxide has been a question of debate taking into account the low affinity of CAT for the substrate. In the red blood cells, CAT and glutathione peroxidase have comparable shares in the removal of intracellularly generated H_2O_2 [232].

The peroxidative activity of CAT in an important source of acetaldehyde in the brain where no other enzyme is known to convert significant amounts of ethanol to acetaldehyde [233, 234]. Treatment of rats with the CAT inhibitor, 3-amino-1,2,4-triazole, decreases voluntary ethanol consumption [235]. It has been suggested that individuals with increased level of CAT activity may be prone to alcoholism [236, 237].

2.8 Pathology

Acatalasemia or hypocalasemia, due to mutations in the gene coding for CAT, have been described in several families. Acatalasemia is due to a genetic deficiency in erythrocyte CAT, in persons homozygous for a mutated gene; hypocatalasmia is the condition in heterozygotes. Both hypocatalasemia and acatalasemia were found not to lead to any clinical effects except for an increased incidence of mouth ulceration in a Japanese family [238, 239].

3 Superoxide Dismutases and Catalase in Longevity Assurance

The free radical theory of aging postulates that free-radical damage is the main course of alterations responsible for aging of cells and organisms [240]. In an apparent agreement with this theory, observations in different organisms point to a relationship between life span and SOD activity. Yeast lacking either CuZnSOD or MnSOD have decreased replicative life span i.e., the number of division a cell can accomplish [241, 242], Interestingly, overexpression of MnSOD increases the "chronologic" life span of the yeast (i.e., the time of survival at the stationary phase) but decreases replicative life span [243].

Results of studies concerning the role of CATs in yeast replicative aging are controversial [242, 244]. In one study, knockout of CAT A decreased life span of yeast grown on glucose media, while on ethanol media both CATs were required for longevity assurance [245].

In *Cenorhabditis elegans*, increased levels of SOD and CAT were found in *age-1* mutant that is characterized by a long life and oxygen resistance. In other mutants of increased life span, *daf-16* and *mev-1*, expression of superoxide dismutases is lowered but that of CATs elevated [246]. Life span of *C. elegans* has been extended using a synthetic SOD/CAT mimetics [247], but this procedure in not effective in *Musca domestica*. Therefore, the effect seems to be species-specific [248].

In *Drosophila melanogaster*, deficiency of CuZnSOD decreases life span and accelerates aging [249, 250] while simultaneous overexpression of CuZnSOD plus CAT [251] or overexpression of CuZnSOD or MnSOD alone increases life span [252, 253, 254]. Lowered (down to 14%) levels of CAT have no effect [255] but knockout of CAT decreases life span [256]; overexpression has no positive effect on the longevity of *Drosophila* [254, 256, 257].

The life-prolonging effect of (-)deprenyl on experimental animals has been ascribed to up-regulation of superoxide dismutase and CAT in the dopaminergic regions of the brain and in other organs [258].

4 Possibilities of Applications of Superoxide Dismutases and Catalase

Unbalanced production of reactive oxygen species, first of all superoxide and hydrogen peroxide, has been postulated as playing a role in the pathogenesis of a number of clinical disorders, such as acute respiratory distress syndrome, ischemia-reperfusion injury, atherosclerosis, neurodegenerative diseases and cancer. As the function of SOD and CAT is the protection against oxidative stress, their application has been proposed in cases where oxidative stress is important in the mechanism of a disease.

One of the first reports on the protective role of administration of SOD concerned alleviation of influenza toxicity in mice by CuZnSOD conjugated to pyran polymers (non-conjugated SOD was ineffective due to the short life time of this low molecular weight protein in the circulation [259].

It has been shown in an experimental guinea pig model that CuZnSOD and MnSOD are effective in treatment of mustard gas burns. Mustard gas, used as a warfare since World War I, causes blistering lesions that are slow to heal, and secondary inflammation. There is no antidote for burns by mustard gas. SODs were found to effectively reduce the burn lesion area when administered intraperitoneally/intralesionally [260].

Exogenous SOD, especially liposome-encapsulated or linked to polyethylene glycol and administered locally by intratracheal insufflation or aerosol inhalation, has been demonstrate to protect experimental animals from pulmonary oxygen toxicity [261–263], although conflicting results have been reported [97]. Lecithinized SOD (phosphatidylcholine-SOD) protected against bleomycin-induced pulmonary fibrosis [264]. Efficacy of SOD treatment was improved by administering liposome-encapsulated CuZnSOD also containing CAT [265]. Polyethylene glycol-attached CuZnSOD + CAT provided better protection against hyperoxia and fiber-induced lung injury than PEG-SOD alone [266].

Animal experiments suggest that SOD/CAT gene therapy may protect against oxidative stress. Rats injected intratracheally with an adenovirus vector containing human CuZnSOD or CAT cDNA showed better survival upon exposure to 100% oxygen (but not against ischemia-reperfusion lung injury) [267]. A clinical trial demonstrated efficacy of CuZnSOD confined in liposomes in the treatment of radiation-induced fibrosis [268]. Numerous studies demonstrated that EC-SOD can protect against ischemia-reperfusion injury and reduce myocardial infarct size in experimental animals [167].

However, generally the use of SODs and CAT as therapeutic agents to attenuate reactive oxygen species-induced injury has mixed success. The main limitations of these enzymes include their large size (which limits cell penetration), short half-life in the circulation, antigenicity, and high costs. Increasing number of low-molecular weight SOD mimetics have been developed to overcome these limitations. Most effective seems to be metalloporphyrins, some of them also possessing CAT activity [269]. These compounds also scavenge lipid peroxyl radicals and peroxynitrite [270]; other classes include salen compounds (also showing SOD-like and CAT-like activity) [271], macrocyclics [272, 273], and nitroxides [274, 275]. SOD and CAT were reported to improve the function of rat heart when present during reperfusion after ischemia [276].

Biotechnological applications of SODs and CATs concern their use in biosensors for determination of concentrations of superoxide and hydrogen peroxide, respectively [277, 278].

References

1. Keller GA, Warner TG, Steimer KS, Hallewell RA (1991) Proc Natl Acad Sci USA 88:7381
2. Crapo JD, Oury T, Rabouille C, Slot JW, Chang LY (1992) Proc Natl Acad Sci USA 89:10405
3. Fridovich I (1995) Annu Rev Biochem 64:97
4. Brouwer M, Brouwer TH, Grater W, Enghild JJ, Thogersen IB (1997) Biochemistry 36:13381
5. Marklund SL (1984) Biochem J 220:269
6. Marklund SL (1984) J Clin Invest 74:1398
7. Marklund SL (1990) Biochem J 266:213
8. Stralin P, Karlsson K, Johansson BO, Marklund SL (1995) Arterioscler Thromb Vasc Biol 15:2032
9. Oury TD, Card JP, Klann E (1999) Brain Res 850:96
10. Zelko IN, Mariani TJ, Folz RJ (2002) Free Radic Biol Med 33:337
11. Marklund SL (1984) Biochem J 222:649
12. Streller S, Wingsle G (1994) Planta 192:195
13. Halliwell B, Gutteridge JMC (1999) Free radicals in biology and medicine. Oxford University Press, Oxford
14. Gabbianelli R, Battistoni A, Polizio F, Carri MT, De Martino A, Meier B, Desideri A, Rotilio G (1995) Biochem Biophys Res Commun 216:841
15. Martin ME, Byers BR, Olson MO, Salin ML, Arceneaux JE, Tolbert C (1986) J Biol Chem 261:9361
16. Kroll JS, Langford PR, Wilks KE, Keil AD (1995) Microbiology 141:2271
17. Youn HD, Kim EJ, Roe JH, Hah YC, Kang SO (1996) Biochem J 318:889
18. Youn HD, Youn H, Lee JW, Yim YI, Lee JK, Hah YC, Kang SO (1996) Arch Biochem Biophys 334:341
19. Kim FJ, Kim HP, Hah YC, Roe JH (1996) Eur J Biochem 241:178
20. Battistoni A, Pacello F, Folcarelli S, Ajello M, Donnarumma G, Greco R, Ammendolia MG, Touati D, Rotilio G, Valenti P (2000) Infect Immun 68:30
21. Scandalios JG (1997) In: Scandalios JG (ed) Oxidative stress and the molecular biology of antioxidant defenses. Cold Spring Harbor Laboratory Press, Cold Spring Harbor, NY, p 527
22. Hunter T, Bannister WH, Hunter GJ (1997) J Biol Chem 272:28652
23. Borgstahl GE, Parge HE, Hickey MJ, Beyer WF, Jr., Hallewell RA, Tainer JA (1992) Cell 71:107
24. Christianson DW (1997) Prog Biophys Mol Biol 67:217
25. Marklund SL (1982) Proc Natl Acad Sci USA 79:7634
26. Fattman CL, Enghild JJ, Crapo JD, Schaefer LM, Valnickova Z, Oury TD (2000) Biochem Biophys Res Commun 275:542
27. Hjalmarsson K, Marklund SL, Engstrom A, Edlund T (1987) Proc Natl Acad Sci USA 84:6340
28. Tainer JA, Getzoff ED, Beem KM, Richardson JS, Richardson DC (1982) J Mol Biol 160:181
29. Sandstrom J, Carlsson L, Marklund SL, Edlund T (1992) J Biol Chem 267:18205
30. Fukai T, Folz RJ, Landmesser U, Harrison DG (2002) Cardiovasc Res 55:239
31. Ookawara T, Kizaki T, Takayama E, Imazeki N, Matsubara O, Ikeda Y, Suzuki K, Li Ji L, Tadakuma T, Taniguchi N, Ohno H (2002) Biochem Biophys Res Commun 296:54
32. Sandstrom J, Karlsson K, Edlund T, Marklund SL (1993) Biochem J 294:853
33. Karlsson K, Marklund SL (1988) Biochem J 255:223

34. Adachi T, Yamada H, Futenma A, Kato K, Hirano K (1995) J Biochem (Tokyo) 117:586
35. Karlsson K, Sandstrom J, Edlund A, Marklund SL (1994) Lab Invest 70:705
36. Bordo D, Djinovic K, Bolognesi M (1994) J Mol Biol 238:366
37. Rodriguez-Trelles F, Tarrio R, Ayala FJ (2001) Proc Natl Acad Sci USA 98:11405
38. Imlay KR, Imlay JA (1996) J Bacteriol 178:2564
39. Tibell L, Aasa R, Marklund SL (1993) Arch Biochem Biophys 304:429
40. Geller BL, Winge DR (1983) Anal Biochem 128:86
41. Bannister JV, Bannister WH, Rotilio G (1987) CRC Crit Rev Biochem 22:111
42. Kardinahl S, Anemuller S, Schafer G (2000) Biol Chem 381:1089
43. Bielski BHJ, Cabelli DE, Arudi RL, Ross AB (1985) J Phys Chem Ref Data 14:1041
44. Klug D, Rabani J, Fridovich I (1972) J Biol Chem 247:4839
45. Tainer JA, Getzoff ED, Richardson JS, Richardson DC (1983) Nature 306:284
46. Salin ML, Wilson WW (1981) Mol Cell Biochem 36:157
47. Getzoff ED, Tainer JA, Weiner PK, Kollman PA, Richardson JS, Richardson DC (1983) Nature 306:287
48. Polticelli F, Bottaro G, Battistoni A, Carri MT, Djinovic-Carugo K, Bolognesi M, O'Neill P, Rotilio G, Desideri A (1995) Biochemistry 34:6043
49. Getzoff ED, Cabelli DE, Fisher CL, Parge HE, Viezzoli MS, Banci L, Hallewell RA (1992) Nature 358:347
50. Folcarelli S, Battistoni A, Falconi M, O'Neill P, Rotilio G, Desideri A (1998) Biochem Biophys Res Commun 244:908
51. Sines J, Allison SA, Wierzbicki A, McCammon JA (1990) J Phys Chem 94:959
52. Lawrence GD, Sawyer DT (1979) Biochemistry 18:3045
53. Verhagen MF, Meussen ET, Hagen WR (1995) Biochim Biophys Acta 1244:99
54. Han WG, Lovell T, Noodleman L (2002) Inorg Chem 41:205
55. Winterbourn CC, Peskin AV, Parsons-Mair HN (2002) J Biol Chem 277:1906
56. Agro' AF, Giovagnoli C, De Sole P, Calabrese L, Rotilio G, Mondovi' B (1972) FEBS Lett 21:183
57. Singh RJ, Goss SP, Joseph J, Kalyanaraman B (1998) Proc Natl Acad Sci USA 95:12912
58. Goss SP, Singh RJ, Kalyanaraman B (1999) J Biol Chem 274:28233
59. Zhang H, Joseph J, Felix C, Kalyanaraman B (2000) J Biol Chem 275:14038
60. Beckman JS, Ischiropoulos H, Zhu L, van der Woerd M, Smith C, Chen J, Harrison J, Martin JC, Tsai M (1992) Arch Biochem Biophys 298:438
61. Crow JP, Sampson JB, Zhuang Y, Thompson JA, Beckman JS (1997) J Neurochem 69:1936
62. MacMillan-Crow LA, Thompson JA (1999) Arch Biochem Biophys 366:82
63. Yamakura F, Taka H, Fujimura T, Murayama K (1998) J Biol Chem 273:14085
64. Jourd'heuil D, Laroux FS, Miles AM, Wink DA, Grisham MB (1999) Arch Biochem Biophys 361:323
65. Auchere F, Rusnak F (2002) J Biol Inorg Chem 7:664
66. Liochev SI, Fridovich I (2000) J Biol Chem 275:38482
67. Liochev SI, Fridovich I (2001) J Biol Chem 276:35253
68. Hempel SL, Buettner GR, O'Malley YQ, Wessels DA, Flaherty DM (1999) Free Radic Biol Med 27:146
69. Zheng M, Storz G (2000) Biochem Pharmacol 59:1
70. Kedziora J, Bartosz G (1988) Free Radic Biol Med 4:317
71. Kim HT, Kim YH, Nam JW, Lee HJ, Rho HM, Jung G (1994) Biochem Biophys Res Commun 201:1526
72. Yoo HY, Chang MS, Rho HM (1999) Mol Gen Genet 262:310
73. Hass MA, Massaro D (1988) J Biol Chem 263:776

74. Inoue N, Ramasamy S, Fukai T, Nerem RM, Harrison DG (1996) Circ Res 79:32
75. Janssen YM, Van Houten B, Borm PJ, Mossman BT (1993) Lab Invest 69:261
76. Frank S, Kampfer H, Podda M, Kaufmann R, Pfeilschifter J (2000) Biochem J 346:719
77. Church SL, Grant JW, Meese EU, Trent JM (1992) Genomics 14:823
78. Zhu CH, Huang Y, Oberley LW, Domann FE (2001) J Biol Chem 276:14407
79. Housset B, Junod AF (1982) Biochim. Biophys Acta 716:283
80. Shiki Y, Meyrick BO, Brigham KL, Burr IM (1987) Am J Physiol 252: C436
81. Warner BB, Stuart L, Gebb S, Wispe JR (1996) Am J Physiol 271: L150
82. Gilks CB, Price K, Wright JL, Churg A (1998) Am J Pathol 152:269
83. Weller BL, Crapo JD, Slot J, Posthuma G, Plopper CG, Pinkerton KE (1997) Am J Respir Cell Mol Biol 17:552
84. Shull S, Heintz NH, Periasamy M, Manohar M, Janssen YM, Marsh JP, Mossman BT (1991) J Biol Chem 266:24398
85. Warner BB, Burhans MS, Clark JC, Wispe JR (1991) Am J Physiol 260:L296
86. Tannahill CL, Stevenot SA, Eaker EY, Sallustio JE, Nick HS, Valentine JF (1997) Am J Physiol 272:G1230
87. Ruetzler CA, Furuya K, Takeda H, Hallenbeck JM (2001) J Cereb Blood Flow Metab 21:244
88. Schaper M, Gergely S, Lykkesfeldt J, Zbaren J, Leib SL, Tauber MG, Christen S (2002) J Neuropathol Exp Neurol 61:605
89. Pani G, Bedogni B, Anzevino R, Colavitti R, Palazzotti B, Borrello S, Galeotti T (2000) Cancer Res 60:4654
90. Martin V, Sainz RM, Mayo JC, Antolin I, Herrera F, Rodriguez C (2003) Endocr Res 29:83
91. Hendrickson DJ, Fisher JH, Jones C, Ho YS (1990) Genomics 8:736
92. Folz RJ, Crapo JD (1994) Genomics 22:162
93. Bowler RP, Arcaroli J, Crapo JD, Ross A, Slot JW, Abraham E (2001) Am J Respir Crit Care Med 164:290
94. Marklund SL (1992) J Biol Chem 267:6696
95. Brady TC, Chang LY, Day BJ, Crapo JD (1997) Am J Physiol 273: L1002
96. Stralin P, Marklund SL (2000) Atherosclerosis 151:433
97. Kinnula VL, Crapo JD (2003) Am J Respir Crit Care Med 167:1600
98. Culotta VC, Klomp LW, Strain J, Casareno RL, Krems B, Gitlin JD (1997) J Biol Chem 272:23469
99. Lyons TJ, Nersissian A, Goto JJ, Zhu H, Gralla EB, Valentine JS (1998) J Biol Inorg Chem 3:650
100. Casareno RL, Waggoner D, Gitlin JD (1998) J Biol Chem 273:23625
101. Schmidt PJ, Ramos-Gomez M, Culotta VC (1999) J Biol Chem 274:36952
102. Marklund S, Beckman G, Stigbrand T (1976) Eur J Biochem 65:415
103. DeCroo S, Kamboh MI, Leppert M, Ferrell RE (1988) Hum Hered 38:1
104. Zhang HJ, Yan T, Oberley TD, Oberley LW (1999) Cancer Res 59:6276
105. Xu Y, Krishnan A, Wan XS, Majima H, Yeh CC, Ludewig G, Kasarskis EJ, St Clair DK (1999) Oncogene 18:93
106. Shimoda-Matsubayashi S, Matsumine H, Kobayashi T, Nakagawa-Hattori Y, Shimizu Y, Mizuno Y (1996) Biochem Biophys Res Commun 226:561
107. Van Landeghem GF, Tabatabaie P, Kucinskas V, Saha N, Beckman G (1999) Hum Hered 49:190
108. Ambrosone CB, Freudenheim JL, Thompson PA, Bowman E, Vena JE, Marshall JR, Graham S, Laughlin R, Nemoto T, Shields PG (1999) Cancer Res 59:602
109. Mitrunen K, Sillanpaa P, Kataja V, Eskelinen M, Kosma VM, Benhamou S, Uusitupa M, Hirvonen A (2001) Carcinogenesis 22:827

110. Van Landeghem GF, Tabatabaie P, Beckman G, Beckman L, Andersen PM (1999) Eur J Neurol 6:639
111. Kimura K, Isashiki Y, Sonoda S, Kakiuchi-Matsumoto T, Ohba N (2000) Am J Ophthalmol 130:769
112. Wang LI, Miller DP, Sai Y, Liu G, Su L, Wain JC, Lynch TJ, Christiani DC (2001) J Natl Cancer Inst 93:1818
113. Hirvonen A, Tuimala J, Ollikainen T, Linnainmaa K, Kinnula V (2002) Cancer Lett 178:71
114. Hong YC, Lee KH, Yi CH, Ha EH, Christiani DC (2002) Toxicol Lett 129:255
115. Hori H, Ohmori O, Shinkai T, Kojima H, Okano C, Suzuki T, Nakamura J (2000) Neuropsychopharmacology 23:170
116. Yamada H, Yamada Y, Adachi T, Goto H, Ogasawara N, Futenma A, Kitano M, Miyai H, Fukatsu A, Hirano K, Kakumu S (1997) Jpn J Hum Genet 42:353
117. Adachi T, Ohta H, Yamada H, Futenma A, Kato K, Hirano K (1992) Clin Chim Acta 212:89
118. Sandstrom J, Nilsson P, Karlsson K, Marklund SL (1994) J Biol Chem 269:19163
119. Yamada H, Yamada Y, Adachi T, Goto H, Ogasawara N, Futenma A, Kitano M, Hirano K, Kato K (1995) Jpn J Hum Genet 40:177
120. Yamada H, Yamada Y, Adachi T, Fukatsu A, Sakuma M, Futenma A, Kakumu S (2000) Nephron 84:218
121. Marklund SL, Nilsson P, Israelsson K, Schampi I, Peltonen M, Asplund K (1997) J Intern Med 242:5
122. Imlay JA (2003) Annu Rev Microbiol 57:395
123. McCord JM, Keele BB, Fridovich I (1971) Proc Natl Acad Sci USA 68:1024
124. Tardat B, Touati D (1993) Mol Microbiol 9:53
125. Flint DH, Allen RM (1996) Chem Rev 96:2315
126. Carlioz A, Touati D (1986) EMBO J 5:623
127. Farr SB, D'Ari R, Touati D (1986) Proc Natl Acad Sci USA 83:8268
128. Bilinski T, Krawiec Z, Liczmanski A, Litwinska J (1985) Biochem Biophys Res Commun 130:533
129. Srinivasan C, Liba A, Imlay JA, Valentine JS, Gralla EB (2000) J Biol Chem 275:29187
130. De Freitas JM, Liba A, Meneghini R, Valentine JS, Gralla EB (2000) J Biol Chem 275: 11645
131. Garay-Arroyo A, Lledias F, Hansberg W, Covarrubias AA (2003) FEBS Lett 539:68
132. Park JI, Grant CM, Davies MJ, Dawes IW (1998) J Biol Chem 273:22921
133. van Loon AP, Pesold-Hurt B, Schatz G (1986) Proc Natl Acad Sci USA 83:3820
134. Costa V, Amorim MA, Reis E, Quintanilha A, Moradas-Ferreira P (1997) Microbiology 143:1649
135. Reaume AG, Elliott JL, Hoffman EK, Kowall NW, Ferrante RJ, Siwek DF, Wilcox HM, Flood DG, Beal MF, Brown RHJ, Scott RW, Snider WD (1996) Nat Genet 13:43
136. Ho YS, Magnenat JL, Gargano M, Cao J (1998) Environ Health Perspect 106(Suppl 5):1219
137. Yoshida T, Maulik N, Engelman RM, Ho YS, Das DK (2000) Circ Res 86:264
138. Kawase M, Murakami K, Fujimura M, Morita-Fujimura Y, Gasche Y, Kondo T, Scott RW, Chan PH (1999) Stroke 30:1962
139. Ho YS, Gargano M, Cao J, Bronson RT, Heimler I, Hutz RJ (1998) J Biol Chem 273:7765
140. McFadden SL, Ohlemiller KK, Ding D, Shero M, Salvi RJ (2001) Noise Health 3:49
141. Zhang X, Han D, Ding D, Dai P, Yang W, Jiang S, Salvi RJ (2002) Chin Med J 115:258
142. Ho YS (1994) Am J Physiol 266: L319
143. Tsan MF (2001) Int J Mol Med 7:13
144. Larsen GL, White CW, Takeda K, Loader JE, Nguyen DD, Joetham A, Groner Y, Gelfand EW (2000) Am J Physiol Lung Cell Mol Physiol 279: L350

145. Wang P, Chen H, Qin H, Sankarapandi S, Becher MW, Wong PC, Zweier JL (1998) Proc Natl Acad Sci USA 95:4556
146. Huang TT, Yasunami M, Carlson EJ, Gillespie AM, Reaume AG, Hoffman EK, Chan PH, Scott RW, Epstein CJ (1997) Arch Biochem Biophys 344:424
147. Liochev SI, Fridovich I (1991) J Biol Chem 266:8747
148. de Haan JB, Cristiano F, Iannello R, Bladier C, Kelner MJ, Kola I (1996) Hum Mol Genet 5:283
149. Epstein CJ, Avraham KB, Lovett M, Smith S, Elroy-Stein O, Rotman G, Bry C, Groner Y (1987) Proc Natl Acad Sci USA 84:8044
150. McCord JM (2002) Methods Enzymol 349:331
151. Gardner R, Salvador A, Moradas-Ferreira P (2002) Free Radic Biol Med 32:1351
152. Offer T, Russo A, Samuni A (2000) FASEB J 14:1215
153. Zielinski R, Pilecki M, Kubinski K, Zien P, Hellman U, Szyszka R (2002) Biochem Biophys Res Commun 296:1310
154. Abramczyk O, Zien P, Zielinski R, Pilecki M, Hellman U, Szyszka R (2003) Biochem Biophys Res Commun 307:31
155. Pinkham JL, Wang Z, Alsina J (1997) Curr Genet 31:281
156. Li Y, Huang TT, Carlson EJ, Melov S, Ursell PC, Olson JL, Noble LJ, Yoshimura MP, Berger C, Chan PH, et al. (1995) Nat Genet 11:376
157. Lebovitz RM, Zhang H, Vogel H, Cartwright JJ, Dionne L, Lu N, Huang S, Matzuk MM (1996) Proc Natl Acad Sci USA 93:9782
158. Melov S, Coskun P, Patel M, Tuinstra R, Cottrell B, Jun AS, Zastawny TH, Dizdaroglu M, Goodman SI, Huang TT, Miziorko H, Epstein CJ, Wallace DC (1999) Proc Natl Acad Sci USA 96:846
159. Lindau-Shepard B, Shaffer JB, Del Vecchio PJ (1994) J Cell Physiol 161:237
160. Hirose K, Longo DL, Oppenheim JJ, Matsushima K (1993) FASEB J 7:361
161. Sun J, Chen Y, Li M, Ge Z (1998) Free Radic Biol Med 24:586
162. Mossman BT, Surinrut P, Brinton BT, Marsh JP, Heintz NH, Lindau-Shepard B, Shaffer JB (1996) Free Radic Biol Med 21:125
163. St Clair DK, Jordan JA, Wan XS, Gairola CG (1994) J Toxicol Environ Health 43:239
164. Ilizarov AM, Koo HC, Kazzaz JA, Mantell LL, Li Y, Bhapat R, Pollack S, Horowitz S, Davis JM (2001) Am J Respir Cell Mol Biol 24:436
165. Wispe JR, Warner BB, Clark JC, Dey CR, Neuman J, Glasser SW, Crapo JD, Chang LY, Whitsett JA (1992) J Biol Chem 267:23937
166. Oury TD, Day BJ, Crapo JD (1996) Lab Invest 75:617
167. Fatman CL, Schaefer LM, Oury TD (2003) Free Radic Biol Med 35:236
168. Carlsson LM, Jonsson J, Edlund T, Marklund SL (1995) Proc Natl Acad Sci USA 92:6264
169. Sheng H, Brady TC, Pearlstein RD, Crapo JD, Warner DS (1999) NeuroSci Lett 267:13
170. Sentman ML, Jonsson LM, Marklund SL (1999) Free Radic Biol Med 27:790
171. Pineda JA, Aono M, Sheng H, Lynch J, Wellons JC, Laskowitz DT, Pearlstein RD, Bowler R, Crapo J, Warner DS (2001) J Neurotrauma 18:625
172. Oury TD, Ho YS, Piantadosi CA, Crapo JD (1992) Proc Natl Acad Sci USA 89:9715
173. Levin ED, Brady TC, Hochrein EC, Oury TD, Jonsson LM, Marklund SL, Crapo JD (1998) Behav Genet 28:381
174. Levin ED, Brucato FH, Crapo JD (2000) Behav Genet 30:95
175. Thiels E, Klann E (2002) Physiol Behav 77:601
176. Suzuki T, Sugino N, Fukaya T, Sugiyama S, Uda T, Takaya R, Yajima A, Sasano H (1999) Fertil Steril 72:720
177. Ookawara T, Haga S, Ha S, Ohishi S, Toshinai K, Kizaki T, Ji LL, Suzuki K, Ohno H (2003) Free Radic Res 37:713

178. Seppanen MM, Fagerstedt K (2000) Physiol Plant 108:279
179. Rosen DR, Siddique T, Patterson D, Figlewicz DA, Sapp P, Hentati A, Donaldson D, Goto J, O'Regan JP, Deng HX, et al. (1993) Nature 362:59
180. Gaudette M, Hirano M, Siddique T (2000) Amyotroph Lateral Scler Other Motor Neuron Disord 1:83
181. Cleveland DW, Liu J (2000) Nat Med 6:1320
182. Bruijn LI, Houseweart MK, Kato S, Anderson KL, Anderson SD, Ohama E, Reaume AG, Scott RW, Cleveland DW (1998) Science 281:1851
183. Huang Y, He T, Domann FE (1999) DNA Cell Biol 18:643
184. Safford SE, Oberley TD, Urano M, St Clair DK (1994) Cancer Res 54:4261
185. Li JJ, Oberley LW, St Clair DK, Ridnour LA, Oberley TD (1995) Oncogene 10:1989
186. Oberley LW (2001) Antioxid Redox Signal 3:461
187. Epperly MW, Bernarding M, Gretton J, Jefferson M, Nie S, Greenberger JS (2003) Exp Hematol 31:465
188. Landmesser U, Merten R, Spiekermann S, Buttner K, Drexler H, Hornig B (2000) Circulation 101:2264
189. Takatsu H, Tasaki H, Kim HN, Ueda S, Tsutsui M, Yamashita K, Toyokawa T, Morimoto Y, Nakashima Y, Adachi T (2001) Biochem Biophys Res Commun 285:84
190. Wang XL, Duarte N, Cai H, Adachi T, Sim AS, Cranney G, Wilcken DE (1999) Atherosclerosis 146:133
191. Wilcken DE, Wang XL, Adachi T, Hara H, Duarte N, Green K, Wilcken B (2000) Arterioscler Thromb Vasc Biol 20:1199
192. Oda A, Bannai C, Yamaoka T, Katori T, Matsushima T, Yamashita K (1994) Horm Metab Res 26:1
193. Adachi T, Ohta H, Hirano K, Hayashi K, Marklund SL (1991) Biochem J 279:263
194. Smith LJ, Shamsuddin M, Sporn PH, Denenberg M, Anderson J (1997) Free Radic Biol Med 22:1301
195. Andreadis AA, Hazen SL, Comhair SA, Erzurum SC (2003) Free Radic Biol Med 35:213
196. Radi R, Turrens JF, Chang LY, Bush KM, Crapo JD, Freeman BA (1991) J Biol Chem 266:22028
197. Zamocky M, Koller F (1999) Prog Biophys Mol Biol 72:19
198. Sichak SP, Dounce AL (1986) Arch Biochem Biophys 249:286
199. Putnam CD, Arvai AS, Bourne Y, Tainer JA (2000) J Mol Biol 296:295
200. Amara P, Andreoletti P, Jouve HM, Field MJ (2001) Protein Sci 10:1927
201. Switala J, Loewen PC (2002) Arch Biochem Biophys 401:145
202. Fita I, Rossmann MG (1985) J Mol Biol 185:21
203. Kirkman HN, Gaetani GF (1984) Proc Natl Acad Sci USA 81:4343
204. Darr D, Fridovich I (1986) Biochem Pharmacol. 35:3642
205. Waldo GS, Fronko RM, Penner-Hahn JE (1991) Biochemistry 30:10486
206. Ivancich A, Jouve HM, Sartor B, Gaillard J (1997) Biochemistry 36:9356
207. Chance B, Sies H, Boveris A (1979) Physiol Rev 59:527
208. Obinger C, Maj M, Nicholls P, Loewen P (1997) Arch Biochem Biophys 342:58
209. Miller MA, Shaw A, Kraut J (1994) Nat. Struct. Biol 1:524
210. Araiso T, Dunford HB (1980) Biochem Biophys Res Commun 94:1177
211. Chance B (1949) J Biol Chem 179:1311
212. Chelikani P, Carpena X, Fita I, Loewen PC (2003) J Biol Chem 278:31290
213. Chance B (1952) J Biol Chem 194:471
214. Kirkman HN, Rolfo M, Ferraris AM, Gaetani GF (1999) J Biol Chem 274:13908
215. Kirkman HN, Galiano S, Gaetani GF (1987) J Biol Chem 262:660
216. Marchler G, Schuller C, Adam G, Ruis H (1993) EMBO J 12:1997

217. Ni W, Trelease RN (1991) Plant Cell 3:737
218. Willekens H, Chamnongpol S, Davey M, Schraudner M, Langebartels C, Van Montagu M, Inze D, Van Camp W (1997) EMBO J 16:4806
219. Quan F, Korneluk RG, Tropak MB, Gravel RA (1986) Nucleic Acids Res 14:5321
220. Bell GI, Najarian RC, Mullenbach GT, Hallewell RA (1986) Nucleic Acids Res 14: 5561
221. Meilhac O, Zhou M, Santanam N, Parthasarathy S (2000) J Lipid Res 41(8):1205
222. Coto-Montes A, Boga JA, Tomas-Zapico C, Rodriguez-Colunga MJ, Martinez-Fraga J, Tolivia-Cadrecha D, Menendez G, Hardeland R, Tolivia D (2001) Free Radic Biol Med 30:785
223. Cao C, Leng Y, Kufe D (2003) J Biol Chem 278:29667
224. Ogata M (1991) Hum Genet 86:331
225. Schellhorn HE (1995) FEMS Microbiol Lett 131:113
226. Loewen P (1996) Gene 179:39
227. Michan S, Lledias F, Baldwin JD, Natvig DO, Hansberg W (2002) Free Radic Biol Med 33:521
228. Smirnoff N (1998) Curr Opin Biotechnol 9:214
229. Durner J, Klessig DF (1996) J Biol Chem 271:28492
230. Chen Z, Silva H, Klessig DF (1993) Science 262:1883
231. Antunes F, Han D, Cadenas E (2002) Free Radic Biol Med 33:1260
232. Gaetani GF, Kirkman HN, Mangerini R, Ferraris AM (1994) Blood 84:325
233. Zimatkin SM, Liopo AV, Deitrich RA (1998) Alcohol Clin Exp Res 22:1623
234. Lands WE (1998) Alcohol 15:147
235. Aragon CM, Amit Z (1992) Neuropharmacology 31:709
236. Koechling UM, Amit Z (1992) Alcohol 27:181
237. Koechling UM, Amit Z, Negrete JC (1995) Alcohol Clin Exp Res 19:1096
238. Hirono A, Sasaya-Hamada F, Kanno H, Fujii H, Yoshida T, Miwa S (1995) Blood Cells Mol Dis 21:232
239. Goth L, Rass P, Madarasi I (2001) Electrophoresis 22:49
240. Harman D (2001) Ann NY Acad Sci 928:1
241. Barker MG, Brimage LJ, Smart KA (1999) FEMS Microbiol Lett 177:199
242. Wawryn J, Krzepilko A, Myszka A, Bilinski T (1999) Acta Biochim Pol 46:249
243. Harris N, Costa V, MacLean M, Mollapour M, Moradas-Ferreira P, Piper PW (2003) Free Radic Biol Med 34:1599
244. Nestelbacher R, Laun P, Vondrakova D, Pichova A, Schuller C, Breitenbach M (2000) Exp Gerontol 35:63
245. Van Zandycke SM, Sohier PJ, Smart KA (2002) Mech Ageing Dev 123:365
246. Yanase S, Yasuda K, Ishii N (2002) Mech Ageing Dev 123:1579
247. Melov S, Ravenscroft J, Malik S, Gill MS, Walker DW, Clayton PE, Wallace DC, Malfroy B, Doctrow SR, Lithgow GJ (2000) Science 289:1567
248. Bayne AC, Sohal RS (2002) Free Radic Biol Med 32:1229
249. Parkes TL, Kirby K, Phillips JP, Hilliker AJ (1998) Genome 41:642
250. Rogina B, Helfand SL (2000) Biogerontology 1:163
251. Orr WC, Sohal RS (1994) Science 263:1128
252. Sun J, Folk D, Bradley TJ, Tower J (2002) Genetics 161:661
253. Tower J (2000) Mech Ageing Dev 118:1
254. Sun J, Tower J (1999) Mol Cell Biol 19:216
255. Orr WC, Arnold LA, Sohal RS (1992) Mech Ageing Dev 63:287
256. Griswold CM, Matthews AL, Bewley KE, Mahaffey JW (1993) Genetics 134:781
257. Mockett RJ, Bayne AC, Kwong LK, Orr WC, Sohal RS (2003) Free Radic Biol Med 34:207

258. Kitani K, Minami C, Isobe K, Maehara K, Kanai S, Ivy GO, Carrillo MC (2002) Mech Ageing Dev 123:1087
259. Oda T, Akaike T, Hamamoto T, Suzuki F, Hirano T, Maeda H (1989) Science 244:974
260. Eldad A, Ben Meir P, Breiterman S, Chaouat M, Shafran A, Ben-Bassat H (1998) Burns 24:114
261. Simonson SG, Welty-Wolf KE, Huang YC, Taylor DE, Kantrow SP, Carraway MS, Crapo JD, Piantadosi CA (1997) J Appl Physiol 83:550
262. Tang G, White JE, Gordon RJ, Lumb PD, Tsan MF (1993) J Appl Physiol 74:1425
263. Jacobson JM, Michael JR, Jafri MHJ, Gurtner GH (1990) J Appl Physiol 68:1252
264. Tamagawa K, Taooka Y, Maeda A, Hiyama K, Ishioka S, Yamakido M (2000) Am J Respir Crit Care Med 161:1279
265. Freeman BA, Turrens JF, Mirza Z, Crapo JD, Young SL (1985) Fed Proc 44:2591
266. White CW, Jackson JH, Abuchowski A, Kazo GM, Mimmack RF, Berger EM, Freeman BA, McCord JM, Repine JE (1989) J Appl Physiol 66:584
267. Danel C, Erzurum SC, Prayssac P, Eissa NT, Crystal RG, Herve P, Baudet B, Mazmanian M, Lemarchand P (1998) Hum Gene Ther 9:1487
268. Delanian S, Baillet F, Huart J, Lefaix JL, Maulard C, Housset M (1994) Radiother Oncol 32:12
269. Patel M, Day BJ (1999) Trends Pharmacol Sci 20:359
270. Ferrer-Sueta G, Vitturi D, Batinic-Haberle I, Fridovich I, Goldstein S, Czapski G, Radi R (2003) J Biol Chem 278:27432
271. Doctrow SR, Huffman K, Marcus CB, Musleh W, Bruce A, Baudry M, Malfroy B (1997) Adv Pharmacol 38:247
272. Samlowski WE, Petersen R, Cuzzocrea S, Macarthur H, Burton D, McGregor JR, Salvemini D (2003) Nat Med 9:750
273. Muscoli C, Cuzzocrea S, Riley DP, Zweier JL, Thiemermann C, Wang ZQ, Salvemini D (2003) Br J Pharmacol 140:445
274. Kwon TH, Chao DL, Malloy K, Sun D, Alessandri B, Bullock MR (2003) J Neurotrauma 20:337
275. Hahn SM, Sullivan FJ, DeLuca AM, Bacher JD, Liebmann J, Krishna MC, Coffin D, Mitchell JB (1999) Free Radic Biol Med 27:529
276. Koke JR, Christodoulides NJ, Chudej LL, Bittar N (1990) Mol Cell Biochem 96:97
277. Tian Y, Mao L, Okajima T, Ohsaka T (2002) Anal Chem 74:2428
278. Yu A, Caruso F (2003) Anal Chem 75:3031

Part C
Secondary Antioxidative Defense, Repair and Removal Systems

The Handbook of Environmental Chemistry Vol. 2, Part O (2005): 151–175
DOI 10.1007/b101150

DNA Repair: Mechanisms and Measurements

Bente Riis · Henrik E. Poulsen (✉)

Dept. of Clinical Pharmacology Q7642, Rigshospitalet, Copenhagen University Hospital, Blegdamsvej 9, 2100 Copenhagen East, Denmark
hepo@rh.dk

Abstract The early findings that significant amounts of modifications are induced to the cellular DNA both spontaneously and as a consequence of metabolism and environmental exposures have led to the discovery of the existence of multiple highly efficient repair mechanisms to maintain the integrity of DNA.

There is increasing evidence of considerable variation between individuals with regard to activity of DNA repair mechanisms and there are examples of diseases that have deficiencies in DNA repair as a risk factor. It is anticipated that detailed understanding of the DNA repair mechanisms will lead to insights for prevention of diseases, for diagnosis, into resistance to chemotherapeutic agents, and for development of therapeutic tools. DNA repair is also considered to be an important factor in aging.

This chapter briefly summarizes the known DNA repair pathways and reviews the approaches for clarifying and measuring DNA repair in vivo and in vitro.

Keywords Direct repair · Base excision repair (BER) · Nucleotide excision repair (NER) · Transcription-coupled repair · HR · NHEJ · DNA repair polymorphisms · DNA repair knockout · Inborn DNA repair defects · Incision assay · Base modifications

1 Introduction

Since DNA is the carrier of genetic information and spontaneous mutations occur only at low frequency, cellular DNA was initially regarded as an essentially stable entity. It is now clear that modifications are induced at high rate to the cellular DNA. This occurs both spontaneously as a consequence of metabolism, and from environmental exposures. During the last decade the existence of efficient repair mechanisms for maintaining the integrity of DNA have been discovered.

By the beginning of 1900 the term gene (derived from genesis) had already been established and the relationship between high-energy radiation and skin cancer discovered. Nobel price winner H. J. Muller (in 1946) established both the concept of induction of mutations and hints of gene repair mechanisms, from studies on Drosophila. This was long before the discovery of DNA as the carrier of genetic information [1]. Undoubtedly, the discovery of the genetic code in DNA moved the focus from the repair mechanisms to transcription and gene function. It took another decade before the importance of DNA repair again attracted attention.

One of the first important insights in the field of DNA repair was made after isolation of *E. coli* mutants with exceptional sensitivity to UV light [2] and subsequent mapping of mutations in genes designated *uvr A*, *B*, and *C* [3]. This led to the model of nucleotide excision repair [4, 5]. Later the base excision repair pathway was exposed when Lindahl and coworkers [6, 7] discovered a series of bacterial enzymes that recognized and removed modified nucleotides by a DNA *N*-glycosylase reaction.

The initial 1966 description of four DNA repair steps [8] still represents the overall model of DNA repair; however, the scheme is now much more complex with at least six repair pathways, each consisting of several proteins (Table 1).

For a period it was believed that DNA repair was confined to the nuclear DNA, as an early finding showed the absence of repair of UV-induced damage in mitochondrial DNA [9]. However, several reports later showed that mitochondria can repair a variety of DNA lesions including strand breaks, alkali-sensitive sites, Fpg-sensitive sites (reviewed by [10]), and mismatches [11]. The significance of mitochondrial DNA repair is not yet clear. Few defects in mitochondrial DNA repair and/or replication have been reported. Two diseases have so far been attributed to alterations in the major mitochondrial DNA repair enzyme polymerase γ: progressive external ophthalmoplegia and Alper's syndrome [12].

Table 1 The first model of DNA repair suggested in 1966 was divided into four general steps. This scheme corresponds very nicely to the present models of DNA repair, though numerous details have been added. From recent reviews we have made a list of the proteins known to be involved in the different repair pathways

	DNA repair in 1966 [8]	DNA repair in 2003 [15–17, 21, 25]
Step 1 Damage recognition and initiation of repair	One of the two DNA strands are interrupted: 1. by radiation induced chain break 2. by enzymatic excision of damaged bases 3. by a recombination enzyme	**BER:** OGG1, NEIL1, NEIL2, MYH, NTH, APE1, UNG, MPG, UV-endonuclease, XRCC1, PARP, PNK **GGR-NER:** XPC, hHR23B, TFIIH, XPG, XPA, RPA **TCR-NER:** Pol II, CSB, CSA, TFIIH, XPG, XPA, RPA **HR:** strand break **NHEJ:** strand break **MMR:** MSH2/6, MSH2/3, MLH1/PMS2, MLH1/PMS1
Step 2 Procession of repair patch	Nucleotides (in a limited distance) are released, presumably through the action of an enzyme	**BER:**DNA pol β, XRCC1, DNA pol δ, DNA pol ε, PCNA, FEN1 **NER:** ERCC1, XPF **HR:** RPA, RAD50/MRE11/NBS1, RAD51B, RAD51C, RAD51D, XRCC2, XRCC3 **NHEJ:** KU70/KU80, DNA-PK_{CS}, Artemis **MMR:** exonuclease 1, FEN1, RPA, PCNA, RFC
Step 3 Synthesis of new DNA	DNA helices are reconstructed by a DNA repair polymerase	**BER:** DNA pol β, XRCC1, DNA pol δ, DNA pol ε **NER:** DNA pol δ, DNA pol ε **HR:** Holliday junction resolvases, BRCA1, BRCA2, **NHEJ:** – **MMR:** DNA pol δ, DNA pol ε

Table 1 (continued)

	DNA repair in 1966 [8]	DNA repair in 2003 [15–17, 21, 25]
Step 4		
Ligation of the DNA strand	Phosphodiester backbone is joined when the last nucleotide is inserted into the gap	**BER**: DNA ligase 1, DNA ligase 3 **NER**: DNA ligase 1 **HR**: DNA ligase 4 **NHEJ**: XRCC4, DNA ligase 4 **MMR**: DNA ligase?

Abbreviations: *APE1* apurinic/apyrimidinic endonuclease 1, *BER* base excision repair, *BRCA1* breast cancer 1, *BRCA2* breast cancer 2, *CSA* Cockayne syndromeA, *CSB* Cockayne syndrome B, *DNA-PK$_{CS}$* DNA-dependent protein kinase catalytic subunit (XRCC7), *ERCC1* excision repair cross-complementing rodent repair deficiency, complementation group 1 *FEN1* flap structure-specific endonuclease 1 (DNase IV), *GGR* global genome repair, *hHR23B* human Rad 23 homologue B, *HR* homologous recombination, *KU70* Ku protein 70 kDa (XRCC6), *KU80* Ku protein 80 kDa (XRCC5), *MLH1* Mut L homologue 1, *MMR* mismatch repair, *MPG N*-methylpurine-DNA glycosylase, *MRE11* meiotic recombination 11 homologue, *MSH2* Mut S homologue 2, *MSH3* Mut S homologue 3, *MSH6* Mut S homologue 6, *MYH* Mut Y homologue, *NBS1* Nijmegen breakage syndrome 1, *NEIL1* Nei endonuclease VIII like, *NEIL2* Nei like 2, *NER* nucleotide excision repair, *NHEJ* non-homologous end joining, *NTH* endonuclease III, *OGG1* 8-oxoguanine DNA N-glycosylase, *PARP* poly (ADP-ribose) polymerase, *PCNA* proliferating cell nuclear antigen, *PMS1* postmeiotic segregation increased 1, *PMS2* postmeiotic segregation increased 2, *PNK* polynucleotide kinase, *Pol* polymerase, *RAD50* human radiation sensitive gene 50 homologue (S. cerevisiae), *RAD51* human radiation sensitive gene 51 homologue (S. cerevisiae), *RAD51B* human RAD51 paralogue B, *RAD51C* human RAD51paralogue C, *RAD51D* human RAD51 paralogue D, *RFC* replication factor C, *RPA* replication protein A, *TCR* transcription coupled repair, *TFIIH* transcription factor IIH, *UNG* uracil-DNA glycosylase, *XPA* Xeroderma pigmentosum complementation group A, *XPC* Xeroderma pigmentosum complementation group C, *XPF* Xeroderma pigmentosum complementation group F, *XPG* Xeroderma pigmentosum complementation group G, *XRCC1* X-ray repair complementing defective repair in Chinese hamster cells 1, *XRCC2* X-ray repair complementing defective repair in Chinese hamster cells 2, *XRCC3* X-ray repair complementing defective repair in Chinese hamster cells 3, *XRCC4* X-ray repair complementing defective repair in Chinese hamster cells 4.

It is anticipated that detailed understanding of the DNA repair mechanisms will lead to insights for prevention of diseases, for diagnosis, into resistance to chemotherapeutic agents, and for development of therapeutic tools. This chapter briefly summarizes the known DNA repair pathways and reviews the approaches for clarifying and measuring DNA repair in vivo and in vitro. Excellent comprehensive reviews have recently been published on repair mechanisms and chromatin structure [13, 14], on the chemistry and biology of DNA repair [15], on the chemistry of glycosylases [16], and on how DNA repair mechanisms may relate to cancer and ageing [17]. In this chapter we also describe the current available methods for estimation of DNA repair, qualitatively and quantitatively.

2 DNA Repair Mechanisms

2.1 Repair of the Precursor Pool

The presence of the oxidized nucleotide triphosphate (e.g., 8-oxodGTP) in the DNA synthesis precursor pool has been shown to be an important part of the potential mutagenic risk of oxidative stress. This is pointed out by the existence of specific 8-oxodGTP triphosphatases (MutT protein in *E.coli* and hMTH1 in humans) that eliminate 8-oxodGTP from the nucleotide precursor pool, and thereby prevent the insertion of the oxidized deoxynucleotide triphosphate opposite dA or dC residues during DNA replication [18, 19]. Both enzymes have also been shown to degrade 8-oxoguanosine triphosphate (8-oxoGTP), which can otherwise be incorporated in RNA opposite C or A [20].

2.2 Direct Repair

Direct repair is a single-step reversal of a lesion in the DNA by a single enzyme. One example is the O^6-methylguanine-DNA methyltransferase (MGMT), which repairs DNA by a "suicide" mechanism involving the transfer of the methyl-group from O^6-methylguanine (or other alkylated substrates) to a specific cysteine residue in the enzyme [21]. This type of repair is especially important with regard to alkylating agents. Another example is the DNA photolyase, which specifically repairs UV-light-induced pyrimidine dimers by the use of energy from blue light [22]. The principal item of the repair mechanism is the transfer of one electron from the photolyase enzyme onto the pyrimidine dimer, which breaks the linkage between the pyrimidine molecules. Subsequently, the electron is returned to the oxidized molecule, whereby the enzyme and the biologically intact DNA are recreated [23]. Photolyase activity has been demonstrated in cell extracts of a large number of higher eukaryotes (reviewed in [24]) and a mammalian DNA photolyase has been cloned [22]; however, it is still unclear whether human cells possess significant photolyase activity [25].

2.3 Base Excision Repair

The mammalian process of base excision repair (BER) deals with most single base damages and abasic sites and acts on both oxidative and non-oxidative base modifications. The modified bases are removed by a subset of repair proteins called DNA glycosylases. Each DNA glycosylase enzyme is specific for one or a few altered bases in the DNA and catalyzes its removal.

The process of BER is divided into two overall biochemical pathways. These are termed "short patch" or "single nucleotide" BER involving replacement of one nucleotide, and "long patch" BER pathway with gap filling of several nucleotides. The repair pathways are not completely clarified as yet, and controversy still exists regarding the involved DNA polymerases, DNA ligases, and accessory proteins. Most of the present work concerning the different BER pathways are in vitro studies of reconstituted pathways with purified proteins, which may not reflect the in vivo mechanisms and do not take into account the possible regulatory mechanisms. Table 1 is based on recent reviews [15–17, 26, 27] and lists the proteins known to be involved in BER.

BER of damaged bases are initiated by a glycosylase that recognizes and removes the damaged base leaving an abasic site. There are two functional types of glycosylases named mono- and bi-functional DNA glycosylases. Mono-functional or simple DNA glycosylases hydrolyze the base-sugar (*N*-1′C) glycosylic bond to release the modified or damaged base from DNA and thereby generate an unmodified abasic (AP) site. Successive incision by an AP endonuclease makes a nick in the DNA strand. Bi-functional DNA glycosylases work by a two-step incision mechanism. Besides hydrolysis of the *N*-glycosylic bonds of their substrates, they also contain a β-lyase activity that cleaves the resulting AP-sites. The completion of the base excision repair process after removal of the damaged base can diverge in at least four pathways. As yet, there is little understanding of the potential mechanisms coordinating the different repair pathways when they are functional in the same cell.

There is strong evidence from studies in mammalian cell extracts that DNA polymerase β (pol β) is the major DNA polymerase participating in single nucleotide gap filling during short patch BER [28–31]). After incision of the DNA strand by an endonuclease or a glycosylase, DNA pol β catalyzes an β-elimination of the 5′ deoxyribose-5-phosphate by an intrinsic dRPase activity and subsequently fills the resultant one-nucleotide gap. The nick is sealed by DNA ligase III. The pathway is dependent on the XXRC1 protein [32]. However, the existence of a pol β-independent single nucleotide repair patch pathway for removal of 8-oxodG in DNA in pol β knock out mouse cell extracts has been established by Dianov et al. [33]. The repair gap in these cells is filled by DNA polymerase δ or ε. It is not clear whether this mechanism operates simultaneous with the pol β-dependent single nucleotide BER pathway in normal cells or only as a back up when pol β is deficient. A DNA pol β-independent long patch (2–6 nucleotides) base excision repair pathway has also been reported as a possible back-up system in pol β knockout mouse cell extracts [34, 35]. Both DNA polymerase δ and ε were able to replace nucleotides at the lesion site.

Long patch BER differ from short patch BER and from NER by both the repair patch size and the enzymes involved. Klungland and Lindahl [36] described the 2–6 nt long repair patch BER after reconstitution of repair of a reduced or oxidized abasic site. Human AP endonuclease, DNA polymerase β, DNA ligase III or I, and the structure-specific FEN1 (flap endonuclease/DNase IV) showed to be essential for the repair process. In addition, PCNA could promote the long patch pathway by stimulation of FEN1. Pol δ could substitute for pol β. Long patch BER has also been described as gap filling of up to 13 nucleotides [32, 34, 37, 38]. This pathway was shown to require AP endonuclease, FEN1, PCNA, DNA polymerase δ or ε, and DNA ligase I.

2.4 Nucleotide Excision Repair

The nucleotide excision repair (NER) pathway is involved in the removal of a wide range of bulky DNA adducts and DNA cross-links. In addition, the NER enzymes may recognize many diverse DNA lesions that do not disturb the DNA helix, including 8-oxodG and other modified bases. Under normal conditions NER does not contribute significantly (maybe as little as 0.1% [39]) to the removal of 8-oxodG in mammalian cell extracts, not even when the pol β-dependent BER pathway was disrupted [33]. It is, however, a common assumption that NER works as a back up system for BER, i.e., when BER is deficient NER can repair the base modifications.

The NER pathway consists of two different sub-pathways known as global genome repair (GGR) and transcription-coupled repair (TCR). Most genes involved contribute to both sub-pathways.

In GGR-NER the initial recognition step is believed to occur via the binding of XPC protein at the site of the lesion [40] in complex with the human homologue of the yeast RAD23B protein (HHR23B) [41]. Subsequent to the recognition of the DNA damage, the basal transcription initiation factor IIH (TFIIH) and the XPA protein are recruited to the lesion where they initiate the local opening of double-stranded DNA around the lesion [42]. XPA and replication protein A (RPA) are then thought to form a pre-incision complex. The first DNA cut is made by the endonuclease XPG five to six nucleotides downstream (3′) from the lesion. Then the XPF-ERCC1 endonuclease complex cleaves the 22nd, 23rd or 24th bond upstream (5′) of the damage and thus releases a 27–29-mer [43]. The gap is filled by DNA polymerase δ and/or ε. DNA replication proteins like PCNA, RPA, and replication factor C (RF-C) support the de novo DNA synthesis in NER. Finally, the 3′ nick is closed by DNA ligase I. Efficient GGR has been shown to require the p53 tumor suppressor protein [44, 45] probably with the p48 protein as the link between p53 and the nucleotide excision repair apparatus [46].

TCR refers to a repair pathway that preferentially and rapidly repairs lesions on the transcribed strand of active genes (less than 1% of human DNA) [47,48]. In TCR-NER the ability of a lesion to block RNA polymerases seems critical. The stalled polymerase recruits at least two TRC-specific factors: CSA and CSB. The remainder of the TCR-NER pathway (i.e., unwinding, incision, excision, DNA synthesis, and ligation) may be identical to the GGR-NER pathway. The breast and ovarian cancer susceptibility gene BRCA1 is involved in TCR [49,50]; however, the exact biological function of the protein remains uncertain (for review see [51]).

2.5 Mismatch Repair

A special repair system designed to recognize and eliminate mismatches from DNA [52] differs from the other repair systems as many of the mismatches recognized are composed of normal nucleotides, but the DNA structure is abnormal because of base mispairing. The mismatch repair in mammalian cells recognizes all possible normal base mismatches, 1–12 base loops, and certain pairs involving modified bases like O^6-methylguanine:T, O^6-methylguanine:C, and O^4-methyladenine:A. The human mismatch repair system consists of the MutS complex, which is thought to bind to a mismatch and bi-directionally searches along the DNA for a nick in one of the strands that will signal the initiation of exonucleolytic degradation [53]. It is presumed that the nicks of Okazaki fragments on the lagging strand and the 3′-terminus of the leading strand provide the necessary signal for initiating the mismatch repair reaction [21]. The resulting single-strand gap that can cover up to 1,000 nucleotides is filled by DNA polymerase and closed by DNA ligase.

2.6 Double Strand Break Repair: Homologous Recombination and End Joining

When strand breaks remain open at a lesion site, or when non-repaired damage blocks the progress of a DNA replication fork to produce a daughter strand gap, a complex cascade of reactions is triggered to stop the cell cycle machinery and recruit repair factors. When, after replication, a second identical DNA copy is available, homologous recombination (HR) seems to be preferred; otherwise cells rely on non-homologous end joining (NHEJ), which is more error-prone [17].

An essential component of the HR repair pathway is the strand-exchange protein, known as RecA in bacteria or Rad51 in yeast. Several mammalian genes have been implicated in recombinational repair on the basis of their sequence homology to yeast Rad51; one of these is XRCC2, which has been shown to be essential for the efficient recombinational repair of DNA double strand breaks

(DSB) between sister chromatids in hamster cells [54]. Human RAD52 and RPA in a physical interaction have also been shown to be essential for homologous recombination [55], and so has the XRCC4 gene, which is required for the repair of DSB in mammalian cells [56].

The NHEJ reaction links ends of a DSB together without any template, using the end binding KU70/80 complex. DNA-dependent protein kinase catalytic subunit (DNA-PK_{CS}) and the nuclease Artemis bind to the KU-complex and trim the overhangs at the double strand break before ligation by the XRCC4-DNA ligase4 complex (reviewed by [57, 58]).

3 Measuring DNA Repair

3.1 Structural Analyses

The structural analyses of DNA repair enzymes are not important in a quantitative aspect but have provided very useful qualitative insights. First of all, their existence is directly proved but the structure is also a foundation for detailing the mechanisms of the repair pathways. Several repair proteins have been identified by various methods and the functional description and assembly of protein complexes have been investigated.

In recent years several high-resolution crystal structures of DNA repair glycosylases bound to their substrates, substrate analogues, or products have been solved. DNA glycosylases are probably the best understood DNA-repair enzymes in terms of damage recognition and catalysis, and several reviews covering this topic have appeared recently [15, 58–61]. Crystal structures of the repair enzymes can show how the substrate bases are exposed to the active site of the enzyme and suggest the mechanisms of catalytic specificity.

Structures of multiple DNA glycosylases have been determined, and though they share little sequence homology they can be placed in two main structural families (reviewed by [59]). The first consists of the uracil DNA glycosylase (UNG), the G:T/U mismatch DNA glycosylase (Mug), the thymine DNA glycosylase (TDG), and the single-strand specific monofunctional DNA glycosylase (SMUG). They share two common motifs, one involved in pyrimidine binding and one involved in glycosidic bond hydrolysis. The second and better-defined family of DNA glycosylases includes human 8-oxoguanine DNA glycosylase (hOGG1), human mutY homologue (MYH), human endonuclease III homologue (hNTH), and methylated DNA-binding domain protein 4(MBD4), along with a number of proteins from other organisms. The crystal structure has revealed that these proteins contain a conserved central α-helical domain. The hallmark is a helix-hairpin-helix (HhH) motif followed at a specified distance by an invariant aspartic acid residue that is essential for the catalytic activity of the enzyme [62].

A general strategy for all glycosylases is to flip the target out of the double helix and into an active site pocket of the protein where catalysis takes place. Recognition of the damaged base in the pocket is accomplished through π-stacking, and through hydrophobic and hydrogen bonding interactions [15]. However, it remains to be shown how these enzymes recognize the damaged base in the DNA before it is flipped out.

The mechanistic steps for eukaryotic NER involve the concerted action of at least six proteins or protein complexes. Recognizing damaged DNA is the first step and is mechanistically highly interesting. Many chemically distinct DNA lesions have to be recognized, and at the same time inappropriate repair must be avoided at other sites. The two sub-pathways, GGR and TCR, employ different methods of damage recognition and different enzymes are involved. For TCR the stalled RNA polymerase seems to be the signal that initiates repair, whereas for GGR the XPC/HR23B protein complex is thought to be responsible for the damage recognition (reviewed by [17]). Scanning force microscopy (SFM) at nanometer resolution has revealed that damage recognition by the XPC/HR23B complex induces a bend in DNA upon binding to a single damaged base at a defined position [63]. The authors expect this distortion to be an important feature required for subsequent assembly of an active NER complex. They also suggest from their experiments that the interaction of XPA-RPA with DNA requires both bending and unwinding of the DNA helix, probably achieved after binding of TFIIH and XPG to the DNA-XPC/HR23B complex.

3.2 Common Polymorphisms

When quantitative assays for specific repair activities in vivo are not available, the "experiments of nature" in the form of genetic polymorphisms can be of value. Very often a change in the genetic sequence leads to decreased activity, provided its function is not essential to the development and survival of the fetus. The study of common genetic polymorphisms can therefore be used to obtain useful information about the DNA repair mechanisms and their importance in vivo.

Polymorphisms in several DNA repair genes have been identified, but little is known about their phenotypic significance. A polymorphism in itself does not uncover a repair defect and cannot be regarded as quantitative. However, the polymorphisms may explain the inter-individual variations in DNA repair capacity observed in humans and may be associated with susceptibility to cancer (for a review, see [64]). Some examples of the work in this area will be summarized below.

A variant allele of XRCC1 in exon 10 (Arg399Gln) may be associated with higher DNA adduct levels in lymphocytes [65], although another study showed that the homozygote variant genotype (399 GG) apparently had a protective effect against bladder cancer relative to carriers of either one or two copies of the common allele [66]. No significant interaction was found with smoking. Different explanations for the discrepancy between the two studies can be found, e.g., another polymorphic gene might be in linkage disequilibrium with XRCC1 or the variants might be more likely to undergo apoptosis or senescence and therefore have a decreased risk of cancer.

The Oggl Ser326Cys polymorphism has been investigated in relation to breast cancer risk in 859 postmenopausal women (425 cases and 434 controls) but no association between genotype and breast cancer risk was found [67].

In DNA ligase I a single nucleotide polymorphism (A to C) in exon 6 was evaluated in a case control study with 530 cases and 570 controls with regard to susceptibility to lung cancer. The polymorphism was found to be very common, but there was no association with the risk of lung cancer, neither in adenocarcinomas nor in squamous cell carcinomas [68].

The XPD polymorphism found in exon 6 (C to A at nucleotide 22541) showed a mild protection against basal cell carcinoma for the CC genotype, whereas the polymorphisms (Asp312Asn) and (Lys751Gln) showed no correlation between genotype and risk of basal cell carcinoma in the study group (N=189) [69].

An XPA polymorphism (A to G) in the 5′ non-coding region of the XPA gene was recently evaluated in 695 lung cancer cases and 695 controls [70]. The presence of one or two copies of the G allele was associated with a reduced lung cancer risk; however, as in the above examples the functional significance is unknown.

A recent study by Mort et al. [71] showed that the NER genes XPD, XPF, XPG, and ERCC1, and the BER gene XRCC1 are not important genetic determinants in colorectal carcinogenesis, while a modest association between colorectal cancer and a polymorphism in the recombination repair gene XRCC3 gene was found.

A thorough review of markers of DNA repair and susceptibility to cancer in humans from 2000 [64] has summarized the results from 64 epidemiologic studies that addressed the association of cancer susceptibility with a putative defect in DNA repair capacity. The authors concluded that the vast majority of studies showed a difference between cancer case subjects and although this observation is compatible with a chromosomal instability due to the cancer itself, it was notable that impaired mutagen sensitivity was also observed in healthy relatives of cancer subjects. There were a variety of functional tests used that only indirectly addressed DNA repair and these showed high variability in their expression. Finally, the issue of confounding was almost totally unexplored.

It is obvious that more studies are warranted, with a precise estimate of the DNA repair capacity in addition to the genotype-cancer association. It is believed that the study of inter-individual variability in DNA repair will greatly contribute to our understanding of human carcinogenesis.

3.3
Naturally Occurring DNA Repair Knockouts

Inborn deficiencies in DNA repair have also provided useful information regarding the importance of DNA repair in vivo. However, as many of the involved proteins are not limited to playing a role in DNA repair, it can be very difficult to determine the contribution of the repair deficit versus the other possible deficits, e.g., insufficient transcription. For example, gene expression profiling by a micro-array technique of primary fibroblast cell cultures from young and old humans and from patients with Werner syndrome has shown that the expression profile of Werner cells closely resembles that of old normal fibroblasts, even though Werner cells only have a single genetic defect that leads to production of a single mutated protein [72].

Known DNA repair-deficient syndromes mainly affect the nucleotide excision repair pathway and the mechanisms for strand break repair. No human disorders caused by inherited BER deficiencies have been identified. The most likely explanations are based on the generated mice knock out models. Deficiency of a single glycosylase may not cause an overt phenotype as the substrates can be repaired by other glycosylases or by other repair systems. In contrast, knock out of BER core proteins often induces embryonic lethality.

Many insights into the importance and the mechanism of human NER have come from the study of cells from patients with hereditary disorders. Xeroderma Pigmentosum (XP) patients are abnormally sensitive to sun exposure, presenting with skin melanomas at an early age, but are also prone to other malignancies caused by known carcinogens due to mutated XP genes [73, 74]. The XP type C patients are only deficient in GGR-NER but proficient in TCR. XP type A patients are completely deficient in NER and display progressive neurological dysfunction in addition.

Cockayne syndrome (CS) patients with mutation in the CSA or CSB gene are completely deficient in TCR-NER, but proficient in GGR. This rare hereditary disease is characterized by postnatal growth failure and early onset of severe neurobiological abnormalities, but no cancers [75]. This may be explained by the fact that the TCR defect causes CS cells to be particularly sensitive to lesion-induced apoptosis, thereby protecting against tumor genesis [17]. This syndrome emphasizes the importance of specific repair of actively transcribed DNA.

Photosensitive trichothiodystrophy (TTD) patients have defects in the XPD or XPB gene and cannot repair cyclobutane dimers (CPD). The phenotype is characterized by many of the symptoms common to the CS patients but with the additional characteristics of brittle hair and nails, and scaly skin. Why mutations in XPD and XPB can give rise to both XP and TTD is explained by the dual functions of the proteins in NER and transcription. The special hallmarks of TTD are thus due to reduced transcription and expression of matrix proteins [76].

Fanconi's anaemia (FA) is an autosomal disorder with bone marrow failure, variable presence of developmental abnormalities, hypersensitivity to DNA cross-linking agents, and a very high incidence of cancer. Different complementation groups have been cloned, but the exact pathway remains uncertain. However, data suggest that the syndrome is linked to mutations in XRCC9 and XRCC11 and therefore to defects in the homologous recombination repair pathway (summarized in [26]).

Nijmegen breakage syndrome (NBS) is associated with defects in NBS1 involved in homologous recombination. The patients display an excessively high risk for the development of lymphatic tumors, immunodeficiency, and chromosomal instability as well as microcephaly and growth retardation [77].

The clinical phenotype of ataxia telangiectasia (AT) syndrome is more striking due to ataxia, cerebral degeneration, and dilated blood vessels in the eyes, but shares many of the characteristics from NBS (chromosome instability, radiation sensitivity, and increased cancer risk). The gene responsible for AT is named ATM and the protein product is a protein kinase, likewise involved in double-strand break repair.

Members of the RECQL gene family are responsible for the syndromes Werner syndrome (mutation in WRN), Bloom syndrome (mutation in BLM), Rothmund–Thomson syndrome (mutation in RECQL4), and RAPADILINO (mutation in RECQL4) (summarized in [78]). All polypeptides encoded by RECQL genes share a central region of seven helicase domains, however, the precise function of the enzymes are unknown. As the patients (at least in Bloom, Werner, and Rothmund–Thomson syndromes) are cancer prone and display chromosomal instability, the most likely affected genome maintenance mechanism is homologous recombination [17].

Hereditary non-polyposis colorectal cancer (HNPCC) is due to defects in mismatch repair (MMR) and thus to deficiencies in removing nucleotides mispaired by DNA polymerases as well as insertion/deletion loops (1–10 bases). This dramatically increases the mutation rate. The affected genes are MLH1 (60%), MSH2 and MSH6.

3.4 Generated Knock Out Mice

The ability to "knock out" a specific gene has proved very useful in elucidating complex DNA repair enzymes and in establishing the role of individual proteins in the suppression or generation of cancer. The conventional target gene replacement technologies used to generate knock out strains are both technically difficult and time consuming. Nevertheless, 148 mice strains are listed in the 5th version of the "Database of mouse strains carrying targeted mutations in genes affecting biological responses to DNA damage" [79]. They are listed in seven tables that represent defects in direct repair, BER, NER, MMR, strand break repair, trans-lesion synthesis, and finally in a table of mouse strains with mutations in genes that influence other cellular responses to DNA damage.

It is beyond the aim of this chapter to review the huge amount of data published regarding these mouse strains. However, a very generalized picture is that many of the knock outs give strikingly normal phenotypes while others are embryonically lethal. The former reflects the overlap in the repair pathways and maybe also a strong selection as the fertility of the knock out mice is often severely lowered. At the other end, the lethal phenotype limits the amount of information gained from the knocked out gene. Still, the method has been useful in revealing the many repair mechanisms.

3.5 RNA Interference

The discovery of small interfering RNAs (siRNA) was in 2002 nominated as the scientific "Breakthrough of the year" by the editors of Science [80]. Providing researchers with a powerful tool for gene regulation, siRNAs have opened a new door for genetic research and therapeutics, and RNA interference (RNAi) is rapidly becoming a chosen technique for down-regulating the expression of a specific gene.

RNA interference (RNAi) was initially proven to be a method of knocking down gene activity by the introduction of a long double-stranded RNA (dsRNA) molecule in invertebrates such as Drosophila and *C.elegans* [81–83]. However, in most mammalian cells the introduction of long dsRNA initiates a cellular interferon response that ultimately causes cell shutdown and leads to apoptosis. Instead, the introduction of small 21–23 nucleotides sequence-specific RNA duplexes was found to initiate post-transcriptional gene knockdown, apparently without triggering this non-specific effect in mammalian cells [84].

The RNAi process can occur in eukaryotes by introducing short (or small) interfering RNA (siRNA) duplexes directly or by expression of long double-stranded RNA (dsRNA) as hairpins or snap-back RNAs and subsequent cleavage into siRNAs, facilitated by an enzyme called Dicer (reviewed by [85]). The siRNA associates with an intracellular multi-protein RNA induced silencing complex (RISC). This complex recognizes and cleaves complementary cellular mRNA. The cleaved mRNA is targeted for degradation, ultimately leading to knock down of post-transcriptional gene expression in the cell [85].

The possibility of engineering cells to provide limiting levels of an essential gene product instead of a complete knock out is a major advantage of this technology. In addition, the RNAi methods are much easier to apply than the conventional gene knock out technology and offer a rapid path for probing gene function and, importantly, the procedure makes it possible to simultaneously suppress several genes. The specificity of RNAi may also make it possible to silence a disease-causing mutant allele specifically and the technology is hoped to become a tool to create RNAi-based therapeutics.

However, pitfalls also exist for this method. Two very recent reports show that a substantial number of siRNA and shRNA vectors can indeed trigger an interferon response and although the effect of this response in vivo is not yet certain, potential non-specific effects not attributable to the gene targeted must be considered [86, 87].

RNAi technology has recently been adapted for investigation of DNA repair processes. Rosenquist et al. have used siRNAs to study Neil1 glycosylase in *E. coli* [88]. An 80% reduction in Neil1 protein was achieved and this led to hypersensitivity to ionizing radiation.

Complete depletions of BRCA1 and BRCA2 proteins, which appear to act in the Fanconi anemia repair pathway of interstrand cross-links and maintenance of genome stability, leads to cell lethality. Thus, instead of gene knock out, siRNA was used to transiently deplete the expression in order to better understand the function of the proteins [89]. It was shown that BRCA2 acts late in the FA response to interstrand cross links, whereas BRCA1 must act early in the FA response and has at least one additional role in interstrand cross-link repair outside the FA pathway.

3.6 Incision Assay

The incision assay [90, 91] is a widely used tool to investigate the mechanisms and activities of DNA base excision repair. It is a functional assay that reveals the repair capacity measured by the incision activity for a synthetic oligonucleotide of 25–30 bases in length with an incorporated base damage (numerous possibilities offered by many biotech companies) at a specific residue. The oligonucleotide is commonly 5′-labeled with [γ-32P]-dATP, but can also be labeled fluorometrically [92]. It is added as a substrate to a reaction mixture containing purified enzymes, or cell or tissue extracts. The volume of the extract is typically normalized to the total protein concentration in the sample. After incubation, the DNA is precipitated by ethanol and separated on a denaturing polyacrylamide gel. Phosphor imaging (fluorescence imaging) reveals the amount of intact and cleaved oligonucleotides. The procedure is relatively simple; however, to use it quantitatively it is very important to work with pure oligonucleotides similar in length and to measure the incision product while the reaction is lineally increasing.

Initially the assay was primarily used qualitatively to investigate the specificity of partly or completely purified repair enzymes from *E. coli* [90, 93] or mammalian cells [91, 94–98]. However, the assay has now been widely used to quantify the repair activity, e.g., [99–106]. Very recently OOG1 activity was measured in protein extracts, prepared from peripheral blood mononuclear cells or lung tissue, from 68 case patients with non-small-cell lung cancer and 68 age- and sex-matched healthy controls [107]. The overall finding was a significantly lower activity of OGG1 in the lung cancer patients, which was believed to be determined by genetic factors. Thus, low OGG1 activity might be a susceptibility factor for lung cancer. Another study suggests that decreased BER activity for lipid peroxidation adducts on adenine and cytosine is associated with inflammatory related lung adenocarcinoma [108].

3.7 Indirect Measurements of DNA Repair Activity

A variety of methods exist for measuring DNA repair. An often-used approach is to measure the amounts of DNA damaged at various time points following the induction of damage.

The alkaline elution assay [109] and the DNA unwinding assay [110] in various modifications can be used to measure single strand breaks (SSB) in DNA. The estimate of repair activity is achieved by measuring SSB immediately after damage induction and after a given incubation period, thus allowing the cells to repair the induced damages. The methods are based on a faster elution through a membrane filter or a faster unwinding of broken DNA strands under alkaline conditions.

The comet assay [111, 112] is also often used to estimate repair activity [113, 114]. Again the principle is the migration patterns of damaged DNA strands versus undamaged DNA under alkaline conditions. The comet assay has the advantage that it requires only small sample quantities.

The three methods above can be combined with enzymatic nicking of the DNA by purified repair proteins to estimate repair of specific base damages or abasic sites.

DNA double strand breaks (DSB) can be measured by the same techniques as the SSB, by changing the alkaline conditions to neutral pH and thereby avoiding the denaturation of the DNA helix. The more DSB, the faster the DNA will migrate whereas SSB will not be detected under neutral conditions. When measured before and after a defined incubation period, the repair capacity of DSB can be estimated.

Various DNA adducts can be measured both in tissue, blood cells, and urine and are widely used as biomarkers of DNA damage and are also used to estimate DNA repair. When estimating repair activity from tissue or blood cells it is important to bear in mind that the amounts of altered bases or bulky adducts reflect steady state levels and that alterations are both dependent on the induction of damage as well as the repair. Thus if a given type of DNA adduct is very efficiently repaired, the levels can possibly remain unaffected by an increased load of damage induction, whereas the levels of more slowly repaired modifications will increase and then normalize. Another important issue when discussing repair kinetics is the complex biology of ROS and thus the possibility that DNA modifications may continue to arise for a long period following induction of a certain type of damage. Thus when one calculates the $T_{1/2}$ of a given type of modification it is necessary to consider this. Conventionally, normal log linear decay is used to calculate the $T_{1/2}$. In our opinion this is not correct and may give biased results, as there is a constant production of damage during the decay. Furthermore, the quantification of adducts may be biased by artificial formation [115] and much of the data must be interpreted with caution [116–120].

Specific DNA adducts are excreted into urine as modified bases and deoxynucleotides [121–123]. As endonucleases and glycosylases repair the damaged DNA, deoxynucleotides and bases respectively are liberated. Oligonucleotides and bulky DNA adducts like benzo[a]pyrene-deoxynucleotides or aflatoxin-B_1-N_7-dG are also excreted. Immuno-affinity chromatography, GC/MS, LC-MS/MS, or HPLE-EC can be used to quantify the wide range of base modifications and thus to estimate the DNA repair activity, whereas ELISA assays do not yet have sufficient proven specificity.

The most widely studied adduct in urine is the oxidation product of guanine either as the nucleoside (8-oxodG), which probably reflects the NER, or as the base (8-OH-Gua), which represents the BER activity. Even though it has been shown that only a small fraction of the 8-oxodG modifications are repaired by the NER system by human cell extracts [124, 125], possibly as low as 0.1% [39], most investigators measure the urinary level of 8-oxodG instead and not the levels of 8-oxoG. Diet has apparently no influence on excretion, either of the nucleobase or the nucleoside in the urine [126], which was the initial reason for not measuring the modified base in urine. It is also much more difficult to measure the modified base in urine because of its much lower electrochemical reactivity. The ratio of excreted modified base to modified nucleoside could be an arbitrary measure of the contribution of NER and BER, however, there are only few data available and it has not yet been clarified whether the ratio of NER:BER changes with increased oxidative stress, or how it is distributed in the population. An estimation of repair capacity in this way is therefore somewhat uncertain.

While the excretion of modified bases and nucleotides relies on functional DNA repair mechanisms, the rate of excretion is a measure of the rate of oxidation, and this rate is not changed by changes in DNA repair. The argument for this is that if there is a change in DNA repair of a given lesion, the levels in tissue DNA will increase/decrease and a new steady state between formation and excretion will occur. In this new steady state the formation and excretion will again be identical and the rate of excretion will therefore be a measure of the rate of formation of the damage, i.e., oxidative stress, independent of repair. This is valid as long as a steady state is actually achieved and as long as there is residual repair capacity [127, 128]. Presently there are no published data on urinary excretion in DNA repair-deficient humans or animals.

A final example of indirect DNA repair measurements is the estimation of DNA repair from the frequency of mutations in a single gene after induction of damage. The hypoxanthine phosphoribosyltransferase (HPRT) gene is excellent as a model gene as mutations inducing loss of function in the gene renders the cell resistant to 6-thioguanine (TG) and thus is easily detectable [129]. Furthermore the gene is carried on the X chromosome and hence only exists in one (expressed) copy in each cell. When estimating repair capacity, part of the cells must be kept in a non-selective media for a given period of time after damage induction to allow the repair systems to work and before the mutation frequencies are determined. With this system it is not possible to detect any silent mutations in the model gene.

3.8 Direct Measurement of DNA Repair Enzyme Activity

The activity of DNA repair enzymes can be estimated from different related assays. The common characteristic is the use of synthetic DNA strands or plasmids with specific incorporated or induced damage.

The repair synthesis assay developed by Wood et al. in 1988 [130] is based on a damaged double-stranded plasmid, which is incubated with cell or tissue extract in the presence of normal dNTP and one ^{32}P-labeled deoxynucleotide. The incorporation of labeled deoxynucleotides by the repair enzymes in the extract corresponds to the repair capacity. The assay, however, is quite laborious and rarely used for quantitative purposes.

3.9 Other DNA Repair Assays

In the host cell reactivation assay [131], cells in culture are transfected with a plasmid containing, e.g., a transcription-blocking damage in a specific gene. Any type of DNA damage that renders the reporter gene product inactive can be assayed. After a given incubation time allowing the cell to repair the damage, the activity of the reporter gene product reflects the repair capacity.

The gene-specific repair assay developed by Bohr and Hanawalt in 1985 [47] is a general technique to study the repair process in individual genes. The principal idea in this approach is to use a DNA repair enzyme to generate a strand break at the site of a lesion, and then to measure the frequency of DNA lesions within specific restriction fragments located within specific genes before and after incubation. The use of Southern probing for quantifying the genes makes it possible to study various genes in the same sample. Until now the assay has only been applied successfully to cell cultures, as a substantial amount of damage is necessary to detect the repair capacity.

Another approach for studying general DNA or gene repair is to detect changes in the genetic code by DNA sequencing and calculate the repair by correlating the number of mutations to the induction of damage. It is important though to notice that a mutational fingerprint is dependent on at least three components: (1) the specificity of DNA repair, (2) the distribution of DNA damage, and (3) the ability of the genetic target to reveal mutation (related to the product and the mutation selection criteria).

4 Conclusions

DNA repair mechanisms, or rather gene repair, had already been suggested a century ago, and about 40 years ago the basic four biochemical steps were suggested (Table 1). Both knowledge and interest in DNA repair have accelerated, particularly during the last decade. It is now clear that a large number of environmental chemicals are transformed in the body, ultimately to carcinogens that act by binding to DNA and thereby inducing adducts, and that such adducts can be repaired to a large extent. Maybe more interesting, it is now clear that products from normal cellular processes, e.g., reactive oxygen species and lipid peroxidation, also are capable of modifying DNA and inducing mutations. Furthermore, it seems that the levels of such endogenously induced DNA modifications occur at much higher rates and result in higher levels of modifications than exogenously derived modifications. Add to this, the fact that endogenous lesions are mutagenic to a large extent. Even more interesting there is increasing evidence of considerable variation between individuals with regard to activity of DNA repair mechanisms and there are examples of diseases that have deficiency in DNA repair as a risk factor. The focus has long been on modification of DNA by endogenous as well as exogenous factors. Now we have to realize that how the DNA repair mechanisms function could be even more important for development of disease and ageing.

References

1. Friedberg EC (2002) The intersection between the birth of molecular biology and the discovery of DNA repair. DNA Repair (Amst) 1:855–67
2. Hill RF (1958) BBA 30:636–7
3. Howard-Flanders P, Boyce RP, Simson E, Theriot L (1962) A genetic locus in E.coli K12 that controls the reactivation of UV-photoproducts associated with thymine in DNA. Proc Natl Acad Sci USA 48:2109–15
4. Setlow RB, Carrier WL (1964) The disappearance of thymine dimers from DNA: An error-correcting mechanism. Proc Natl Acad Sci USA 51:226–31
5. Boyce RP, Howard-Flanders P (1964) Release of ultraviolet light-induced thymine dimers from DNA in E.coli K-12. Proc Natl Acad Sci USA 51:293–300
6. Lindahl T (1974) An *N*-glycosidase from Escherichia coli that releases free uracil from DNA containing deaminated cytosine residues. Proc Natl Acad Sci USA 71:3649–53
7. Riazuddin S, Lindahl T (1978) Properties of 3-methyladenine-DNA glycosylase from Escherichia coli. Biochemistry 17:2110–8
8. Howard-Flanders P, Boyce RP (1966) DNA repair and genetic recombination: studies on mutants of Escherichia coli defective in these processes. Radiat Res (Suppl)
9. Clayton DA, Doda JN, Friedberg EC (1974) The absence of a pyrimidine dimer repair mechanism in mammalian mitochondria. Proc Natl Acad Sci USA 71:2777–81
10. Croteau DL, Bohr VA (1997) Repair of oxidative damage to nuclear and mitochondrial DNA in mammalian cells. J Biol Chem 272:25409–12
11. Mason PA, Matheson EC, Hall AG, Lightowlers RN (2003) Mismatch repair activity in mammalian mitochondria. Nucleic Acids Res 31:1052–8
12. Copeland WC, Longley MJ (2003) DNA polymerase gamma in mitochondrial DNA replication and repair. Sci World J 3:34–44
13. Lambert MW, Lambert WC (1999) DNA repair and chromatin structure in genetic diseases. Prog Nucleic Acid Res Mol Biol 63:257–310
14. Wilson DM, III, Sofinowski TM, McNeill DR (2003) Repair mechanisms for oxidative DNA damage. Front Biosci 8:d963–d981
15. Schärer OD (2003) Chemistry and biology of DNA repair. Angew Chem Int Ed Engl 42:2946–74
16. Stivers JT, Jiang YL (2003) A mechanistic perspective on the chemistry of DNA repair glycosylases. Chem Rev 103:2729–59
17. Hoeijmakers JH (2001) Genome maintenance mechanisms for preventing cancer. Nature 411:366–74

18. Maki H, Sekiguchi M (1992) MutT protein specifically hydrolyses a potent mutagenic substrate for DNA synthesis. Nature 355:273–5
19. Mo JY, Maki H, Sekiguchi M (1992) Hydrolytic elimination of a mutagenic nucleotide, 8-oxodGTP, by human 18-kilodalton protein: sanitization of nucleotide pool. Proc Natl Acad Sci USA 89:11021–5
20. Taddei F, Hayakawa H, Bouton M-F et al. (1997) Counteraction by MutT protein of transcriptional errors caused by oxidative damage. Science 278:128–30
21. Chaney SG, Sancar A (1996) DNA repair: enzymatic mechanisms and relevance to drug response. J Natl Cancer Inst 88:1346–60
22. Yasui A, Eker AP, Yasuhira H, Kobayashi T, Takao M, Oikawa A (1994) A new class of DNA photolyases present in various organisms including aplacental mammals. EMBO J 13:6143–51
23. Sancar A (1994) Mechanisms of DNA excision repair. Science 266:1954–6
24. Yasui A, Eker AP (1999) DNA photolyases. In: Nickoloff JA, Hoekstra MF (eds) DNA damage and repair, vol II: DNA repair in higher eucaryotes. Humana, Totowa, NJ, pp 9–32
25. Whitmore SE, Potten CS, Chadwick CA, Strickland PT, Morison WL (2001) Effect of photoreactivating light on UV radiation-induced alterations in human skin. Photodermatology, Photoimmunology and Photomedicine 17:213–7
26. Thacker J, Zdzienicka MZ (2003) The mammalian XRCC genes: their roles in DNA repair and genetic stability. DNA Repair (Amst) 2:655–72
27. Sinha RP, Hader DP (2002) UV-induced DNA damage and repair: a review. Photochem Photobiol Sci 1:225–36
28. Sobol RW, Horton JK, Kuhn R et al. (1996) Requirement of mammalian DNA polymerase-beta in base-excision repair. Nature 379:183–6
29. Nealon K, Nicholl ID, Kenny MK (1996) Characterization of the DNA polymerase requirement of human base excision repair. Nucleic Acids Res 24:3763–70
30. Dianov GL, Price A, Lindahl T (1992) Generation of single nucleotide repair patches following excision of uracil residues from DNA. Mol Cell Biol 12:1605–12
31. Singhal RK, Prasad R, Wilson SH (1995) DNA polymerase β conducts the gap-filling step in uracil-initiated base excision repair in a bovine testis nuclear extract. J Biol Chem 270:949–57
32. Cappelli E, Taylor R, Cevasco M, Abbondandolo A, Caldecott K, Frosina G (1997) Involvement of XRCC1 and DNA ligase III gene products in DNA base excision repair. J Biol Chem 272:23970–5
33. Dianov GL, Bischoff C, Piotrowski J, Bohr VA (1998) Repair pathways for processing of 8-oxoguanine in DNA by mammalian cell extract. J Biol Chem 273:33811–6
34. Fortini P, Pascucci B, Parlanti E, Sobol RW, Wilson SH, Dogliotti E (1998) Different DNA polymerases are involved in the short- and long-patch base excision repair in mammalian. Biochemistry 37:3575–80
35. Stucki M, Pascucci B, Parlanti E et al. (1998) Mammalian base excision repair by DNA polymerases δ and ε. Oncogene 71:835–43
36. Klungland A, Lindahl T (1997) Second pathway for completion of human DNA base excision repair: reconstitution with purified proteins and requirement for DNase IV (FEN1). EMBO J 16:3341–8
37. Matsumoto Y, Kim K, Bogenhagen DF (1994) Proliferating cell nuclear antigen-dependent abasic site repair in Xenopus laevis oocytes: an alternative pathway of base excision DNA repair. Mol Cell Biol 199414:6187–97
38. Frosina G, Fortini P, Rossi O et al. (1996) Two pathways for base excision repair in mammalian cells. J Biol Chem 271:9573–8

39. Reardon JT, Bessho T, Kung HC, Bolton PH, Sancar A (1997) In vitro repair of oxidative DNA damage by human nucleotide excision repair system: possible explanation for neurodegeneration in Xeroderma Pigmentosum patients. Proc Natl Acad Sci USA 94: 9463–8
40. Sugasawa K, Ng JM, Masutani C et al. (1998) Xeroderma pigmentosum group C protein complex is the initiator of global genome nucleotide excision repair. Mol Cell 2:223–32
41. Masutani C, Sugasawa K, Yanagisawa J, et al. (1994) Purification and cloning of a nucleotide excision repair complex involving the xeroderma pigmentosum group C protein and a human homologue of yeast RAD23. EMBO J 13:1831–43
42. Evans E, Moggs JG, Hwang JR, Egly JM, Wood RD (1997) Mechanism of open complex and dual incision formation by human nucleotide excision repair factors. EMBO J 16:6559–73
43. Sancar A, Tang M-S (1993) Nucleotide excision repair. Photochem Photobiol 57:905–21
44. Ford JM, Hanawalt PC (1997) Expression of wild-type p53 is required for efficient global genomic nucleotide excision repair in UV-irradiated human fibroblasts. J Biol Chem 272:28073–80
45. Ford JM, Hanawalt PC (1995) Li-Fraumeni syndrome fibroblasts homozygous for p53 mutations are deficient in global DNA repair but exhibit normal transcription-coupled repair and enhanced UV resistance. Proc Natl Acad Sci USA 92:8876–80
46. Hwang BJ, Ford JM, Hanawalt PC, Chu G (1999) Expression of the *p48* xeroderma pigmentosum gene is p53-dependent and is involved in global genomic repair. Proc Natl Acad Sci USA 96:424–8
47. Bohr VA, Smith CA, Okumoto DS, Hanawalt PC (1985) DNA repair in an active gene: removal of pyrimidine dimers form the DHFR gene of CHO cells is much more efficient than in the genome overall. Cell 40:359–69
48. Mellon I, Spivak G, Hanawalt PC (1987) Selective removal of transcription-blocking DNA damage from the transcribed strand of the mammalian DHFR gene. Cell 51:241–9
49. Gowen LC, Avrutskaya AV, Latour AM, Koller BH, Leadon SA (1998) BRCA1 required for transcription-coupled repair of oxidative DNA damage. Science 281:1009–12
50. Le Page F, Randrianarison V, Marot D et al. (2000) BRCA1 and BRCA2 are necessary for the transcription-coupled repair of the oxidative 8-oxoguanine lesion in human cells. Cancer Res 60:5548–52
51. Venkitaraman AR (2001) Functions of BRCA1 and BRCA2 in the biological response to DNA damage. J Cell Sci 114:3591–8
52. Modrich P, Lahue R (1996) Mismatch repair in replication fidelity, genetic recombination, and cancer biology. Annu Rev Biochem 65:101–33
53. Lehmann AR, Bridges BA, Hanawalt PC et al. (1996) Workshop on processing of DNA damage. Mut Res 364:245–70
54. Johnson RD, Liu N, Jasin M (1999) Mammalian XRCC2 promotes the repair of DNA double-strand breaks by homologous recombination. Nature 401:397–9
55. Park MS, Ludwig DL, Stigger E, Lee S-H (1996) Physical interaction between human RAD52 and RPA is required for homologous recombination in mammalian cells. J Biol Chem 271:18996–9000
56. Grawunder U, Zimmer D, Kulesza P, Lieber MR (1998) Requirement for an interaction of XRCC4 with DNA ligase IV for wild-type V(D)J recombination and DNA double-strand break repair in vivo. J Biol Chem 273:24708–14
57. Rassool FV (2003) DNA double strand breaks (DSB) and non-homologous end joining (NHEJ) pathways in human leukemia. Cancer Lett 193:1–9
58. Lieber MR, Ma Y, Pannicke U, Schwarz K (2003) Mechanism and regulation of human non-homologous DNA end-joining. Nat Rev Mol Cell Biol 4:712–20

59. Schärer OD, Jiricny J (2001) Recent progress in the biology, chemistry and structural biology of DNA glycosylases. Bioessays 23:270–81
60. Hollis T, Lau A, Ellenberger T (2001) Crystallizing thoughts about DNA base excision repair. Prog Nucleic Acid Res Mol Biol 68:305–14
61. Hosfield DJ, Daniels DS, Mol CD, Putnam CD, Parikh SS, Tainer JA (2001) DNA damage recognition and repair pathway coordination revealed by the structural biochemistry of DNA repair enzymes. Prog Nucleic Acid Res Mol Biol 68:315–47
62. Nash HM, Bruner SD, Schärer OD et al. (1996) Cloning of a yeast 8-oxoguanine DNA glycosylase reveals the existence of a base-excision DNA-repair protein superfamily. Curr Biol 6:968–80
63. Janicijevic A, Sugasawa K, Shimizu Y et al. (2003) DNA bending by the human damage recognition complex XPC-HR23B. DNA Repair 2:325–36
64. Berwick M, Vineis P (2000) Markers of DNA repair and susceptibility to cancer in humans: an epidemiologic review. JNCI Cancer Spectrum 92:874–97
65. Duell EJ, Wiencke JK, Cheng TJ, et al. (2000) Polymorphisms in the DNA repair genes XRCC1 and ERCC2 and biomarkers of DNA damage in human blood mononuclear cells. Carcinogenesis 21:965–71
66. Stern MC, Umbach DM, van Gils CH, Lunn RM, Taylor JA (2001) DNA repair gene XRCC1 polymorphisms, smoking, and bladder cancer risk. Cancer Epidemiol Biomarkers Prev 10:125–31
67. Vogel U, Nexo BA, Olsen A et al. (2003) No association between OGG1 Ser326Cys polymorphism and breast cancer risk. Cancer Epidemiol Biomarkers Prev 12:170–1
68. Shen H, Spitz MR, Qiao Y, Zheng Y, Hong WK, Wei Q (2002) Polymorphism of DNA ligase I and risk of lung cancer – a case-control analysis. Lung Cancer 36:243–7
69. Vogel U, Hedayati M, Dybdahl M, Grossman L, Nexo BA (2001) Polymorphisms of the DNA repair gene XPD: correlations with risk of basal cell carcinoma revisited. Carcinogenesis 22:899–904
70. Wu X, Zhao H, Wei Q et al. (2003) XPA polymorphism associated with reduced lung cancer risk and a modulating effect on nucleotide excision repair capacity. Carcinogenesis 24:505–9
71. Mort R, Mo L, McEwan C, Melton DW (2003) Lack of involvement of nucleotide excision repair gene polymorphisms in colorectal cancer. Br J Cancer 89:333–7
72. Kyng KJ, May A, Kolvraa S, Bohr VA (2003) Gene expression profiling in Werner syndrome closely resembles that of normal aging. Proc Natl Acad Sci USA 100:12259–64
73. ap Rhys CMJ, Bohr VA (1996) Mammalian DNA repair responses and genomic instability. In: Feige U, Morimoto RI, Yahara I, Polla B (eds) Stress-inducible cellular responses. Birkhäuser, Basel, Switzerland, pp 289–305
74. Lehmann AR (1995) Nucleotide excision repair and the link with transcription. TIBS 20:402–5
75. Cooper PK, Nouspikel T, Clarkson SG, Leadon SA (1997) Defective transcription-coupled repair of oxidative base damage in Cockayne syndrome patients from XP group G. Science 275:990–3
76. Vermeulen W, Rademakers S, Jaspers NG, et al. (2001) A temperature-sensitive disorder in basal transcription and DNA repair in humans. Nat Genet 27:299–303
77. Digweed M, Reis A, Sperling K (1999) Nijmegen breakage syndrome: consequences of defective DNA double strand break repair. Bioessays 21:649–56
78. Siitonen HA, Kopra O, Kaariainen H et al. (2003) Molecular defect of the RAPADILINO syndrome expands the phenotype spectrum of RECQL diseases. Hum Mol Genet: ddg306
79. Friedberg EC, Meira LB (2003) Database of mouse strains carrying targeted mutations in genes affecting biological responses to DNA damage. Version 5. DNA Repair 2:501–30

80. Couzin J (2002) Breakthrough of the year: small RNAs make big splash. Science 298: 2296–7
81. Grishok A, Mello CC (2002) RNAi (Nematodes: Caenorhabditis elegans). Adv Genet 46:339–60
82. Caplen NJ, Fleenor J, Fire A, Morgan RA (2000) dsRNA-mediated gene silencing in cultured Drosophila cells: a tissue culture model for the analysis of RNA interference. Gene 252:95–105
83. Parrish S, Fleenor J, Xu S, Mello C, Fire A (2000) Functional anatomy of a dsRNA trigger: differential requirement for the two trigger strands in RNA interference. Mol Cell 6:1077–87
84. Elbashir SM, Harborth J, Lendeckel W, Yalcin A, Weber K, Tuschl T (2001) Duplexes of 21-nucleotide RNAs mediate RNA interference in cultured mammalian cells. Nature 411:494–8
85. Hannon GJ (2002) RNA interference. Nature 418:244–51
86. Bridge AJ, Pebernard S, Ducraux A, Nicoulaz AL, Iggo R (2003) Induction of an interferon response by RNAi vectors in mammalian cells. Nat Genet 34:263–4
87. Sledz CA, Holko M, de Veer MJ, Silverman RH, Williams BR (2003) Activation of the interferon system by short-interfering RNAs. Nat Cell Biol 5:834–9
88. Rosenquist TA, Zaika E, Fernandes AS, Zharkov DO, Miller H, Grollman AP (2003) The novel DNA glycosylase, NEIL1, protects mammalian cells from radiation-mediated cell death. DNA Repair 2:581–91
89. Bruun D, Folias A, Akkari Y, Cox Y, Olson S, Moses R (2003) siRNA depletion of BRCA1, but not BRCA2, causes increased genome instability in Fanconi anemia cells. DNA Repair 2:1007–13
90. Chung MH, Kasai H, Jones DS et al. (1991) An endonuclease activity of *Escherichia coli* that specifically removes 8-hydroxyguanine from DNA. Mut Res 254:1–12
91. Yamamoto F, Kasai H, Bessho T et al. (1992) Ubiquitous presence in mammalian cells of enzymatic activity specifically cleaving 8-hydroxyguanine-containing DNA. Jpn J Cancer Res 83:351–7
92. Kreklau EL, Limp-Foster M, Liu N, Xu Y, Kelley MR, Erickson LC (2001) A novel fluorometric oligonucleotide assay to measure O(6)-methylguanine DNA methyltransferase, methylpurine DNA glycosylase, 8-oxoguanine DNA glycosylase and abasic endonuclease activities: DNA repair status in human breast carcinoma cells overexpressing methylpurine DNA glycosylase. Nucleic Acids Res 29:2558–66
93. Tchou J, Kasai H, Shibutani S et al. (1991) 8-Oxoguanine (8-hydroxyguanine) DNA glycosylase and its substrate specificity. Proc Natl Acad Sci USA 88:4690–4
94. Bessho T, Tano K, Kasai H, Ohtsuka E, Nishimura S (1993) Evidence for two DNA repair enzymes for 8-hydroxyguanine (7,8-dihydro-8-oxoguanine) in human cells. J Biol Chem 268:19416–21
95. Nagashima M, Sasaki A, Morishita K et al. (1997) Presence of human cellular protein(s) that specifically binds and cleaves 8-hydroxyguanine containing DNA. Mut Res 383: 49–59
96. Rosenquist TA, Zharkov DO, Grollman AP (1997) Cloning and characterization of a mammalian 8-oxoguanine DNA glycosylase. Proc Natl Acad Sci USA 94:7429–34
97. Croteau DL, ap Rhys CMJ, Hudson EK, Dianov GL, Hansford RG, Bohr VA (1997) An oxidative damage-specific endonuclease from rat liver mitochondria. J Biol Chem 272: 27338–44
98. Bjørås M, Luna L, Johnsen B et al. (1997) Opposite base-dependent reactions of a human base excision repair enzyme on DNA containing 7,8-dihydro-8-oxoguanine and abasic sites. EMBO J 16:6314–22

99. Asami S, Hirano T, Yamaguchi R, Tomioka Y, Itoh H, Kasai H (1996) Increase of a type of oxidative DNA damage, 8-hydroxyguanine, and its repair activity in human leukocytes by cigarette smoking. Cancer Res 56:2546–9
100. Hirano T, Higashi K, Sakai A et al. (2000) Analyses of oxidative DNA damage and its repair activity in the livers of 3′-methyl-4-dimethylaminoazobenzene-treated rodents. Jpn J Cancer Res 91:681–5
101. Lin LH, Cao S, Yu L, Cui J, Hamilton WJ, Liu PK (2000) Up-regulation of base excision repair activity for 8-hydroxy-2′-deoxyguanosine in the mouse brain after forebrain ischemia-reperfusion. J Neurochem 74:1098–105
102. Asami S, Hirano T, Yamaguchi R, Tsurudome Y, Itoh H, Kasai H (2000) Increase in 8-hydroxyguanine and its repair activity in the esophagi of rats given long-term ethanol and nutrition-deficient diet. Jpn J Cancer Res 91:973–8
103. Sorensen M, Jensen BR, Poulsen HE et al. (2001) Effects of a Brussels sprouts extract on oxidative DNA damage and metabolising enzymes in rat liver. Food Chem Toxicol. 39:533–40
104. Souza-Pinto N, Hogue BA, Bohr VA (2001) DNA repair and aging in mouse liver: 8-oxodG glycosylase activity increase in mitochondrial but not in nuclear extracts. Free Radic Biol Med 30:916–23
105. Riis B, Risom L, Loft S, Poulsen HE (2002) Increased rOGG1 expression in regenerating rat liver tissue without a corresponding increase in incision activity. DNA Repair 1:419–24
106. Riis B, Risom L, Loft S, Poulsen HE (2002) OGG1 mRNA expression and incision activity in rats are higher in foetal tissue than in adult liver tissue, while 8-oxodeoxyguanosine levels are unchanged. DNA Repair 1:709–17
107. Paz-Elizur T, Krupsky M, Blumenstein S, Elinger D, Schechtman E, Livneh Z (2003) DNA Repair activity for oxidative damage and risk of lung cancer. JNCI Cancer Spectrum 95:1312–9
108. Speina E, Zielinska M, Barbin A et al. (2003) Decreased repair activities of 1,*N*(6)-ethenoadenine and 3,*N*(4)-ethenocytosine in lung adenocarcinoma patients. Cancer Res 63:4351–7
109. Kohn KW, Erickson LC, Ewig RA, Friedman CA (1976) Fractionation of DNA from mammalian cells by alkaline elution. Biochemistry 15:4629–37
110. Kanter PM, Schwartz HS (1982) A fluorescence enhancement assay for cellular DNA damage. Mol Pharm 22:145–51
111. Ostling O, Johanson KJ (1984) Microelectrophoretic study of radiation-induced DNA damages in individual mammalian cells. Biochem Biophys Res Commun 123: 291–8
112. Singh NP, McCoy MT, Tice RR, Schneider EL (1988) A simple technique for quantation of low levels of DNA damage in individual cells. Exp Cell Res 175:184–91
113. Collins AR, Dusinská M, Horvathova E, Munro E, Savio M, Stetina R (2001) Inter-individual differences in repair of DNA base oxidation, measured in vitro with the comet assay. Mutagenesis 16:297–301
114. Olsen AK, Bjortuft H, Wiger R et al. (2001) Highly efficient base excision repair (BER) in human and rat male germ cells. Nucleic Acids Res 29:1781–90
115. Collins A, Cadet J, Epe B, Gedik C (1997) Problems in the measurement of 8-oxoguanine in human DNA. Report of a workshop, DNA oxidation, Aberdeen, UK, 19–21 January, 1997. Carcinogenesis 18:1833–6
116. ESCODD (2000) Comparison of different methods of measuring 8-oxoguanine as a marker of oxidative DNA damage. ESCODD (European Standards Committee on Oxidative DNA Damage). Free Radic Res 32:333–41

117. ESCODD (2002) Comparative analysis of baseline 8-oxo-7,8-dihydroguanine in mammalian cell DNA, by different methods in different laboratories: an approach to consensus. Carcinogenesis 23:2129–33
118. ESCODD (2002) Inter-laboratory validation of procedures for measuring 8-oxo-7,8-dihydroguanine/8-oxo-7,8-dihydro-2′-deoxyguanosine in DNA. Free Radic Res 36:239–45
119. ESCODD (2003) Measurement of DNA oxidation in human cells by chromatographic and enzymic methods. Free Radic Biol Med 34:1089–99
120. Riis B (2002) Comparison of results from different laboratories in measuring 8-oxo-2′-deoxyguanosine in synthetic oligonucleotides. Free Radic Res 36:649–59
121. Ames BN (1989) Endogenous DNA damage as related to cancer and aging. Mutat Res 214:41–6
122. Weimann A, Belling D, Poulsen HE (2001) Measurement of 8-oxo-2-deoxyguanosine and 8-oxo-2-deoxyadenosine in DNA and human urine by high performance liquid chromatography-electrospray tandem mass spectrometry. Free Radic Biol Med 30:757–64
123. Weimann A, Belling D, Poulsen HE (2002) Quantification of 8-oxo-guanine and guanine as the nucleobase, nucleoside and deoxynucleoside forms in human urine by high-performance liquid chromatography-electrospray tandem mass spectrometry. Nucleic Acids Res 30:E7
124. Jaiswal M, Lipinski LJ, Bohr VA, Mazur SJ (1998) Efficient in vitro repair of 7-hydro-8-oxodeoxyguanosine by human cell extracts: involvement of multiple pathways. Nucleic Acids Res 26:2184–91
125. Fraga CG, Shigenaga MK, Park JW, Degan P, Ames BN (1990) Oxidative damage to DNA during aging: 8-hydroxy-2′-deoxyguanosine in rat organ DNA and urine. Proc Natl Acad Sci USA 87:4533–7
126. Gackowski D, Rozalski R, Roszkowski K, Jawien A, Foksinski M, Olinski R (2001) 8-Oxo-7,8-dihydroguanine and 8-oxo-7,8-dihydro-2′-deoxyguanosine levels in human urine do not depend on diet. Free Radic Res 35:825–32
127. Poulsen HE, Loft S (1998) Interpretation of oxidative DNA modification: Relation between tissue levels, excretion of urinary repair products and single cell gel electrophoresis (comet assay). In: Aruoma OI, Halliwell B (eds) DNA & free radicals: Techniques, mechanisms and applications. OICA International, London, pp 261–70
128. Poulsen HE, Loft S, Weimann A (2000) Urinary measurement of 8-oxodG (8-oxo-2′-deoxyguanosine). In: Lunec J, Griffiths HR (eds) Measuring in vivo oxidative damage: a practical approach. Wiley, London, pp 69–80
129. Maher VM, McCormick JJ (1996) The HPRT gene as a model system for mutation analysis. In: Pfeifer GP (ed) Technologies for detection of DNA damage and mutations. Plenum, New York, pp 381–90
130. Wood RD, Robins P, Lindahl T (1988) Complementation of the Xeroderma Pigmentosum DNA repair defect in cell free extracts. Cell 53:97–106
131. Athas WF, Hedayati MA, Matanoski GM, Farmer ER, Grossman L (1991) Development and field-test validation of an assay for DNA repair in circulating human lymphocytes. Cancer Res 51:5786–93

The Handbook of Environmental Chemistry Vol. 2, Part O (2005): 177–201
DOI 10.1007/b101151

Protein Repair and Degradation

Diana Poppek · Tilman Grune (✉)
Research Institute of Environmental Medicine, Heinrich Heine University,
Auf'm Hennekamp 50, 40225 Düsseldorf, Germany
Tilman.Grune@uni-duesseldorf.de

Abstract Protein oxidation is one of the important processes taking place during oxidative stress. Numerous amino acids can be modified within proteins resulting in a large variety of oxidatively modified proteins. These oxidation reactions are accompanied by secondary unfolding of the proteins and modification by non-protein oxidation products. For maintaining the protein pool two processes must be taken into account: protein repair and protein degradation. The known protein repair mechanisms are limited to a few amino acid modifications. However, special removal systems exist in all cellular compartments for degrading oxidized proteins. The major removal system for oxidized proteins in the nucleus and the cytosol is the proteasomal system. The structure, composition, and function of this proteolytic system will be reviewed here. Severe oxidation, however, leads to the formation of insoluble, non-degradable aggregates. These aggregates accumulate within cells, especially during neurodegenerative diseases and the aging process.

Keywords Proteasome · Protein oxidation · Protein degradation · Protein repair

1 Introduction

Oxidative stress is a harmful condition that occurs due to an excess of free radical and oxidant production over the primary antioxidant defense systems. Consequently the undefeated free radicals and oxidants are able to react and modify cellular components including proteins.

Since the primary source of oxygen species (ROS) is the superoxide radical anion $O_2^{\bullet -}$ [1] and the most reactive free radical is the hydroxyl radical ($HO^{\bullet}$) a wide spectrum of various oxidants damage the cellular components. The hydroxyl radical, for example, reacts very quickly with all major components of cells, e.g., proteins, hydrocarbons, nucleic acids, and lipids [1].

An oxidative attack on proteins results in site-specific amino acid modifications, fragmentation of the polypeptide chain, aggregation and cross-linking, altered electrical charge and surface properties, and changed susceptibility to proteolysis. Since the persistent presence of high concentrations of oxidized proteins in the intracellular protein pool might have adverse effects, the worst outcome can be cell death and, therefore, it is absolutely necessary for the cell to degrade oxidized proteins. These processes become of increasing importance since it is known that numerous diseases, like Alzheimer's disease, Parkinson's disease, arteriosclerosis, and the physiological aging process, are combined with the accumulation of oxidized protein material.

Therefore, this chapter of the "Handbook of Environmental Chemistry – Oxidants and Antioxidative Defense Systems" aims to summarize current knowledge of the formation of oxidized proteins and the removal of oxidized protein forms from biological tissue. It will therefore concentrate on the part of the secondary antioxidative defense system aimed at maintaining an intact intracellular protein pool.

2 Protein Oxidation

Reactions of free radicals with proteins result in the oxidation of various amino acid side chains, often leading to a loss of function of the protein [2].

The variety of protein oxidation products of whole polypeptide chains is therefore the result of the variety of 20 amino acids present in the proteins and the various free radicals interacting with the protein. In more complex biological systems the term "secondary oxidative protein modification" indicates the modification of proteins by a compound resulting from oxidation processes in other parts of the cell. For example, the oxidant attack on poly-unsaturated fatty acids in cells produces a number of reactive aldehydes that readily modify proteins. Therefore, the number of oxidative protein modifications in proteins or more complex biological matrices is very often more numerous than for free amino acids [3–5].

The oxidation of free aliphatic amino acids very often leads to the formation of NH_4^+ and a ketoacid. In order to modify the amino acid side-chains the presence of oxidation-susceptible groups is necessary. In a structured protein these groups should also fulfill the criteria of steric availability for the oxidant attack. The oxidation of both the sulfur-containing amino acids is well described – cysteine and methionine are highly susceptible to oxidative damage [2]. On the other hand, aromatic amino acids may be regarded as preferred targets for attack by reactive oxygen species and are readily oxidized to various hydroxyl derivates [6]. To describe all amino acid modifications by various oxidants is beyond the scope of this chapter. It is extensively reviewed in the literature [1, 7, 8].

It is worth mentioning that some oxidative structural modifications are strongly dependent on the interacting oxidant. Oxidative modification of tyrosine by chlor-related oxidants or peroxynitrite results, for example, in clorinated tyrosine forms or nitrotyrosine. Under certain conditions this gives the possibility of making conclusions about the oxidizing agent.

Furthermore, it should be mentioned that some of the amino acid side-chain modifications are actually reversible, including thiolation and disulfide bond formation, whereas others are irreversible, like methionine sulfone formation or nitration.

Oxidative amino acid side-chain modifications do not result in a stable "end" product of the oxidation process, but very often highly reactive intermediates are formed. These include chemically reactive groups, like ketones and aldehydes, or the formation of protein hydroperoxides. The presence of such protein hydroperoxides leads to a process called protein peroxidation. Here secondary reactions occur if the protein hydroperoxide decomposes and initiates further oxidative reactions, again forming oxidized protein forms.

All the changes to proteins that have been described so far are based on the modification of amino acid side-chains in proteins, but of course the protein structure is also influenced by oxidation processes.

One of the most prominent changes in the primary protein structure is oxidative polypeptide chain fragmentation. Here an alpha carbon-centered radical is formed and interacts with oxygen to form first a peroxyl species and a hydroperoxide. Decomposition of such alpha-carbon hydroperoxides result in peptide chain cleavage with formation of ketoacyl (carbonyl) and amide derivates of the carboxy- and amino-terminal amino acids. Polypeptide backbone fragmentation leads to formation of protein fragments with derivatized terminal amino acids (Fig. 1). The formation of these so-called protein fragments largely depends on the oxidant. Hypochloric acid is one of the oxidants leading to a high yield of protein fragments.

Besides the primary protein structure, the higher protein structures are also affected by oxidative processes. The normal folding of proteins is disrupted by the formation of covalently modified amino acids. Therefore, a partial unfolding of the protein takes place and, in a number of cases, the hydrophobic amino acids of the inner protein core are exposed to the surface (Fig. 2). The unfolding of the protein seems to be a crucial step in the further fate of the oxidized

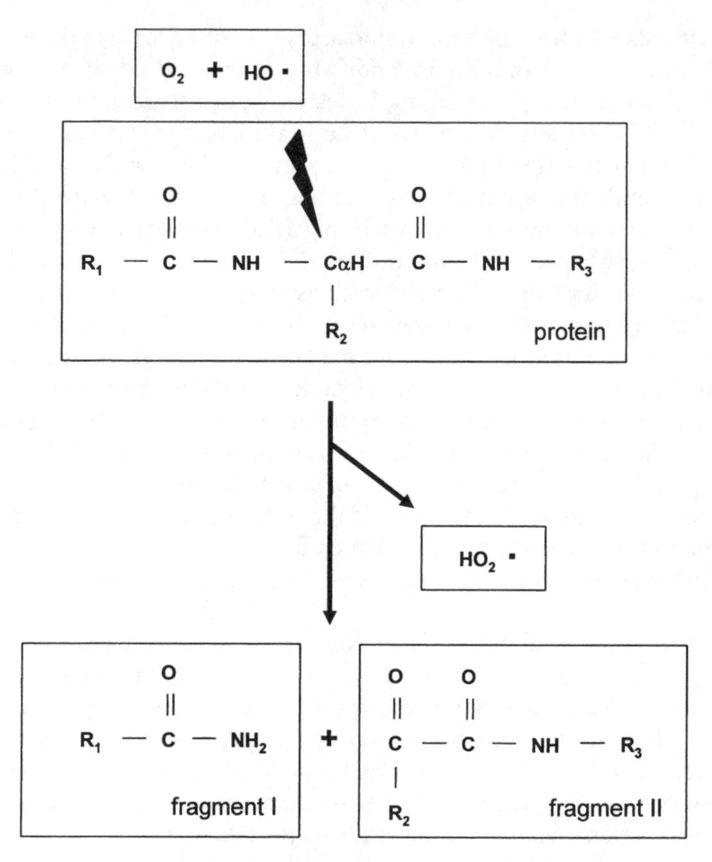

Fig. 1 Proposed mechanisms of protein fragmentation by oxidants

protein. Although "pure" oxidation processes normally increase the hydrophility of the products due to hydroxylation, chlorination etc. it is the unfolding that contributes to the formation of hydrophobic surface structures. These hydrophobic surface patches tend to interact with other hydrophobic domains resulting (in the case of interaction with another hydrophobic protein domain) in the formation of protein–protein aggregates (Fig. 2). Changes in the electric net charge and surface charges can facilitate and stabilize this interaction. Therefore, non-covalent protein aggregates are formed. The term "protein aggregates" describes oligomeric complexes of normally non-interacting protein molecules that can be either structured or amorphous. These protein aggregates are often insoluble and metabolically stable.

Within the protein aggregates numerous chemical reactions take place and often result in the formation of cross-linked protein aggregates. Some of the cross-linking reactions are reversible, e.g., if the cross-linking reaction forms a disulfide bond. On the other hand, covalent bounds are formed due to the interaction of two carbon-centered radicals. These cross-links are very often

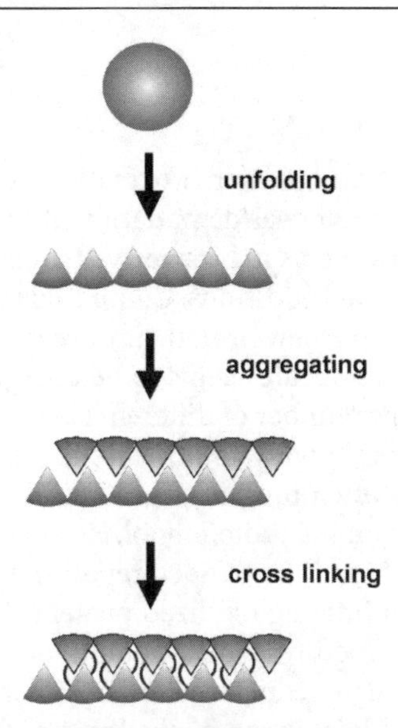

Fig. 2 Oxidation-induced unfolding of a globular protein. Due to unfolding the inner hydrophobic core of the protein is exposed to the surface. These hydrophobic patches tend to aggregate. The aggregates become cross-linked in secondary reactions

irreversible. Such an irreversible formation of cross-links is the formation of a 2,2-biphenyl cross-link between two tyrosyl radicals. Other covalent cross-linking reactions are the formation of Schiff's bases between amino groups of proteins and oxidation-driven ketones within the protein.

Interestingly, the generation of cross-linked aggregates is supported by various non-proteinogenous agents like lipid peroxidation products, including the most abundant lipid peroxidation-driven aldehydes malondialdehyde and 4-hydroxynonenal. Ongoing cross-linking reactions and interaction with cellular components leads to the formation of a fluorescent material. One of the widest definitions for this is the age pigment-like fluorophore [9]. Other authors use the termini lipofuscin, ceroid, protein inclusion bodies, aggresomes, or plaques, often in connection with some special cellular condition for the formation of this material.

Thus numerous changes in the protein pool occur due to oxidation, depending on the acting oxidant and presence and accessibility of oxidizable amino acid residues in the protein molecule. Very often these protein modifications are accompanied by a loss of normal protein function and in some case the acquiring of new pathological functions (gain of function). Oxidative modifications result in modification of the susceptibility to protein degradation and, in the case of extracellular proteins, in altered uptake rates. One has to take into account that the oxidation state of the protein pool is the result of two components: protein oxidation and the repair or removal of these oxidized protein forms.

3 Protein Repair

To maintain the cellular protein pool after oxidative stress two principle ways exist: repair of the damage or removal/degradation of the damaged molecule. The repair of oxidized proteins seems to be a very efficient way for maintaining the protein pool under oxidizing conditions. Compared to the high amount of ATP required for the synthesis of a new protein after removal of a damaged one, the direct repair of a protein molecule seems to be energetically desirable. On the other hand, due to the large number of different chemical modifications it seems to be chemically and evolutionally impossible to develop the required set of enzymes. Therefore, very often protein degradation is the only way to remove such damaged proteins from the protein pool. However, for some oxidative protein modifications evolutionally developed repair mechanisms.

One possibility for minimizing oxidized protein damage is the thiol repair (Fig. 3). This repair system requires either glutathione or the thioredoxin system. The thioredoxin/thioredoxin reductase repair system [10] is able to reduce disulfide bonds. It can dethiolate protein disulfides and thus is an extremely important regulator for redox homeostasis in the cells. Thioredoxin is a small ubiquitous protein that contains a pair of cysteines that undergo reversible oxidation and are re-reduced by the enzyme thioredoxine reductase. The thioredoxin reductase transfers electrons from NADPH to thioredoxin via a flavin.

A thiol/disulfide-regulating enzyme thiol transferase has recently been found [11]. The enzyme is extremely oxidation resistant. It is thought to prevent the accumulation of oxidation-induced protein–protein disulfide bonds within the protein pool. This cytosolic enzyme is GSH-dependent and has a low molecular weight (11.8 kDa). The enzyme repairs oxidatively damaged proteins through its unique catalytic site containing a cysteine moiety, which can specifically dethiolate protein-S-S-glutathione and restore protein-free thiol groups for a proper enzyme or protein function. Most importantly, it has been demonstrated that this thiol transferase has a remarkable resistance to hydrogen peroxide induced oxidation in cultured human and rabbit lens epithelial cells and remains active even under conditions when glutathione peroxidase and glutathione reductase are severely inactivated.

Besides thiol repair there also exists direct repair for one of the oxidation products of methionine, methionine sulfoxide. The enzyme peptide methionine sulfoxide reductase reduces the methionine sulfoxide formed in proteins due to oxidation and is therefore able to reconstitute the normal protein (Fig. 3). Besides methionine sulfoxide there exists a further oxidation product of methionine, methionine sulfone, which can not be repaired. The cycle of methionine oxidation and efficient methionine sulfoxide repair, and the early and easy oxidation of the methionine in proteins, led some authors to hypothesize that methionine acts as an intramolecular antioxidant for some proteins and so protects other amino acids from oxidation [12]. Besides the peptide methionine sulfoxide reductases, there also exists methionine reductases able to

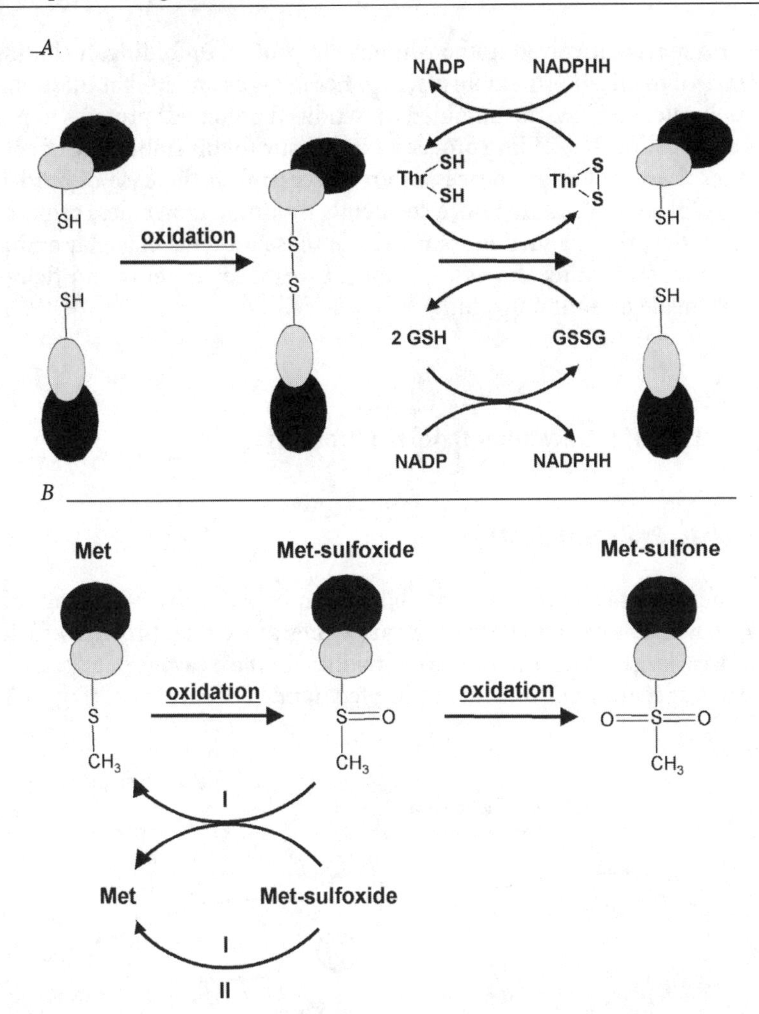

Fig. 3A, B Protein repair mechanisms. *Panel A* demonstrates the oxidation of thiol groups, leading to the formation of an intermolecular cross-link. This modification can be repaired by two mechanisms, one involving gluathione and the other thioredoxin. *Panel B* demonstrates the methionine oxidation, leading to methionine sulfoxide and methionine sulfone. Methionine sulfoxide can be repaired either in its protein-bound or soluble form by two types of enzymes (I and II). For more detailed description see the text

reduce the free (not protein-bound) methionine sulfoxide to methionine. Only the peptide methionine sulfoxide reductases are able to reduce the oxidized form of methionine in both peptides and proteins.

Besides these direct repair mechanisms a number of cellular proteins are involved in the re-folding activity of oxidized proteins. The protein disulfide isomerase exhibits a chaperone-like activity and facilitates disulfide exchange reactions in large inactive protein substrates [13, 14]. A set of heat shock and stress proteins is involved in the response to protein unfolding, including the presence of oxidized protein forms. In general it is assumed that these chaperone molecules stabilize the unfolded (or oxidized unfolded) proteins to prevent their aggregation. Re-folding processes occur due to this stabilizing effect [15]. Whether these re-folding processes are successful in the case of oxidatively damaged proteins (and therefore covalently modified molecules) remains obscure. On the other hand, it is assumed that these proteins may play an important role in preventing the aggregation of oxidized proteins and delivering them to the degradation machineries of the cell.

4 Degradation of Intracellular Oxidized Proteins

4.1 Intracellular Proteolytic Systems

Cells, and in particular mammalian cells, are equipped with a wide range of proteases that enable them to degrade a variety of intracellular proteins efficiently. There are a large number of proteases involved in the cleavage of a specific substrate or a specific motif present in a limited number of substrates (Fig. 4). These

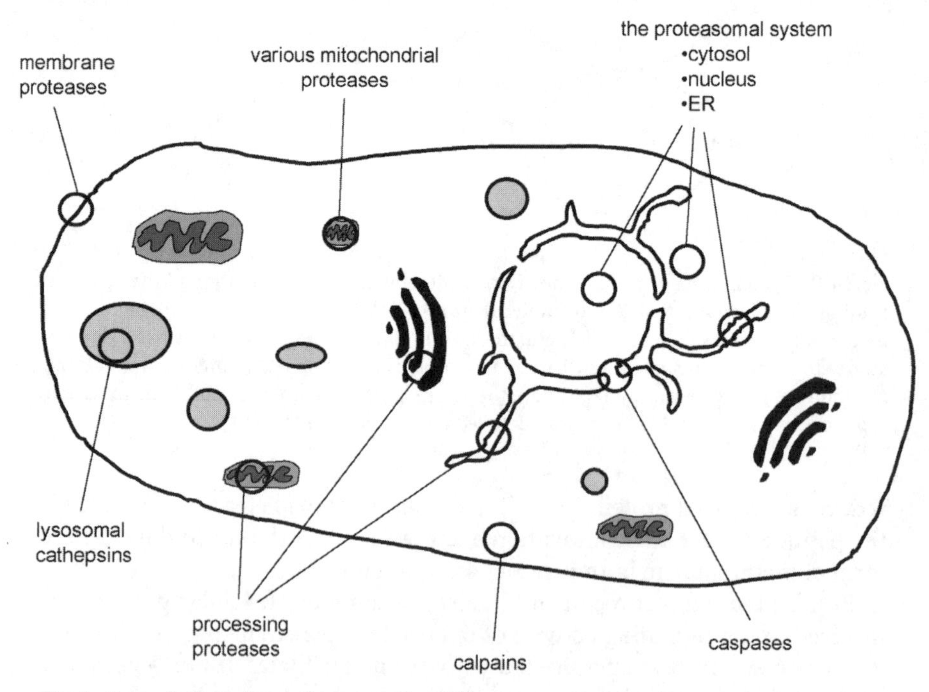

Fig. 4 Proteolytic systems of mammalian cells

include the processing proteases, proteases involved in the regulation of several hormones, and to a certain extent the caspases of the proteolytic cascade of the apoptosis. On the other hand, there exist several proteolytic systems with a wide range of substrates. These include the lysosomal proteases, the calcium-dependent calpains, and the proteasomal system. Besides these proteolytic systems, cells contain a whole array of peptidases involved in the cleavage of smaller peptides or in degrading polypeptides step-by-step as exopeptidases. Since this chapter deals with the removal of oxidized proteins we will not include the various peptidases. One should be aware of the fact that numerous proteases are also able to degrade peptides, but in a living cell peptidases very often do so more efficiently.

The proteolytic systems of the cell do have some overlapping substrate specificity. Most of the extracellular proteins taken up by cells are degraded within the endosomal/lysosomal compartment. Additionally, some (preferentially long-lived?) cytosolic proteins and proteins of cellular organelles taken up by autophagocytosis are also lysosomally degraded. Targeting to the lysosomes is either realized by direct transport of cytosolic proteins through the lysosomal membrane [16] or by the processes of autophagy [17].

Calpains are preferentially involved in the degradation of proteins of the cytoskeleton. Although other proteolytic systems are also important for the removal of cytoskeletal proteins, the bulk of intracellular proteins are degraded by the proteasomal system [18]. This system is therefore the major proteolytic system involved in the degradation of proteins and in particular of oxidized proteins. The proteasomal system is often referred to as the ubiquitin-proteasome system due to its tight interaction with the ubiquitination machinery of the cell.

4.2 The Ubiquitin-Proteasome System

Proteasome research began in the late 1960s [19] when Harris discovered a barrel-shaped particle in erythrocyte extracts. From then on knowledge about the proteasome and the regulating factors of this protease has been added to step by step. Today it is generally accepted that the ubiquitin-proteasome system is responsible for the turnover of the bulk of intracellular cytosolic and nuclear proteins [20]. It is known that this protease is located in the cytosol, the nucleus, and that it is attached to the ER and other cell membranes.

The core particle of the ubiquitin-proteasome system is the 20S "core" proteasome. This protein complex consists of 28 subunits forming a barrel-like cylinder (Fig. 5). The 28 subunits are arranged in four rings, each of them containing seven subunits. Although in mammalian cells all the subunits in the outer and the inner rings are different (so the particle consists of two sets of 14 different subunits) the subunits of the outer rings are referred to as α-subunits and the subunits of the inner rings are called β-subunits. Evolutionally the various α-subunits developed from a single α-subunit forming the outer proteasomal rings in Archaeobacter proteasome, whereas the β-subunits are the

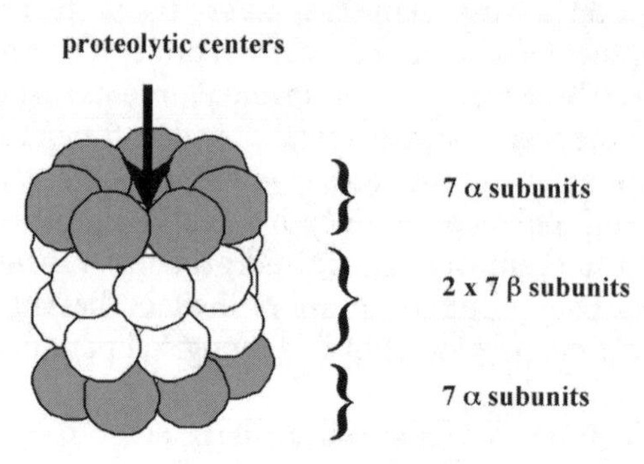

Fig. 5 Structure of the 20S core proteasome

result of the evolution of a single β-subunit. The α-subunits and β-subunits are highly homologous. All subunits of the proteasome are within the molecular weight range of 18–35 kDa. The entire 20S proteasome has a molecular weight of 670–700 kDa [20].

The 20S proteasome is a multicatalytic protease containing several active centers. These are located in the hole of the cylinder and are encoded by the β-subunits. The 20S proteasome acts independently of ATP or any other factor. It is able to degrade unfolded proteins since those can enter the active centers of the proteasome through the opening formed by the α-subunits. This opening is usually covered by domains of the α-subunits and the active center is therefore only accessible after a certain activation of the proteasome. The 20S proteasome is able to cleave proteins on the carboxyl side of basic, hydrophobic, and acidic amino acids, described as a trypsin-like, chymotrypsin-like, and peptidylgluamyl-peptide hydrolase-like proteolytic activity.

The 20S proteasome is the proteolytic core of the ubiquitin-proteasome system. Two major regulators of the 20S proteasome have been described. One is referred to as the 19S regulator (or PA700) and the other as the 11S regulator (or PA28). Each of the regulators is able to bind to either of the ends of the cylindrical 20S proteasomal particle [20]. Thus, a number of variants of proteasomes are formed (Fig. 6). Each 19S regulator adds an additional 15–20 subunits to the proteasome, some of them exerting enzymatic activities. The 19S regulator has a molecular weight of 700 kDa giving the entire complex (19S–20S–19S) a molecular weight of 2,000 kDa (or 2 MDa). This protease is also referred to as 26S proteasome. Basically the 19S regulator is a particle containing a "base" and a "lid". The base contains several ATPases. The binding of the 19S regulator enables the proteasome to recognize and selectively degrade ubiquitinated proteins. The polyubiquitin is removed from the ubiquitin-targeted substrate and the substrate protein is degraded. This process is ATP-stimulated. Besides several other functions in the cell, ubiquitin is one of the major targeting mechanisms for protein degradation. To act as a protein degradation signal multiple

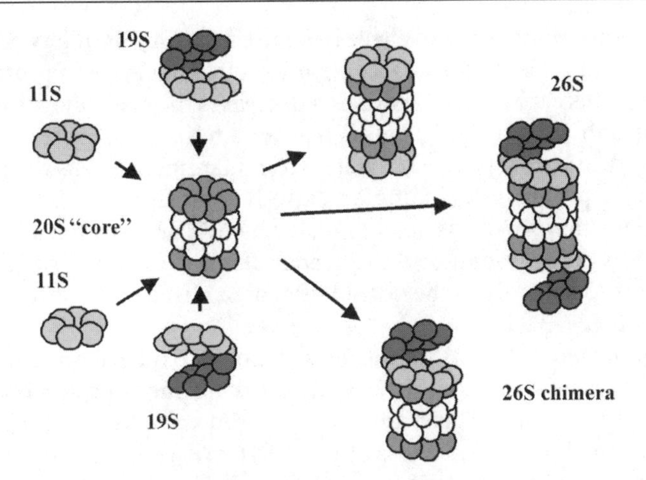

Fig. 6 Binding of the 11S and 19S regulators to the 20S core proteasome forms new proteasomal particles

ubiquitin molecules have to be transferred onto the substrate protein. This is a highly regulated process catalyzed by the ubiquitination system [21].

The other well-described regulator of the proteasome is the 11S regulator (Fig. 6). It consists of the PA28α and the PA28β subunits. A PA28γ subunit also exists in the nucleus. The whole complex has a molecular weight of 200 kDa and forms a heptameric ring. The addition of the 11S regulator to the proteasome results in a dramatic increase in the degradation of model peptides.

Other mechanisms of proteasomal regulation are known such as the expression of three "new" subunits in the immune response, several phosphorylation sites for the 20S proteasome as well for the proteasomal regulators [20], the activation by poly-ADP-ribose polymerase [22], and some authors have suggested the presence of proteasome inhibitors.

The whole ubiquitin-proteasome system is involved in multiple cellular functions like antigen processing, cell division and cell cycle progression, or cellular differentiation. These processes require the selective and targeted degradation of regulatory proteins, like cyclins, cyclin-dependent kinases, p53, p21 or the degradation of regulatory factors like IRP-2, or the IkB degradation in the activation of NF-kB. The proteasome is involved in the degradation of structural intracellular proteins and malfolded or unfolded proteins from the cytosol and ER. Furthermore, it is known today that the proteasome degrades most of the oxidatively modified proteins of the cytosol and the nucleus.

4.3
Degradation of Cytosolic Oxidized Protein Substrates

The structural changes introduced into proteins due to oxidation are connected with changes in proteolytic susceptibility of the oxidized protein. A large number of various model proteins have been used to test this. It has been demonstrated that mild oxidation is accompanied by an increase in proteolytic susceptibility whereas excessive oxidation decreases the proteolytic susceptibility of protein substrates (Fig. 7). This is true for a number of proteases, including the isolated 20S proteasome. In lysates from mammalian cells the proteasome is the major protease degrading oxidatively modified proteins [4, 5, 23–35]. Interestingly this process is not stimulated by ATP or any other known factors. ATP inhibits the degradation of oxidized proteins to a certain extent. In fact the presence of ATP decreases the degradation of oxidized proteins moderately but consistently [29–32,35].

One of the requirements for the degradation of oxidized proteins by the 20S proteasome is its activation, very often occurring due to proteasome isolation or by low levels of SDS. This feature of increased proteolytic susceptibility due to moderate oxidation and a loss of proteolytic susceptibility due to excessive oxidation seems to be independent of the protein substrate and the oxidant used.

The large number of globular, soluble proteins that have been tested raises questions about the general features of protein oxidation processes that target the oxidized protein towards proteasomal degradation. Several groups have presented data showing that the general future of oxidized proteins, which are recognized by the proteasome, is unfolding or at least local unfolding and the exposure of hydrophobic moieties to the protein surface [29, 32, 36]. A clear dependence of the recognition by the 20S proteasome and the unfolding of the protein or the hydrophobicity of the substrate protein could be demonstrated.

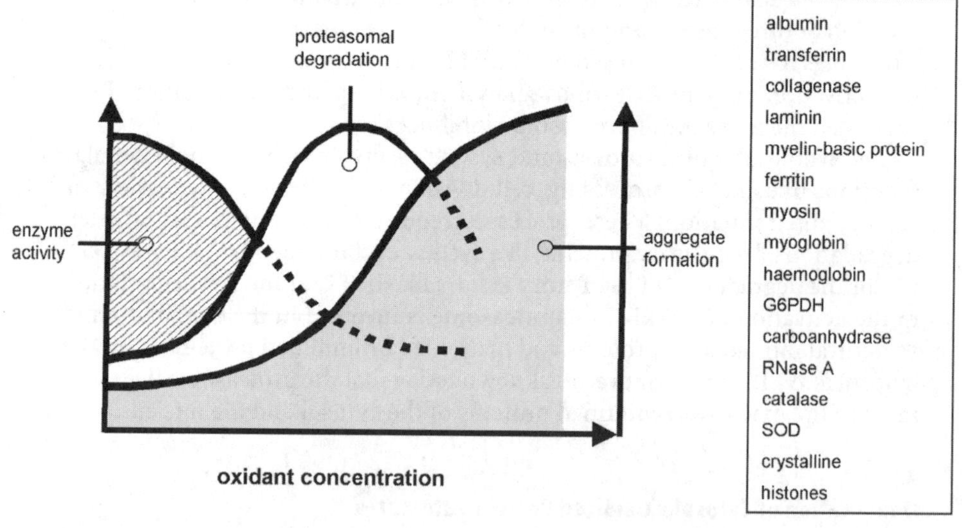

Fig. 7 Enzyme activity, proteasomal degradation, and protein aggregate formation are dependent on the oxidation status of a given protein. These dependencies were tested for the listed proteins

As described above, it is likely that highly oxidized proteins tend to aggregate and that the recognition by the proteasome is therefore decreased.

Over the last few years we and others have been able to demonstrate that the same processes occur in a large variety of mammalian cells. Different forms of mild oxidative stress significantly increase the intracellular degradation of both short- and long-lived proteins [30, 31]. This increase in overall protein turnover is accompanied by the selective removal of oxidized proteins from the intracellular protein pool. It has been demonstrated that this removal of oxidized proteins and the increase in the overall protein turnover is strongly proteasome-dependent. These processes are now being investigated during the exposure of cells to a large variety of oxidants. Interestingly, the oxidative burst of macrophages and microglial cells is also able to oxidize the intracellular protein pool, leading to an increase in the overall protein turnover in these cells [37].

There is still a question about which constituents of the proteasomal system in cells are involved in the recognition and degradation of oxidized proteins. Data available have so far concentrated on the role of the 20S proteasome, although not all regulatory mechanisms of the proteasome have been investigated during oxidative stress. Present data suggests that during oxidative stress several of the ubiquitinating enzymes are inactivated [38] and the 26S proteasome activity is decreased [39]. On the other hand, the 20S proteasome remains stable under oxidizing conditions. The role of ubiquitination in the targeting of oxidized proteins to proteasomal degradation has been looked at in a series of studies. Although it could be shown that oxidation leads to some accumulation of ubiquitinated proteins in the recovery phase, it could not be demonstrated that this ubiquitinated protein fraction contained oxidized proteins. On the other hand, Shringarpure et al. [40] clearly demonstrated that cells lacking an active ubiquitination machinery were still able to degrade oxidized proteins. In further oxidation studies, these authors could not demonstrate an ubiquitination of oxidized proteins in in vitro systems.

It has been clearly demonstrated that oxidized proteins are also degraded in the nucleus by the proteasomal system. Here the proteasome is activated by the PARP-1 enzyme [41], probably to facilitate the removal of oxidized protein forms from the nucleus.

To conclude, the bulk of the oxidized proteins in the cytosol and the nucleus of mammalian cells are degraded by the proteasome (Fig. 8). There are no known co-factors or regulators involved in the intracellular recognition and degradation of oxidized proteins. Furthermore, the process of removal of oxidized proteins from the intracellular protein pool does not require ubiquitin and seems to be independent of the activity of the 26S proteasome.

4.4 Lysosomal Turnover of Oxidized Proteins

Lysosomal proteases are thought to be involved in the degradation of autophagocytosed intracellular proteins or of phagocytosed extracellular pro-

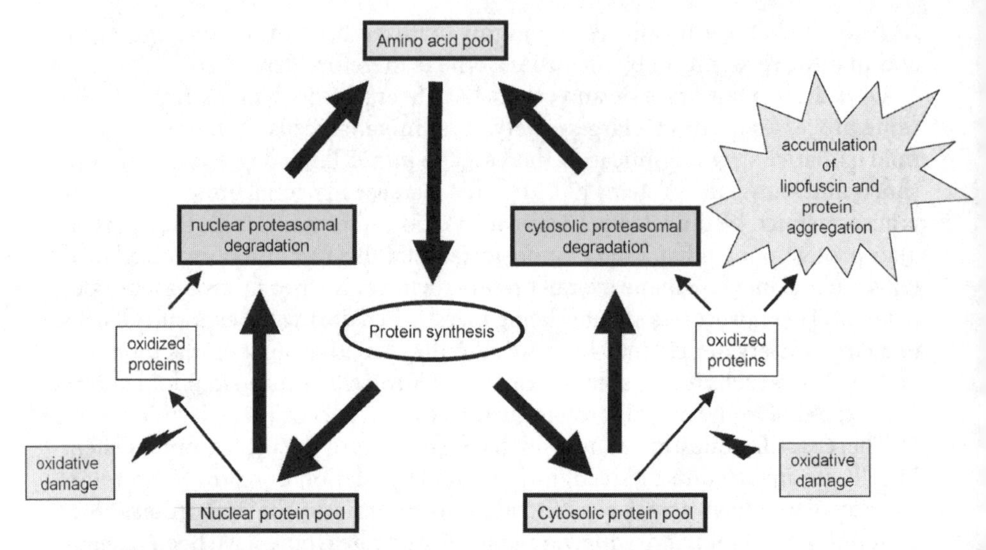

Fig. 8 Cellular protein turnover. Due to protein synthesis the nuclear and cytosolic protein pools are formed. A part of these proteins is oxidized and degraded, whereas the bulk of the proteins are degraded due to normal protein turnover. Only a small amount of oxidized proteins accumulates in aggregates

teins. No clear preferential degradation and recognition of oxidized intracellular proteins via the lysosomal pathway has been demonstrated up to now. However, oxidized extracellular proteins are internalized with a slightly higher activity (see Sect. 5 "Degradation of extracellular proteins"). Considering the fact that a large number of membrane proteins are turned over via the lysosomal degradation pathway it is remarkable how few studies have been performed on oxidative damage and the subsequent degradation and turnover of these proteins. Originally it was thought that membrane proteins are shielded from proteases by the surrounding lipid bilayer. However, today it is assumed that there exist proteases that are also able to cleave proteins within the lipid bilayer, e.g., the presenilins [42]. In the case of severe oxidative stress membrane proteins undergo fragmentation, driven by oxidants and lipid-radicals, as shown for monoamino oxidase in the mitochondrial outer membrane [43]. One should keep in mind, however, that many membrane proteins have large domains in the hydrophilic environment that are clearly accessible to soluble proteases. On the other hand, oxidized membrane proteins (or the remains of those after fragmentation or initial proteolytic processing) are possibly degraded within lysosomes. The same seems to be true for soluble oxidized proteins within the endosomal compartment. However, Stolzing et al. [44] demonstrated an increased importance of the proteasomal system in the degradation of up-taken extracellular and formerly endosomal localized oxidized proteins. Therefore, the role of the lysosomes in the degradation of oxidized proteins still remains somewhat obscure.

4.5 Role of Calpains in the Degradation of Oxidized Proteins

Calpains are cytosolic Ca^{2+}-dependent cysteine proteases [45]. They are located close to the cytoskeleton and mostly degrade cytoskeletal proteins, protein kinases, phosphatases, membrane receptors, transport proteins, and regulatory proteins [46, 47]. Stimulation of calpain activity was shown following oxidative stress [48–55] in combination with an increase in intracellular Ca^{2+} concentration [50, 54]. No studies exist that point towards a selective recognition of oxidized protein substrates by calpains [56]. However, it has been demonstrated that the oxidized cytoskeletal protein ezrin is degraded by the proteasome [57].

4.6 Degradation of Oxidized Proteins in Mitochondria

Mitochondria contain their own proteolytic systems homologous to those of bacteria. It has been known for several years that mitochondria are able to degrade oxidized proteins selectively [58]. However, it was discovered only recently that the mitochondrial LON-protease is responsible. This enzyme is also able to recognize oxidized proteins (oxidized aconitase has been used in the study) and degrade them selectively [59, 60]. The role of this protease was verified by using various inhibitors and anti-sense strategies. Interestingly the LON-protease requires an ATP-stimulation, in contrast to the cytosolic proteasomal degradation. The targeting or recognition mechanism of the oxidized substrate by the LON-protease remains to be discovered.

5 Degradation of Extracellular Proteins

It is assumed that extracellular proteins are either taken up by cells and degraded in the lysosomal compartment or that some of these proteins are degraded by extracellular proteases. As stated already, little is known about the ability of lysosomal proteases to selectively recognize and degrade oxidatively damaged proteins. Stolzing et al. [44] were able to demonstrate an increased degradation of oxidized extracellular proteins. It was clearly demonstrated that this is an intracellular process. However, with increasing oxidation of the used substrate proteins an increasing role of the proteasomal system was verified by the use of inhibitors [44]. Interestingly, endocytosed Apo B from LDL accumulates in the lysosomes of macrophages [61]. Grant et al. demonstrated a longer half-life of endocytosed oxidized albumin in comparison to non-oxidized [61]. It was concluded that the oxidation of proteins produces some local resistance to proteolytic degradation. If one assumes that Apo B [62] and albumin [62] were extensively oxidized, these reports agree well with the data of Stolzing et al. [44], demonstrating a low degradation rate of extensively oxidized extracellular proteins. Interestingly, the studies of Stolzing et al. also demonstrated, for more complex material such as apoptotic bodies, a preferential intracellular degradation and some role of the proteasomal system [63]. This might be important since some oxidation of the protein pool has been demonstrated in apoptotic material [63].

6 Accumulation of Cross-Linked Proteins

As already mentioned: oxidation of proteins and an insufficient degradation of these proteins leads to the accumulation of malfolded protein forms, which tend to aggregate. These protein aggregates are in general insoluble and poor substrates for proteolytic degradation. A fairly broad spectrum of protein aggregate formation occurs not due to the formation of covalent cross-links but due to hydrophobic and electrostatic interactions [5, 8, 23, 64]. However, a number of various covalent cross-links are involved in protein aggregation. The most thoroughly investigated cross-links are the formation of a 2,2′-biphenyl cross-link by two tyrosyl radicals [8, 65] and the formation of -S-S- bonds. In addition to those cross-links directly involving amino acids, numerous "natural" cross-linking reagents form due to oxidative stress. Especially, during lipid peroxidation numerous products form, which act as protein cross-linkers. For the most abundant lipid peroxidation products – malondialdehyde and 4-hydroxynonenal – protein cross-linking effects have been demonstrated by several authors [66, 67]. It is discussed whether the final step of protein cross-linking is the formation of an insoluble, fluorescent material that accumulates within the cells. This material is called lipofuscin, ceroid, or age pigment-like fluorophores by various authors, indicating the involvement of carbohydrates in the final fluorophore formation [9, 68]. It is widely accepted that all these pigments have the same principal origin [9] although there might be tissue-specific or disease-specific differences [9, 69–73]. The involvement of cross-linked protein oxidation products in lipofuscin formation has been demonstrated by the immunohistochemical detection of dityrosine within these pigments [74]. It has been shown by various authors that lipofuscin accumulates in postmitotic cells during aging [75–78]. The involvement of free radicals as one of the initial steps in the formation of fluorescent oxidized/cross-linked aggregates has been postulated [77, 79–83]. Additionally, the role of catalytic iron was underlined [75], which might be the result of the release of catalytically active iron from autophagocytosed mitochondria as postulated by Brunk and Terman [75]. Large amounts of the fluorescent oxidized/cross-linked aggregates are found in lysosomes; a role of non-function of the lysosomal degradation and turnover machinery was postulated [75, 84, 85]. Besides the accumulation of oxidized/cross-linked material in lysosomes a cytosolic accumulation of this material is also assumed. The accumulation of oxidized/cross-linked proteins in the cytosol inhibits proteasomal protein degradation [86, 87] and therefore interferes with the degradation of newly oxidized proteins. This seems to be a self-accelerating process. Therefore, several diseases (most importantly age-related diseases) and the aging process itself are accompanied by the accumulation of oxidized/cross-linked proteins within certain cell types. It is more and more accepted that these cross-linked proteins progressively disturb the cellular function and several of these diseases are sometimes referred to as "gain of function" diseases (Fig. 9).

Furthermore, these protein aggregates not only accumulate within the intracellular compartment, but also in the extracellular space. This is the result of oxidation of extracellular proteins or due to the release of intracellular protein aggregates from dead cells.

Regardless of the localization of the accumulated oxidized/cross-linked proteins it remains rather unclear whether this is a result of malfunction of one or other of the proteolytic systems or the result of an excessive modification of the

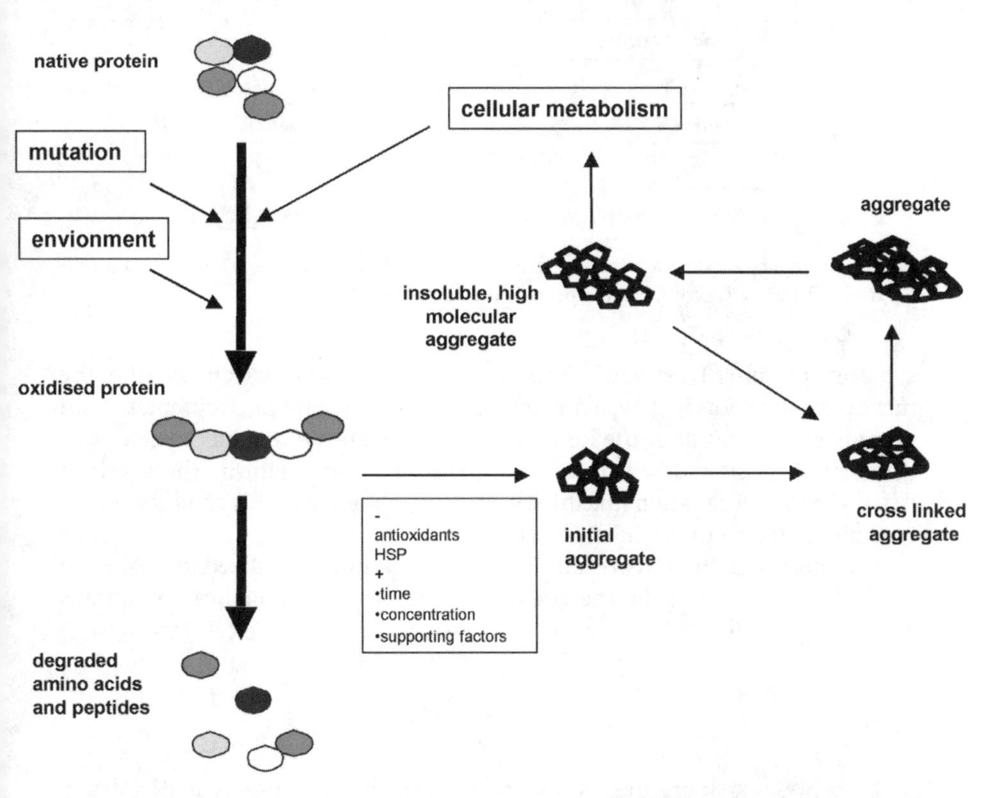

Fig. 9 Formation of protein aggregates. Proteins are oxidized due to environmental and metabolic influences. The part of the oxidized proteins escaping degradation form initial aggregates. These become cross-linked and insoluble due to intramolecular reactions. The chemically highly reactive surface of the aggregates tends to attract more proteins, in particular oxidized ones, and a process of enlarging of the aggregates takes place. It is important to note that these aggregates disturb cellular metabolism

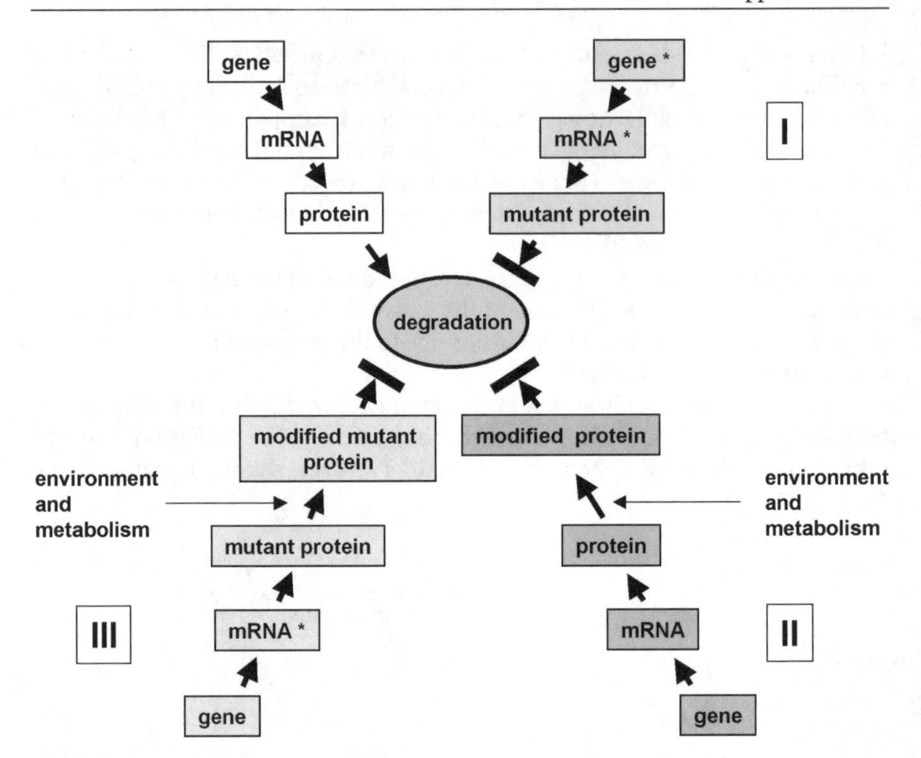

Fig. 10 Protein aggregates are formed due to mutations (*I*), due to posttranslational modifications (*II*), or to the combination of both mechanisms (*III*)

substrate proteins. However, it is also unclear in several situations whether the turnover of proteins is slowed down and therefore the protein accumulates and is oxidized as a result of the long existence of the protein in an oxygen environment. This slow-down of protein turnover might again be the result of defects in the degradation machinery or the result of metabolic changes, or due to some mutation in the substrate (Fig. 10).

Therefore, although the result – the accumulation of oxidized/cross-linked proteins – is comparable, the reason for the pathophysiological symptoms might be quite different.

6.1 Aging

A large body of literature exists concerning the measurement of protein oxidation during aging in different models. Most of the studies measure the formation of protein-bound carbonyls [88–108]. Other authors suggested or used other amino acid modifications, like tyrosine oxidation products such as dityrosine, *o*-tyrosine or nitrotyrosine [109–112], the formation of 5-hydroxyl-2-amino valeric acid [113], or methionine sulfoxide formation [112]. Changes in the protein thiol content were also taken as a measure of the age-related changes in protein structure [114]. Additionally, measurements were made using parameters of protein structure or changes in protein aggregation to demonstrate age-related changes in protein oxidation [9, 77, 115, 116].

It could be demonstrated that the protein carbonyl content of various animals and humans increases with age [78, 89, 93–98, 100–103, 108, 114, 117, 118]. Furthermore, the increase in the amount of oxidized protein in the liver, lens, lymphocytes, heart, skin, and skeletal muscle of various species was reported [88, 103, 105–107, 109, 112, 115, 119]. For various rodent and fly species a correlation between life expectancy and protein oxidation was found [88, 104]. The level of oxidized proteins increases in fly muscle mitochondria [100, 120] and mice synaptic mitochondria during aging [90, 99], underlining the special role of mitochondria in age-related free radical generation. It is interesting to note that the accumulation of oxidized proteins seems not to be a random process and that certain proteins are accumulated at higher rates in an oxidized form, as demonstrated for high molecular weight mitochondrial proteins [120], for mitochondrial aconitase [121], and for cytosolic carbonic anhydrase III [122]. Whether this accumulation is due to an increased targeting of these proteins for oxidation or whether it is the result of a decline in the degradation of the oxidized proteins still remains unclear.

Since the age-related accumulation of oxidized proteins may be the result of both increased protein oxidation and/or declined protein breakdown, it is interesting to investigate the activity of the intracellular proteolytic systems. It has been demonstrated by various authors [5, 78, 93, 105, 117, 118, 123–125] that the proteasome (often referred to as neutral alkaline protease) activity declines in the brain and liver during aging. These experiments were performed under various conditions using different substrates, including oxidized proteins or artificial fluorogenic substrates. Using these substrates it was demonstrated in several studies that the peptidylglutamyl hydrolyzing activity of the proteasome was preferentially affected during aging [126–128]. Although the trypsin-like activity is also oxidation-dependent, it seems to be protected during aging due to the interaction of the proteasome with HSP90 [127]. Since it was demonstrated that the proteasome is responsible for degradation of the bulk of intracellular proteins [20] and oxidatively modified proteins [5, 30, 31, 129] almost nothing is known about the oxidation- and age-related changes of lysosomal proteolytic systems, which seem to remain constant [116, 130].

6.2 Protein Accumulation in Neurodegeneration

One of the common outcomes of several neurodegenerative diseases is the accumulation of proteins in the extracellular environment, in neurons, or in both. Since the early sixties the fact that protein aggregates could be found in a wide variety of neurodegenerative diseases gained recognition [131, 132]. In familiar forms of these diseases, a mutation in the gene of a specific aggregated proteins is found. Thus, each aggregated protein is intimately connected to a different disease.

Protein aggregates often contain abnormally phosphorylated proteins. The aggregate action within the cell is normally unrelated to the original function of the protein, but interferes with the normal function of neurons and therefore may lead to neurologic impairment. Furthermore, it has been reported that several protein aggregates are able to nucleate the accumulation of more protein and therefore reduce the amount of normal functioning proteins in the cell. This nucleate function was extensively described for the tau-protein.

Since the process of aggregation is slow due to the complex process of specific intermolecular interactions required, the process of formation of disturbing amounts of protein aggregates within a cell and in a significant number of neurons might take many years or even decades. This results in a very complex (and often varying) pattern of molecular-affected brain regions and most prominent clinical observations. Interestingly the aggregating disease-specific protein is not often restricted to the malfunctioning brain areas, e.g., polyglutamine-containing protein aggregates, typical for the trinucleotide repeat diseases, are also expressed in non-affected brain regions and even outside of the brain. The specificity of the neurodegenerative processes for a definite brain area may therefore be determined by associated binding proteins that accelerate or even start the formation of aggregates or resulting functional impairment [133]. Besides mutations, changes in the turnover of a protein might be the reason for the accumulation of one or other protein in neurodegenerative diseases.

Protein oxidation is certainly one of the most predominant protein modifications in neurodegeneration. On the other hand, a possible way by which a protein aggregate could lead to neurotoxicity would be via mediation of radical formation due to metabolic impairment of affected cells. This process again leads to secondary protein oxidation and aggregate modification. One of the best-studied proteins is the amyloid peptide, which seems capable of inducing free radical production in Alzheimer's disease [134]. Amyloid peptide is able to bind metals and these are able to produce radicals through the Fenton reaction [135, 136].

A further way to elevate oxidative stress in the brain in neurodegenerative diseases is the activation of microglial cells, the resident macrophages of the brain. These cells are responsible for the removal of aggregated proteins from the brain and are activated by these. Activation of microglia produces large amounts of free radicals [137] that contribute to the burden of oxidative stress in the affected brain.

It is likely that oxidative stress acts as a promoting factor for neuronal cell death. Oxidative stress and protein oxidation seems to alter the onset and the development of the disease, but are not necessarily the cause for the disease. However, if a particular protein does not provoke the effect by itself, the formation of protein aggregates is able to create secondary changes that might be harmful.

7
Conclusions and Future Problems

Protein oxidation is considered to be one of the important processes during oxidative stress. Once oxidative stress occurs protein oxidation takes place. The measurable changes in the protein pool are always the result of the protein oxidation process itself, protein repair, and the degradation of oxidized proteins (Fig. 11). Cells are able to recover from protein oxidation, mostly by a selective degradation of the oxidized proteins. However, during the lifetime and in certain diseases the removal of oxidized proteins is sub-optimal and therefore proteins accumulate and interfere with the cellular metabolism.

Unfortunately, no established mechanisms or therapeutic strategies are available to selectively prevent protein oxidation. It is possible that site-specific antioxidants exist, which are able to prevent selective oxidation of protein moieties. One further approach could be the stimulation of proteasomal activity and perhaps specificity. On the other hand it seems to be possible to prevent the formation of aggregates through induction of chaperone genes that help to prevent protein aggregation. Furthermore, the induction of other small molecules to inhibit the formation and seeding of aggregates at an early stage could be a strategy to prevent the accumulation of oxidized proteins.

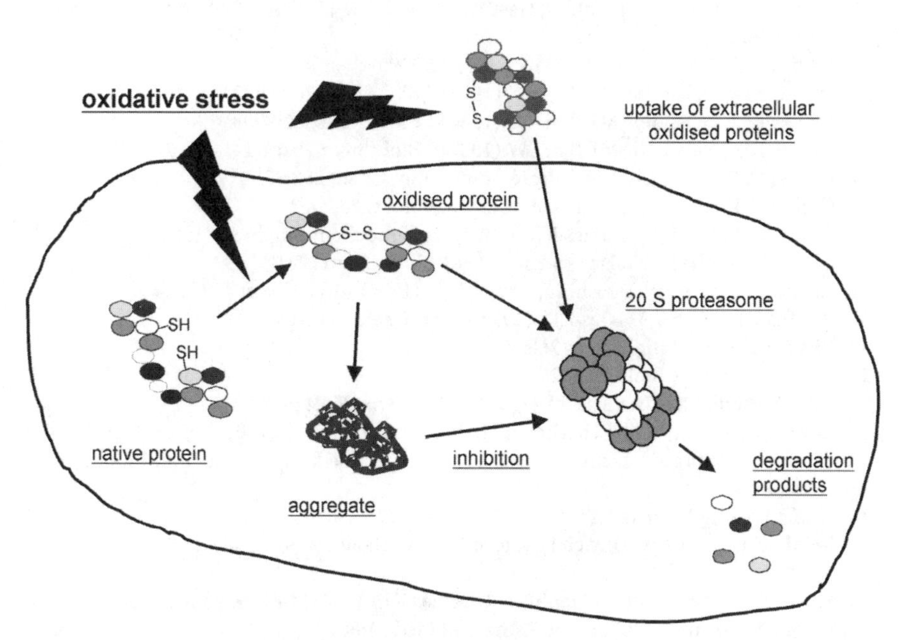

Fig. 11 Fate of oxidized proteins. Intracellular oxidized proteins are degraded by the proteasome to prevent aggregation. If aggregation occurs, these aggregates are able to inhibit the proteasome. A part of the extracellular oxidized proteins is taken up by cells and also degraded by the proteasome

References

1. Naskalski JW, Bartosz G (2001) Adv Clin Chem 35:161
2. Shringapure R, Davies KJA (2002) 32:1084
3. Grune T (1999) Biogerontology 1:31
4. Grune T, Davies KJA (1997) Biofactors 6:165
5. Grune T, Reinheckel T, Davies KJA (1997) FASEB J 11:526
6. Shringapure R, Grune T, Davies KJA (2001) Cell Mol Life Sci 58:1442
7. Dean RT, Fu S, Stocker R, Davies M (1997) Biochem J 324:1
8. Stadtman ER (1993) Annu Rev Biochem 62:797
9. Yin D (1996) Free Radic Biol Med 21:871
10. Masutani H, Yodoi J (2002) Methods Enzymol 347:279
11. Lou MF (2003) Prog Retin Eye Res 22:657
12. Levine RL, Mosoni L, Berlett BS Stadtman ER (1996) Proc Natl Acad Sci USA 93:15036
13. Kregel KC (2002) J Appl Physiol 92:2177
14. Puig A, Gilbert HF (1994) Biol Chem 264:13963
15. Houry WA (2001) Curr Protein Pept Sci 2:227
16. Chiang HL, Dice FJ (1988) J Bio Chem 263:6797
17. Mortimore GE, Miotto G, Venerando R, Kadowaki M (1996) Autophagy Biochem 27:93
18. Rock KL, Gramm C, Rothstein L, Clark K, Stein R, Dick L, Hwang D, Goldberg AL (1994) Cell 78:761
19. Harries JR (1968) Biochim Biophys Acta 150:534
20. Coux O, Tanaka K, Goldberg AL (1996) Annu Rev Biochem 65:801
21. Hershko A, Ciechanover A (1998) Annu Rev Biochem 67:425
22. Ullrich O, Reinheckel T, Sitte N, Hass R, Grune T, Davies KJA (1999) Proc Natl Acad Sci USA 96:6223
23. Davies KJA (1987) Protein J Biol Chem 262:9895
24. Davies KJA, Delsignor ME, Lin SW (1987) J Biol Chem 262:9902
25. Davies KJA, Delsignor ME, Lin SW (1987) J Biol Chem 262:9908
26. Davies KJA, Delsignor ME, Lin SW (1987) J Biol Chem 262:9914
27. Giulivi C, Davies KJA (1993) Proteasome J Biol Chem 268:8752
28. Giulivi C, Davies KJA (1994) Methods Enzymol C 233:363
29. Giulivi C, Pacifici RE, Davies KJA (1994) Arch Biochem Biophys 311:329
30. Grune T, Reinheckel T, Davies KJA (1996) J Biol Chem 271:15504
31. Grune T, Reinheckel T, Joshi M, Davies KJA (1995) J Biol Chem 270:2344
32. Pacifici RE, Kono Y, Davies KJA (1993) J Biol Chem 268:15405
33. Rivett AJ (1985) J Biol Chem 260:300
34. Rivett AJ (1985) J Biol Chem 260:1260
35. Salo D, Pacifici RE, Davies KJA (1990) J Biol Chem 265:11919
36. Lasch P, Petras T, Ullrich O, Backmann J, Naumann D, Grune T (2001) J Biol Chem 276:9492
37. Gieche J, Mehlhase J, Licht A, Zacke T, Sitte N, Grune T (2001) Biochim Biophys Acta 1538:321
38. Shang F, Gong X, Taylor A (1997) J Biol Chem 272:23086
39. Reinheckel T, Sitte N, Ullrich O, Kuckelkorn U, Davies KJA, Grune T (1998) Biochem J 335:637
40. Shringarpure R, Grune T, Mehlhase J, Davies KJA (2003) J Biol Chem 278:311
41. Ullrich O, Grune T (2001) Free Radic Biol Med 31:887
42. Kimberly WT, Wolfe MS (2003) J Neurosci Res 74:353
43. Dean RT, Thomas SM, Garner A (1986) Biochem J 240:489
44. Stolzing A, Wegner A, Grune T (2002) Arch Biochem Biophys 400:171

45. Goll DE, Thompson VF, Li H, Wei W, Cong J (2003) Physiol Rev 83:731
46. Chan SL, Mattson MP (1999) J Neurosci Res 58:167
47. Sun A, Cheng J (1999) Clin Neuropharmacol 22:164
48. Anderssonet M, Sjostrand J, Petersen A, Karlsson JO (1998) Ophthalmic Res 30:157
49. Dare E, Gotz ME, Zhivotovski B, Manzo L, Ceccatelli S, (2000) J Neurosci Res 62:557
50. Ishihara I, Minami Y, Nishizaki T, Matsuoka T, Yamamura H (2000) Neurosci Lett 279:97
51. Kishimoto S, Sakon M, Umeshita K, Miyoshi H, Taniguchi K, Meng W, Nagano H, Dono K, Ariyosi H, Nakamori S, Kawasaki T, Gotoh M, Imajoh-Ohmi S (2000) Transplantation 69:2314
52. McCracken E, Dewar D, Hunter AJ (2001) Brain Res 892:329
53. Miyoshi H, Umeshita K, Sakon M, Imajoh-Ohmi S, Fujitani K, Gotoh M, Oiki E, Kambayashi J, Monden M (1996) Gastroenterology 110:1897
54. Ray SK, Fidan M, Nowak MW, Wilford GG, Hogan EL, Banik NL (2000) Brain Res 852:326
55. Ray SK, Wilford GG, Crosby CV, Hogan EL, Banik NL (1999) Brain Res 829:18
56. Mehlhase J, Grune T (2002) Biol Chem 383:559
57. Grune T, Reinheckel T, North JA, Li R, Bescos PB, Shringarpure R, Davies KJA (2002) FASEB J 16:1602
58. Marcillat O, Zhang Y, Lin SW, Davies KJA (1988) Biochem J 254:225
59. Bota D, Davies KJA (2002) Nat Cell Biol 4:674
60. Bota D, Van Remmen H, Davies KJA (2002) FEBS Lett 532:103
61. Jessup W, Mander EI, Dean RT (1992) Biochem Biophys Acta 1126:167
62. Grant AJ, Jessup W, Dean RT (1993) Free Radic Res Commun 18:259
63. Stolzing A, Grune T (2004) FASEB J FJE 03–0374
64. Sommerburg O, Ullrich O, Sitte N, von Zglinicki D, Grune T (1998) Free Radic Biol Med 24:1369
65. Heinecke JW, Li W, Daehnke HL, Goldstein JA (1993) Biol Chem 268:4069
66. Friguet B, Stadtman ER, Sweda LI (1994) J Biol Chem 269:21639
67. Friguet B, Szweda LI (1997) FEBS Lett 405:21
68. Yin D (1995) In: Kitani K, Ivy GO, Shimasaki H (eds) Lipofuscin and ceroid pigments. Karger, Basel, Switzerland, p 159
69. Fearnley IM, Walker JE, Martinus RD, Jolly RD, Kirkland KB, Shaw GJ, Palmer DN (1990) J Biochem 268:751
70. Hunt JV, Wolff SP (1991) Free Radic Res Commun 12–13:115
71. Jolly RD, Douglas BV, Davey PM, Roiri JE (1995) In: Kitani K, Ivy GO, Shimasaki H (eds), Lipofuscin and ceroid pigments. Karger, Basel, Switzerland, p 283
72. Katz ML, Christianson JS, Norbury NE, Gao C-L Koppang N (1994) J Biol Chem 269:9906
73. Klikugawa K, Beppu M, Sato A (1995) In: Kitani K, Ivy GO, Shimasaki H (eds) Lipofuscin and ceroid pigments. Karger, Basel, Switzerland, p 1
74. Kato Y, Maruyama W, Naoi M, Hashizume Y, Osawa T (1998) FEBS Lett 439:231
75. Brunk UT, Terman A (1998). In: Cadena E, Packer L (Ed), Understanding the process of aging. Dekker, Basel, Switzerland, p 229
76. Marzabadi MR, Yin D, Brunk UT (1992) EXS 62:78
77. Nakano M, Oenzil F, Mizuno T, Gotoh S (1995) In: Kitani K, Ivy GO, Shimasaki H (eds) Lipofuscin and ceroid pigments. Karger, Basel, Switzerland, p 69
78. Sitte N, Merker K, von Zglinicki T, Davies KJA, Grune T (2000) FASEB J 14:2503
79. Aloj-Totaro E, Cuomo V, Pisanti FA (1986) Arch Gerontol Geriatr 5:343
80. Brunk UT, Jones CB, Sohal RS (1992) 275:395
81. Carpenter KLH, van der Veen, C, Taylor SE, Hardwick SJ, Clare K, Hegyi L, Mitchinson MJ (1995) In: Kitani K, Ivy GO, Shimasaki H (eds), Lipofuscin and ceroid pigments. Karger, Basel, Switzerland, p 53

82. Sohal RS, Brunk UT (1989) Adv Exp Med Biol 266:17
83. Yin DZ (1992) Mech Ageing Dev 62:35
84. Rattan SIS (1996) Exp Gerontol 31:33
85. Terman A (1995) In: Kitani K, Ivy GO, Shimasaki H (eds) Lipofuscin and ceroid pigments. Karger, Basel, Switzerland, p 319
86. Bence NF, Sampat RM, Kopito RR (2001) Science 292:1552
87. Sitte N, Huber M, Grune T, Ladhoff A, Doecke W-D, von Zglinicki T, Davies KJA (2000) FASEB J 14:1490
88. Agarwal S, Sohal RS (1996) Exp Gerontol 31:387
89. Aksenova MV, Aksenov MY, Carney JM, Butterfield DA (1998) Mech Aging Dev 100:157
90. Banaclocha MM, Hernandez AI, Martinez N, Ferrandiz ML (1997) Brain Res 762:256
91. Bradley MO, Dice J, Hayflick L, Schimke RT (1975) Exp Cell Res 96:103
92. Butterfield DA, Howard BJ, Yatin S, Allen KL, Carney JM (1997) Proc Natl Acad Sci USA 94:674
93. Carney JM, Starke-Reed PE, Oliver CN, Landum RW, Cheng MS, Wu JF, Floyd RA (1991) Proc Natl Acad Sci USA 88:3633
94. Cini M, Moretti A (1995) Neurobiol Aging 16:53
95. de la Cruz CP, Revilla E, Venero JL, Ayala A, Cano J, Machado A (1996) Free Radic Biol Med 20:53
96. Dubey A, Forster MJ, Lal H, Sohal RS (1996) Arch Biochem Biophys 333:189
97. Dubey A, Forster MJ, Sohal RS (1995) Arch Biochem Biophys 324:249
98. Forster MJ, Dubey A, Dawson KM, Stutts WA, Lal H, Sohal RS (1996) Proc Natl Acad Sci USA 93:4765
99. Martinez M, Hernandez AI, Martinez N, Ferrandiz ML (1996) Brain Res 731:246
100. Smith CD, Carney JM, Starke-Reed PE, Oliver CN, Stadtman ER, Floyd RA, Markesbery WR (1991) Proc Natl Acad Sci USA 88:10540
101. Sohal RS, Agarwal S, Dubey A, Orr WC (1993) Proc Natl Acad Sci USA 90:7255
102. Sohal RS, Dubey A (1994) Free Radic Biol Med 16:612
103. Sohal RS, Ku H-H, Agarwal S, Forster MJ, Lal H (1994) Mech Aging Dev 74:121
104. Sohal RS, Sohal BH, Orr WC (1995) Free Radic Biol Med 19:499
105. Stadtman ER, Starke-Reed PE, Oliver CN, Carney JM, Floyd RA (1992) In: Emerit I, Chance B (eds) Free radicals and aging. Birkhäuser, Basel, Switzerland, p 72
106. Tian L, Cai Q, Bowen R, Wie H (1995) Free Radic Biol Med 19:859
107. Youngman LD, Park J-YK, Ames BN (1992) Proc Natl Acad Sci USA 89:9112
108. Agarwal S, Sohal RS (1993) Relationship Biochem Biophys Res Comm 194:1203
109. Leeuwenburgh C, Hansen P, Shaish A, Holloszy JO, Heinecke JW (1998) Am J Physiol 274:R453
110. Leeuwenburgh C, Wagner P, Holloszy JO, Sohal RS, Heinecke JW (1997) Arch Biochem Biophys 346:74
111. Meucci E, Mordente A, Martorana GE J (1991) Biol Chem 266:4692
112. Wells-Knecht MC, Lyons TJ, Mc Cance DR, Thorpe SR, Baynes JW (1997) J Clin Invest 100:839
113. Ayala A, Cutler RG (1996) Free Radic Biol Med 21:65
114. Agarwal S, Sohal RS (1994) Mech Aging Dev 75:11
115. Chao C-C, Ma Y-S, Stadtman ER (1997) Proc Natl Acad Sci USA 94:2969
116. Porta E, Llesuy S, Monserrat AJ, Benavides S, Travacio M (1995) In: Kitani K, Ivy GO, Shimasaki H (eds) Lipofuscin and ceroid pigments. Karger, Basel, Switzerland, p 81
117. Sitte N, Merker K, von Zglinicki T, Grune T (2000) Free Radic Biol Med 28:701
118. Sitte N, Merker K, von Zglinicki T, Grune T, Davies KJA (2000) FASEB J 14:2495

119. Wells-Knecht MC, Huggins TG, Dyer DG, Thorpe SR, Baynes JW (1993) J Biol Chem 268:12348
120. Agarwal S, Sohal RS (1995) Mech Aging Dev 85:55
121. Yan L-J, Levine RL, Sohal RS (1997) Proc Natl Acad Sci USA 94:11168
122. Cabiscol E, Levine RL (1995) J Biol Chem 270:14742
123. Agarwal S, Sohal RS (1994) Arch Biochem Biophys 309:24
124. Berlett BS, Stadtman ER (1997) J Biol Chem 272:20313
125. Starke-Reed PE, Oliver CN (1989) Arch Biochem Biophys 275:559
126. Anselmi B, Conconi M, Veyrat-Durebex C, Turlin E, Biville F, Alliot J, Friguet B (1998) J Gerontol 53:B173
127. Conconi M, Friguet B (1997) Mol Biol Rep 24:45
128. Conconi M, Szweda LI, Levine RL, Stadtman ER, Friguet B (1996) Arch Biochem Biophys 331:232
129. Grune T, Blasig IE, Sitte N, Roloff B, Haseloff R, Davies KJA (1998) J Biol Chem 273:10857
130. Benuck M, Banay-Schwartz M, Lajtha A (1992) Life Sci 52:877
131. Dichgans J and Schulz JB (1999) Nervenarzt 70 1072
132. Orgel L E (1963) Proc Natl Acad Sci USA 49:517
133. Harper JD, Lansbury PT Jr (1997) Annu Rev Biochem 66:385
134. Floyd RA (1999) Proc Soc Exp Biol Med Dec 222:236
135. Baldwin AS (1996) Ann Rev Immunol 14:649
136. Kaltschmidt B, Uherek M, Volk B, Baeuerle PA, Kaltschmidt C (1997) Proc Natl Acad Sci USA 94:2642
137. Lezoualc'h F, Sagara Y, Holsboer F, Behl C (1998) J Neuroscience 18:3224

The Handbook of Environmental Chemistry Vol. 2, Part O (2005): 203–218
DOI 10.1007/b101152

Modulation of Cellular Signaling Processes by Reactive Oxygen Species

Lars-Oliver Klotz (✉)

Institut für Biochemie und Molekularbiologie I, Heinrich-Heine-Universität Düsseldorf, Universitätsstrasse 1, Geb. 22.03.E02, 40225 Düsseldorf, Germany
LarsOliver.Klotz@uni-duesseldorf.de

Abstract Exposure of cells to reactive oxygen species (ROS) may result not only in cell death by excessive oxidation of biomolecules but also cause the activation of cellular stress signaling pathways. The modes of activation of these signaling cascades include the direct interaction of ROS with signaling proteins, i.e., their oxidation, or the indirect action of ROS via biomolecule oxidation products that, in turn, modulate the activities of signaling proteins. The oxidation of ROS target molecules such as glutathione, tyrosine phosphatases, and thioredoxin may result in modulation of signaling cascades, including the mitogen-activated protein kinase (MAPK) pathways. The biological consequences of ROS-induced signaling depend upon which subtypes of MAPK (ERK, JNK, or p38) are predominantly activated.

Keywords MAPK · Oxidative stress · PTPase · EGFR

Abbreviations

EGFR Epidermal growth factor receptor
ERK Extracellular signal-regulated kinase
GSH Glutathione
JNK c-Jun N-terminal kinase
MAPK Mitogen-activated protein kinase
MEK MAPK/ERK kinase
MKK MAPK kinase

MKKK MKK kinase
PDGFR Platelet-derived growth factor receptor
PTPase Phosphotyrosine phosphatase/protein tyrosine phosphatase
ROS Reactive oxygen species
RTK Receptor tyrosine kinase

1 Introduction

The generation of reactive oxygen species (ROS) in cells has been widely regarded as an adverse effect that a cell has to be prepared to handle in order to avoid its death. Indeed, there are physiological and pathophysiological conditions that coincide with, and maybe in part result from, the excessive generation of ROS. Cell death, however, is at the upper end of the spectrum of possible cellular reactions towards a stressful stimulus. Regarding the lower end, it has even been demonstrated that a certain basic "tone" and cellular steady-state concentration of ROS is essential for proper cellular proliferation [1]. For example, the activation of certain cell membrane-bound receptors, such as the epidermal growth factor receptor (EGFR) or the platelet-derived growth factor receptor (PDGFR), is known to not only coincide with the endogenous generation of ROS but also to, in part, rely on ROS formation as a costimulatory factor for downstream signaling events [2–4]. Yet the reactions of cells towards an excessive endogenous ROS formation or the direct exposure to sources of ROS are somewhere between these extremes. They depend on the extent of oxidative stress, i.e., the extent to which the balance between steady-state concentrations of cellular antioxidants and prooxidants is shifted in favor of the latter [5]. The cellular reactions range from an adaptation to the new situation with its higher steady-state concentrations of ROS, to a stop of normal proliferation, to an induction of a regulated, apoptotic, rather than a passive, necrotic, cell death. It is the signaling events involved in these cellular stress responses that will be discussed in this chapter.

Following a rough overview on the modulation of cellular signaling events by ROS and the possible modes of action of ROS as initiators of signaling processes, a group of kinases crucial for the regulation of cellular proliferation and survival, the mitogen-activated protein kinases (MAPK), will serve as an example for signaling molecules targeted by ROS.

2 Principles of the Regulation of Metabolic and Signaling Pathways by ROS

In the 1980s it was established that there are cellular adaptation mechanisms at the level of gene expression that enable cells to cope with oxidative stress. In particular, prokaryotic redox-sensitive gene-expression regulatory units controlling the expression of antioxidant enzymes were identified. In *E. coli*, these include the transcription factors OxyR, which acts as a hydrogen peroxide sensor, and SoxR, which is sensitive to changes in concentrations of superoxide and nitrogen monoxide (for review, see [6]). Redox-sensitive transcription factors were then also identified in mammalian cells, including activating protein AP-1 and nuclear factor NF-κB [7, 8]. Not only may transcription factors be directly targeted by oxidants and their activity modulated thereby, but redox regulation may also aim at upstream events, such as the modulation of kinase activities that are linked to transcriptional and translational regulation of gene expression.

The modulation of biochemical metabolic pathways by ROS may be either by direct interaction of ROS with the pathway, i.e., by oxidation of an enzyme or a substrate, or in an indirect manner, e.g., by the action of an oxidation product of a biological target of ROS on the pathway. For example, the glycolytic enzyme, glyceraldehyde 3-phosphate dehydrogenase (GAPDH), due to its essential cysteine residue at the active site, is very sensitive toward oxidation and has been described as being directly inactivated by ROS such as peroxynitrite [9]. The blockade of a metabolic pathway may be accomplished indirectly as well, such as by the inactivation of an enzyme by an oxidized form of a biomolecule, the concentration of which relates to levels of ROS: an example is the regulation of enzymes by the glutathione (GSH)/glutathione disulfide (GSSG) ratio [10], i.e., by thiol/disulfide exchange and by glutathionylation (see below). Products of lipid peroxidation, 4-hydroxynonenal or malondialdehyde, may also react with proteins and thereby modulate metabolic flux, establishing another example for the indirect modulation of metabolic pathways by ROS. Both direct and indirect modes of interaction may be inhibitory or stimulatory in nature, resulting either in the blockade of a metabolic pathway or its stimulation.

The same principles apply for the activation of signaling pathways by ROS: both direct and indirect as well as inhibitory and stimulatory interactions of ROS with signaling molecules are known. Some examples are listed in Table 1.

However, there is an obvious problem in talking about inhibitory and stimulatory interactions. If the enhanced generation of hydrogen peroxide induces cellular lipid peroxidation, resulting in the formation of 4-hydroxynonenal, which, in turn, inactivates an enzyme, this may clearly be regarded as an indirect inhibitory interaction. If, however, this enzyme is a negative regulator of a signaling pathway, the net result of this indirect *inhibitory* interaction is the *stimulation* of a pathway. Table 1 must be read with this caveat in mind.

A feature common, and somewhat special, to redox regulation of signal transduction and gene expression is that a major part of it is at the level of covalent modification of the direct target proteins. Different from ligand/receptor interactions, there are targets, rather than receptors, for oxidants – one notable exception being the interaction between nitrogen monoxide and the heme prosthetic group of guanylyl cyclases – the targets usually being oxidized and covalently modified upon interaction.

Table 1 Examples for the interaction of reactive oxygen species (ROS) with signaling pathways

Mode of interaction	Signalling event
Direct	
Inhibitory	Oxidation and inhibition of zinc-finger transcription factors by hydrogen peroxide, peroxyl radicals, singlet oxygen, nitrogen monoxide, peroxynitrite [60]
Stimulatory	Activation of c-Src by peroxynitrite-induced nitration of inhibitory domain [58]
Indirect	
Inhibitory	Inhibition of protein tyrosine phosphatase-1B by glutathionylation [15]
Stimulatory	Activation of transcription factor AP-2 by ceramides, the generation of which is initiated by singlet oxygen, acting on cellular membranes [61]; activation of p53-dependent signaling by oxidative DNA damage [62]

Much of what is known about oxidant-induced signaling was found in experiments using hydrogen peroxide as oxidant, often applied as a bolus yielding concentrations in the high micromolar to low millimolar range. H_2O_2 is a popular subject of studies regarding oxidant signaling because it is commercially available, reasonably stable, and thus easy to work with. It has to be kept in mind, however, that not all ROS share the same biological targets and they vary significantly as to their biochemical reactivity. Hence, findings from experiments with hydrogen peroxide as a model oxidant should not be inappropriately generalized and taken as a fixed template for signaling events induced by all ROS or by oxidative stress, irrespective of the way it is brought about.

2.1 Cellular Targets in ROS-Regulated Signaling Events

The interaction of ROS with signal transduction cascades may occur at several levels. Considering a model signaling cascade consisting of:

1. Messenger (a hormone, transmitter, cytokine)
2. Receptor
3. Second messenger (Ca^{2+}, cAMP, cGMP, phosphoinositides, and others)
4. Target proteins for receptor or second messenger (such as adapter proteins, kinases, phosphatases)
5. Effector molecules, including transcription factors and others,

it is obvious that there are a multitude of potential interactions with ROS in terms of a direct oxidation of almost any of the involved molecules. Even a number of possible indirect interactions are apparent when one considers ROS-initiated processes, such as lipid peroxidation, that may result in a loss of membrane integrity and so affect the proper orientation of membrane-bound receptors and of membrane-anchored signaling molecules, such as Ras. Furthermore, the release of Ca^{2+} from the endoplasmic reticulum (ER) may be triggered by lipid peroxidation of the ER membrane system.

Hence, to obtain an overview, the regulators most susceptible to oxidative modification in a certain signaling cascade must be identified. Some targets common to the activation of several signaling pathways by ROS are mentioned below.

2.1.1 Glutathione

Oxidation of the major low-molecular-weight thiol in eukaryotic cells, glutathione (γ-glutamylcysteinylglycine, GSH), results in the formation of mixtures of the corresponding sulfenic acid, sulfinate, sulfonate, disulfide, and disulfide S-oxide, depending on the nature of the oxidant. The two latter, as well as glutathione sulfenic acid and S-nitrosoglutathione, are capable of glutathionylating protein thiols [11, 12]. Glutathionylation, i.e., the formation of mixed protein/glutathione disulfides, reversibly alters the reactivity of several kinases and phosphatases, including protein kinase A [13], multiple protein kinase C isoforms [14], protein kinase D [14], PTP1B [15], PP2a [16], as well as of the transcription factor c-Jun [17].

Glutathionylation may inhibit proteins, yet there is an important difference from inactivation by oxidation of cysteine residues by ROS, forming sulfinic or sulfonic acids. Both cysteine oxidation states are generally regarded as not reducible under cellular conditions although the cellular reduction and reactivation of a peroxiredoxin cysteine sulfinic acid has recently been described;

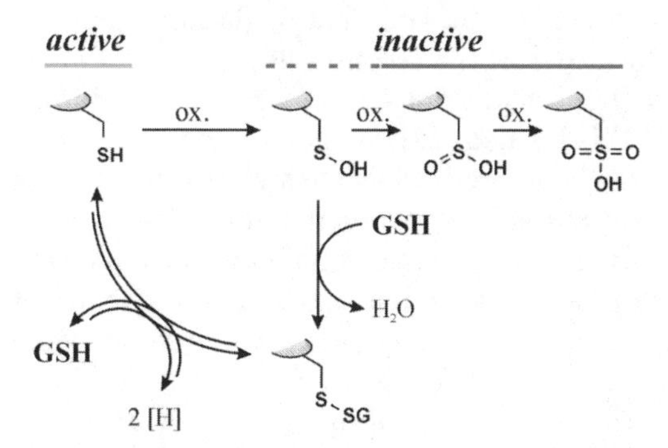

Fig. 1 Glutathionylation as reversible inactivation of signaling proteins. A protein cysteine residue may be oxidized to a sulfenic acid by reactive oxygen species. This, in turn, may either be irreversibly oxidized further to a sulfinic or sulfenic acid or react with glutathione (GSH) to form a mixed protein/glutathione disulfide, which may be reduced to regenerate the thiol

the mechanism, however, remains to be elucidated [18, 19]. Glutathionylation, on the other hand, is reversible (Fig. 1) and may, by transiently forming protein/glutathione mixed disulfides, prevent protein thiols from becoming irreversibly oxidized.

2.1.2
Phosphotyrosine Phosphatases (PTPases)

Receptor tyrosine kinases (RTKs) are activated (phosphorylated) by inhibition of a negatively regulating phosphatase upon treatment with UV (A, B, or C), hydrogen peroxide, or iodoacetamide. The phosphatase activity, (i.e., dephosphorylation and inactivation of RTKs) is restored upon the addition of thiol-regenerating agents, if not inhibited irreversibly by iodoacetamide [20]. H_2O_2 not only inactivates membrane-bound phosphatases but also diminishes cytosolic general protein tyrosine phosphatase activity in mouse fibroblasts [21]. Further, the activation of JNK by sodium arsenite, which is reactive towards thiols (especially vicinal dithiols), is by inactivation of a JNK phosphatase [22].

All phosphotyrosine phosphatases (also protein tyrosine phosphatases, PTPases) known so far harbor an essential cysteine thiolate in their active site, which serves as a nucleophile accepting the phosphate moiety of the phosphatase substrate, forming an intermediate phosphocysteine [23, 24]. An oxidation of this cysteine residue (e.g., by peroxynitrite or by singlet oxygen) would lead to the inactivation of the phosphatase, concomitantly allowing kinase activities to become predominant. For example, peroxynitrite is known to oxidize thiols to preferably form the respective disulfides (RSSR) or sulfenic acids (RSOH) and, at more acidic pH, thiyl radicals [25]. Peroxynitrite was further shown to efficiently inhibit both cellular protein tyrosine phosphatase activity [26] and various isolated tyrosine phosphatases [27]. Different from PTPase inactivation by hydrogen peroxide (vide supra), this latter inhibition is not reversible with DTT, indicating that oxidation is beyond the disulfide – which could be formed only if a second sulfhydryl group were available at the active site – or sulfenic acid state, and that probably sulfinic (RSO_2H) or sulfonic acid (RSO_3H) is formed [27]. Singlet oxygen is also known to efficiently react with thiols, the main product also being the respective disulfide, followed by sulfinic acid, sulfonic acid, and even the disulfide-S-oxide (R-S-SO-R) [28]. Hence, a mechanism for activation of signaling pathways by peroxynitrite and singlet oxygen may be envisaged that is based upon the oxidative inactivation of tyrosine phosphatases. Yet this cannot account for all oxidant-induced signaling. Peroxynitrite-induced activation of Akt (protein kinase B, see above) is an example of activation of a signaling pathway by an oxidant that does not solely rely on the oxidative inactivation of a phosphatase: prior peroxynitrite or singlet oxygen treatment renders cells refractory to subsequent Akt activation by growth factors, indicating that phosphatase inhibition cannot be the sole mechanism responsible for activation of the kinase [29]. If the inhibition of a regulating phosphatase by peroxynitrite were responsible for the increased Akt phosphorylation, then phosphorylation would not be expected to be inhibited by peroxynitrite pre-treatment, but rather would be expected to be enhanced due to the loss of negative regulation. Similarly, it has been demonstrated that the activation of Src-family kinases in erythrocytes does not strictly depend on the inactivation of a tyrosine phosphatase [30]: In the presence of carbon dioxide, the inhibition of membrane-bound PTPases elicited by treatment of erythrocytes with peroxynitrite was attenuated; nevertheless, activation of Src kinases did not appear to be affected.

2.1.3 Thioredoxin

The 12 kDa protein thioredoxin (Trx) is a redox-sensitive protein disulfide reductase with a pair of cysteines at its active site that is kept in the reduced state by the action of the selenoenzyme thioredoxin reductase at the expense of NADPH (for a review, see [31]). Trx was proven to bind to, and thereby inhibit, apoptosis signal-regulating kinase 1 (ASK1), which acts as a MAPK kinase kinase (MKKK, see below) for the JNK and p38 pathways [32]. This binding is abrogated upon oxidation of the above-mentioned critical pair of cysteines, such as by exposure to oxidants, thereby activating ASK1. Trx is also known to modulate the activity of transcription factors such as NF-κB and AP-1 [33], thus establishing a direct link between oxidant treatment of cells and transcriptional regulation.

3 Activation of Mitogen-Activated Protein Kinases by ROS: Mechanisms and Consequences

Mitogen-activated protein kinase (MAPK) cascades were among the first signaling pathways demonstrated to be activated by ROS. These effects have since been extensively investigated. Here, a brief overview on MAPK properties will be followed by a summary of the known physiological roles of MAPK and a paragraph on the mechanisms and consequences of activation of MAPK cascades by ROS.

3.1 Mitogen-Activated Protein Kinases

The classical MAPK were discovered in the late 1980s/early 1990s as 42–44 kDa proteins that are tyrosine phosphorylated in cells exposed to insulin or mitogenic stimuli such as growth factors [34–36]. They were found to be Ser/Thr kinases that phosphorylate their substrates at residues that are C-terminally flanked by proline, i.e., they are proline-directed kinases. Their activation not only requires tyrosine phosphorylation but rather a dual tyrosine and threonine phosphorylation in a Thr-Glu-Tyr motif that is catalyzed by MAPK kinases (MAPKK or MKK). This dual phosphorylation results in drastic structural changes of the protein and a more than 1000-fold higher kinase activity, which is not seen in monophosphorylated MAPK protein [37].

Two very similar groups of kinases with apparent molecular masses in the 46–57 kDa and 38 kDa ranges, respectively, were soon discovered in experiments with stressful stimuli such as the application of protein biosynthesis inhibitors or exposure of cells to (unphysiological) ultraviolet (UV) radiation (UVC <280 nm) [38] or lipopolysaccharide [39, 40]. These stress-activated protein kinases (SAPK) are also proline-directed kinases that are activated by dual phosphorylation. The phosphorylation motif, however, differs from that of the classical MAPK, and is Thr-Pro-Tyr and Thr-Gly-Tyr, respectively, in the two SAPK groups.

All the mentioned kinases are now called MAPK, with the classical MAPK, the p44 and p42 kinases (Thr-Glu-Tyr phosphorylation motif), being termed extracellular signal-regulated kinases ERK-1 and ERK-2, respectively, because of the extracellular nature of the stimuli that lead to their activation. The 46–54 kDa proteins (Thr-Pro-Tyr phosphorylation motif) are the c-Jun N-terminal kinases (JNK), named after their phosphorylation of the transcription factor c-Jun. Ten JNK isoforms exist, coded by three genes (jnk-1 through jnk-3) with the isoforms resulting from differential splicing [41]. The p38-MAPK group of kinases (Thr-Gly-Tyr phosphorylation motif) comprises four isoforms, α through δ, with 38–43 kDa [42]. ERK-1/2, JNK and p38 are the three major and best known MAPK groups. Further MAPK have been described, the most prominent being ERK-5 (big MAPK-1, BMK-1), which is similar to ERK-1/2 in its mode of activation.

In summary, MAPK are proline-directed Ser/Thr kinases that are activated by dual phosphorylation of a T-X-Y motif (with the exception of the unusual MAPK ERK-3 and ERK-4; Table 2). Their activation is catalyzed by dual specificity (Thr/Tyr) kinases, the MKK, which, in turn, are activated by MKK kinases (MKKK) that are often triggered by interaction with small GTP-binding proteins.

As can be seen in Fig. 2, a simplified summary of cascades leading to activation of the MAPK, multiple MKK, and MKKK exist. The respective three-component cascades (MKKK-MKK-MAPK) are activated in many ways, e.g., by interaction with small GTP-binding proteins, such as Ras, Rac or Cdc42. In the case of ERK-1/2, a well-known cascade starting at the level of a growth factor receptor tyrosine kinase is depicted: the MKK for ERK-1/2 are MKK-1 and MKK-2, which are also termed MEK-1/2 (for MAPK/ERK kinase). An MKKK activating MEK-1/2 is Raf, a Ser/Thr kinase which, in turn, is activated by interaction with the active (GTP-bound) form of Ras.

Activation of Ras may be brought about by stimulation of receptor tyrosine kinases such as the epidermal growth factor receptor (EGFR). The activation (i.e., dimerization and multiple tyrosine phosphorylation) of the EGFR is caused by binding of the epidermal growth factor and results in activation of Ras via adaptor protein-mediated recruitment of a guanine nucleotide exchange factor.

But what are the consequences of ERK activation? From the mere fact that a growth factor induces activation of EGFR, leading to the activation of ERK-1/2, it can be concluded that substrates of ERK must include proteins involved in the regulation of cell growth and proliferation, some of which are dealt with below.

Table 2 Terminology of MAPKs [63, 64]

	Aliases	Phosphorylation motif	Hugo nomenclature*
MAPK			
ERK		-T-E-Y-	
ERK1	MAPK 1, $p44^{MAPK}$		MAPK3
ERK2	MAPK 2, $p42^{MAPK}$		MAPK1
ERK5	BMK1		MAPK7
JNK		-T-P-Y-	
JNK1	SAPK1c; SAPKγ		MAPK8 (four splice variants)
JNK2	SAPK1a; SAPKα		MAPK9 (four splice variants)
JNK3	SAPK1b; SAPKβ		MAPK10 (two splice variants)
p38		-T-G-Y-	
p38α	SAPK2a; CSBP; RK		MAPK14
p38β	SAPK2b		MAPK11
p38γ	SAPK3; ERK6		MAPK12
p38δ	SAPK4		MAPK13
Further MAPKs			
ERK3	$p97^{MAPK}$	-S-E-G-	MAPK6
ERK4	$p63^{MAPK}$	-S-E-G-	MAPK4

* MAPK nomenclature as recommended by The Human Genome Organisation (Hugo). The recommendations can be found under http://www.gene.ucl.ac.uk/nomenclature/genefamily/prkm.html. *Abbreviations*: *BMK* big MAPK, *CSBP* cytokine-suppressive antiinflammatory drug binding protein, *ERK* extracellular regulated kinase, *JNK* c-Jun-N-terminal kinase, *MAPK* mitogen-activated protein kinase, *RK* reactivating kinase, *SAPK* stress-activated protein kinase.

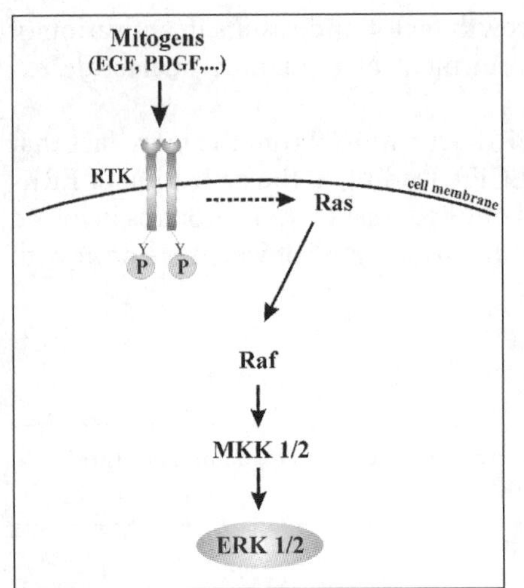

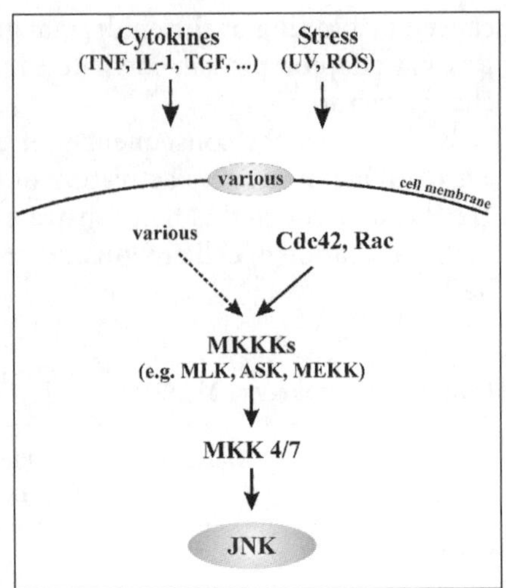

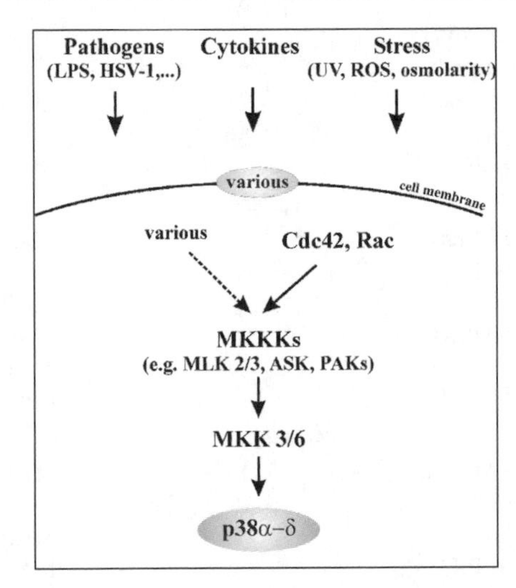

Fig. 2 Cascades leading to activation of ERK-1/2, JNK and p38 mitogen-activated protein kinases. *Abbreviations*: *ASK* apoptosis signal-regulating kinase, *cdc* cell division cycle, *EGF* epidermal growth factor, *ERK* extracellular signal-regulated kinase, *HSV* herpes simplex virus, *IL* interleukin, *JNK* c-Jun N-terminal kinase, *LPS* lipopolysaccharide, *MEKK* MAPK/ERK kinase kinase, *MKK* mitogen-activated protein kinase kinase, *MKKK* MKK kinase, *MLK* mixed-lineage kinase, *PAK* p21-activated kinase, *PDGF* platelet-derived growth factor, *ROS* reactive oxygen species, *RTK* receptor tyrosine kinase, *TGF* transforming growth factor, *TNF* tumor necrosis factor, *UV* ultraviolet (radiation)

3.2 Consequences of MAPK Activation

In 1995, Xia et al. [43] proposed a crucial role of JNK and p38 as pro-apoptotic stimuli in PC-12 cells, whereas ERK activation seemed to be antiapoptotic. Indeed, taking into account the substrates of the kinases (Fig. 3), it is clear that ERK-1/2 play a major role in the regulation of cell growth and proliferation, whereas JNK and p38 are involved in the regulation of apoptosis/differentiation and in the inflammatory response, respectively.

For example, ERK-1/2 phosphorylate and thereby activate various transcription factors, including Elk-1 and Sap-1, thus regulating the expression of genes involved in the regulation of cellular proliferation, such as cyclin D1 or

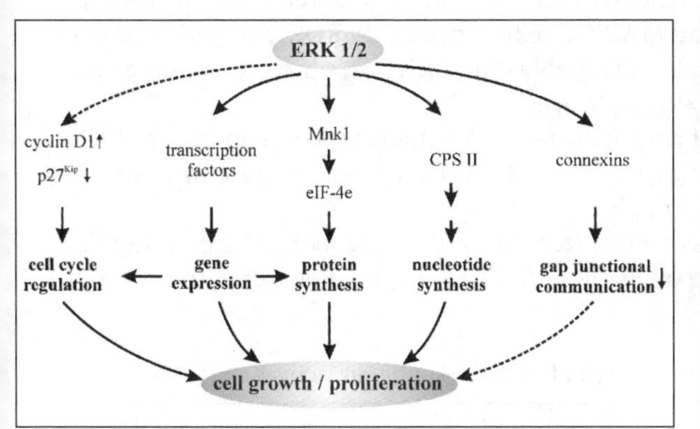

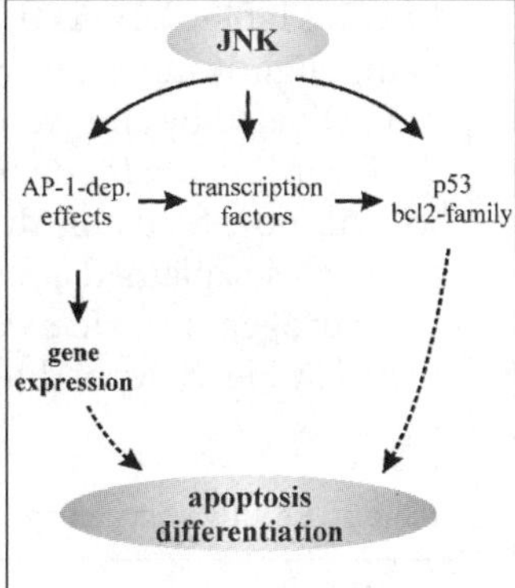

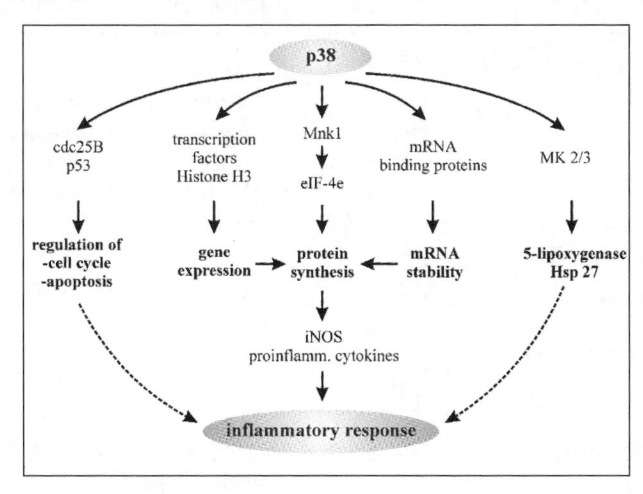

Fig. 3 Physiological significance and downstream targets of ERK, JNK ,and p38. *Abbreviations*: *CPS* carbamoyl phosphate synthetase, *eIF* eukaryotic initiation factor, *hsp* heat shock protein, *iNOS* inducible *NO* synthase, *MK* MAPK-activated protein kinase, *Mnk* MAPK signal-integrating kinase

the inhibitor of cyclin-dependent kinases, $p27^{kip1}$. Further, translation may be positively affected by ERK-1/2, as well as – at the level of pyrimidine biosynthesis – nucleotide synthesis as a preparatory step for DNA biosynthesis. Both protein synthesis and DNA replication are prerequisites for cell growth.

Although the importance of JNK and p38 in mediating apoptosis following stress has been well documented in a variety of model systems, there are clear exceptions to this generality. In fact, cases have been identified in which JNK and/or p38 activations appear to play a role in protecting cells against apoptosis [44] and cases in which *activation* of ERKs is a prerequisite for apoptosis [45].

3.3 Activation of MAPK by ROS

Many different ROS and oxidative stimuli lead to the activation of cellular MAPK (Table 3). Indeed, all MAPK can be activated by ROS, with JNK and p38 being slightly more generally activatable than ERK-1/2, which, for instance, are not activated by singlet oxygen [46, 47].

Regarding the mode of activation of MAPK upon exposure of cells to ROS, we will now focus on the activation of ERK-1/2 to demonstrate how far the principles explained above apply.

Hydrogen peroxide was shown to activate ERK-1/2 along the pathway depicted in Fig. 2, depending on EGFR and MKK-1/2 [48]. Regarding the effect of

Table 3 Reactive oxygen species (ROS) and oxidative conditions activating MAPK

ROS	Reference	Comment
Hydrogen peroxide, H_2O_2	[65, 66]	
Superoxide, $O_2^{\bullet-}$	[52, 67]	Superoxide generated by redox cycling
Singlet oxygen, $^1O_2$	[46, 68]	Only intracellular generation of singlet oxygen leads to activation of JNK and p38; no ERK activation (human skin fibroblasts)
Peroxynitrite, $ONOO^-$	[69]	
NO/NOx	[70]	NO-saturated PBS; NO-donors; endogenous NO
Hypoxia/reperfusion	[71–73]	
Photodynamic therapy	[74]	
UV		
UVA (320–400 nm)	[46, 68, 75]	MAPK activation pattern similar to singlet oxygen
UVB (280–320 nm)	[76–78]	
UVC (<280)	[38, 79, 80]	

NOx nitric oxiderelated species, *UV* ultraviolet.

another ROS, peroxynitrite, however, there are reports as diverse as the cell types used for the studies. According to the literature, there is activation of ERKs (i) via EGFR and downstream targets [49], (ii) independent from EGFR and Raf, but dependent on activation of MKK-1/2 [48], and (iii) activation independent from EGFR and even MKK-1/2, but partly coming from a Ca^{2+}-dependent PKC-isoform [50]. Thus, more than one point of attack is possible for ROS to achieve ERK activation. Regarding the activation of the EGFR as an upstream event in ROS-induced ERK activation, the classical hypothesis of how a species like hydrogen peroxide can activate a growth factor receptor tyrosine kinase in the absence of ligand is that a PTPase negatively regulating the receptor is inactivated [20, 51], resulting in a net increase in receptor tyrosine phosphorylation (see Sect. 2.1.2).

Using an alkylating and redox-cycling quinone, menadione (2-methyl-1,4-naphthoquinone), Abdelmohsen et al. [52] similarly demonstrated that activation of EGFR (resulting in activation of MKK-1/2 and ERK-1/2) was most likely by inactivation of a PTPase. Interestingly, however, two other quinones, a strong alkylator, *p*-benzoquinone, and an exclusive redox-cycler, 2,3-dimethoxy-1,4-naphthoquinone (DMNQ), also activated EGFR and ERK-1/2, independent of any PTPase. Rather, the mere depletion of GSH (see Sect. 2.1.1) by benzoquinone was sufficient for activation, pointing to thiol-containing regulators other than PTPases as important players.

There is another way of activating EGFR-dependent signaling by ROS. The non-receptor tyrosine kinase, c-Src, has been postulated to be involved in EGFR activation in some cases (for review, see [53]), and the phosphorylation of the EGFR in endothelial cells exposed to H_2O_2 was indeed demonstrated to be inhibitable by addition of PP2, a Src tyrosine kinase inhibitor [54], which also blocked the activation of Akt by H_2O_2 – known to rely upon EGFR activation [55] – in HeLa cells [56]. Like hydrogen peroxide, peroxynitrite activates Src and Src-family kinases in various human cell types [30, 57, 58]. The role of Src in receptor tyrosine kinase activations by peroxynitrite, however, remains to be resolved. The activation of c-Src by peroxynitrite was postulated to be by interference with posttranslational modification of Src: phosphorylation of a C-terminal tyrosine helps c-Src fold into an inactive conformation by way of an intramolecular phosphotyrosine/SH2-domain interaction (for review, see [59]). This might be prevented by peroxynitrite-induced nitration of this tyrosine, rendering it unavailable for phosphorylation [58]. In summary, RTK like the EGFR appear to be central players in the activation of ERK-1/2 by ROS.

4
Conclusions

Exposure of cells to ROS may result in the activation of cellular signal transduction pathways such as the MAPK. Hence, the outcome of ROS-induced signaling is in part defined by the substrates and effector molecules of the pathways. The physiological consequences of the activation of such processes largely depends on the extent/intensity and duration of activation as well as on type of cell, type of ROS, and the pattern of signaling pathways activated. For example, ROS may both enhance proliferation (probably via ERK-1/2) and induce apoptosis (possibly via activation of JNK) in target cells (see Fig. 3).

Acknowledgment Work in the author's laboratory is supported by Deutsche Forschungsgemeinschaft (Bonn, Germany): SFB 503/B1, SFB 575/B4, GRK 320.

References

1. Finkel T, Holbrook NJ (2000) Nature 408:239
2. Sundaresan M, Yu ZX, Ferrans VJ, Irani K, Finkel T (1995) Science 270:296
3. Sundaresan M, Yu ZX, Ferrans VJ, Sulciner DJ, Gutkind JS, Irani K, Goldschmidt-Clermont PJ, Finkel T (1996) Biochem J 318:379
4. Bae YS, Kang SW, Seo MS, Baines IC, Tekle E, Chock PB, Rhee SG (1997) J Biol Chem 272:217
5. Sies H (1985) Oxidative stress: introductory remarks. In: Sies H (ed) Oxidative stress. Academic Press, London, p 1
6. Zheng M, Storz G (2000) Biochem Pharmacol 59:1
7. Abate C, Patel L, Rauscher FJ, III, Curran T (1990) Science 249:1157
8. Schreck R, Rieber P, Baeuerle PA (1991) EMBO J 10:2247
9. Buchczyk DP, Grune T, Sies H, Klotz LO (2003) Biol Chem 384:237
10. Sies H (1986) Angew Chem Int Ed Engl 25:1058
11. Klatt P, Lamas S (2000) Eur J Biochem 267:4928
12. Huang KP, Huang FL (2002) Biochem Pharmacol 64:1049
13. Humphries KM, Juliano C, Taylor SS (2002) J Biol Chem
14. Chu F, Ward NE, O'Brian CA (2001) Carcinogenesis 22:1221
15. Barrett WC, DeGnore JP, Konig S, Fales HM, Keng YF, Zhang ZY, Yim MB, Chock PB (1999) Biochemistry 38:6699
16. Rao RK, Clayton LW (2002) Biochem Biophys Res Commun 293:610
17. Klatt P, Molina EP, De Lacoba MG, Padilla CA, Martinez-Galesteo E, Barcena JA, Lamas S (1999) FASEB J 13:1481
18. Woo HA, Kang SW, Kim HK, Yang KS, Chae HZ, Rhee SG (2003) J Biol Chem 278:47364
19. Woo HA, Chae HZ, Hwang SC, Yang KS, Kang SW, Kim K, Rhee SG (2003) Science 300:653
20. Knebel A, Rahmsdorf HJ, Ullrich A, Herrlich P (1996) EMBO J 15:5314
21. Sullivan SG, Chiu DT, Errasfa M, Wang JM, Qi JS, Stern A (1994) Free Radic Biol Med 16:399
22. Cavigelli M, Li WW, Lin A, Su B, Yoshioka K, Karin M (1996) EMBO J 15:6269
23. Fauman EB, Saper MA (1996) Trends Biochem Sci 21:413
24. Kolmodin K, Aqvist J (2001) FEBS Lett 498:208
25. Quijano C, Alvarez B, Gatti RM, Augusto O, Radi R (1997) Biochem J 322 (Pt 1):167
26. Mallozzi C, Di Stasi AM, Minetti M (1997) FASEB J 11:1281
27. Takakura K, Beckman JS, MacMillan-Crow LA, Crow JP (1999) Arch Biochem Biophys 369:197
28. Devasagayam TP, Sundquist AR, Di Mascio P, Kaiser S, Sies H (1991) J Photochem Photobiol B 9:105
29. Klotz LO, Schieke SM, Sies H, Holbrook NJ (2000) Biochem J 352:219

30. Mallozzi C, Di Stasi MA, Minetti M (2001) Free Radic Biol Med 30:1108
31. Arner ES, Holmgren A (2000) Eur J Biochem 267:6102
32. Saitoh M, Nishitoh H, Fujii M, Takeda K, Tobiume K, Sawada Y, Kawabata M, Miyazono K, Ichijo H (1998) EMBO J 17:2596
33. Schenk H, Klein M, Erdbrugger W, Dröge W, Schulze-Osthoff K (1994) Proc Natl Acad Sci USA 91:1672
34. Boulton TG, Yancopoulos GD, Gregory JS, Slaughter C, Moomaw C, Hsu J, Cobb MH (1990) Science 249:64
35. Boulton TG, Nye SH, Robbins DJ, Ip NY, Radziejewska E, Morgenbesser SD, DePinho RA, Panayotatos N, Cobb MH, Yancopoulos GD (1991) Cell 65:663
36. Gomez N, Cohen P (1991) Nature 353:170
37. Chen Z, Gibson TB, Robinson F, Silvestro L, Pearson G, Xu B, Wright A, Vanderbilt C, Cobb MH (2001) Chem Rev 101:2449
38. Derijard B, Hibi M, Wu IH, Barrett T, Su B, Deng T, Karin M, Davis RJ (1994) Cell 76:1025
39. Han J, Lee JD, Tobias PS, Ulevitch RJ (1993) J Biol Chem 268:25009
40. Han J, Lee JD, Bibbs L, Ulevitch RJ (1994) Science 265:808
41. Gupta S, Barrett T, Whitmarsh AJ, Cavanagh J, Sluss HK, Derijard B, Davis RJ (1996) EMBO J 15:2760
42. Ono K, Han J (2000) Cell Signal 12:1
43. Xia Z, Dickens M, Raingeaud J, Davis RJ, Greenberg ME (1995) Science 270:1326
44. Assefa Z, Vantieghem A, Declercq W, Vandenabeele P, Vandenheede JR, Merlevede W, de Witte P, Agostinis P (1999) J Biol Chem 274:8788
45. Wang X, Martindale JL, Holbrook NJ (2000) J Biol Chem 275:39435
46. Klotz LO, Pellieux C, Briviba K, Pierlot C, Aubry JM, Sies H (1999) Eur J Biochem 260:917
47. Klotz LO, Kröncke KD, Sies H (2003) Photochem Photobiol Sci 2:88
48. Zhang P, Wang YZ, Kagan E, Bonner JC (2000) J Biol Chem 275:22479
49. Jope RS, Zhang L, Song L (2000) Arch Biochem Biophys 376:365
50. Bapat S, Verkleij A, Post JA (2001) FEBS Lett 499:21
51. Herrlich P, Böhmer FD (2000) Biochem Pharmacol 59:35
52. Abdelmohsen K, Gerber PA, Von Montfort C, Sies H, Klotz LO (2003) J Biol Chem 278:38360
53. Zwick E, Hackel PO, Prenzel N, Ullrich A (1999) Trends Pharmacol Sci 20:408
54. Chen K, Vita JA, Berk BC, Keaney JF, Jr. (2001) J Biol Chem 276:16045
55. Wang X, McCullough KD, Franke TF, Holbrook NJ (2000) J Biol Chem 275:14624
56. Ostrakhovitch EA, Lordnejad MR, Schliess F, Sies H, Klotz LO (2002) Arch Biochem Biophys 397:232
57. Mallozzi C, Di Stasi AM, Minetti M (1999) FEBS Lett 456:201
58. MacMillan-Crow LA, Greendorfer JS, Vickers SM, Thompson JA (2000) Arch Biochem Biophys 377:350
59. Brown MT, Cooper JA (1996) Biochim Biophys Acta 1287:121
60. Kröncke KD, Klotz LO, Suschek CV, Sies H (2002) J Biol Chem 277:13294
61. Grether-Beck S, Bonizzi G, Schmitt-Brenden H, Felsner I, Timmer A, Sies H, Johnson JP, Piette J, Krutmann J (2000) EMBO J 19:5793
62. Meek DW (1999) Oncogene 18:7666
63. Cohen P (1997) Trends Cell Biol 7:353
64. Widmann C, Gibson S, Jarpe MB, Johnson GL (1999) Physiol Rev 79:143
65. Guyton KZ, Liu Y, Gorospe M, Xu Q, Holbrook NJ (1996) J Biol Chem 271:4138
66. Wang X, Martindale JL, Liu Y, Holbrook NJ (1998) Biochem J 333:291
67. Baas AS, Berk BC (1995) Circ Res 77:29
68. Klotz LO, Briviba K, Sies H (1997) FEBS Lett 408:289

69. Schieke SM, Briviba K, Klotz LO, Sies H (1999) FEBS Lett 448:301
70. Lander HM, Jacovina AT, Davis RJ, Tauras JM (1996) J Biol Chem 271:19705
71. Pombo CM, Bonventre JV, Avruch J, Woodgett JR, Kyriakis JM, Force T (1994) J Biol Chem 269:26546
72. Müller JM, Krauss B, Kaltschmidt C, Baeuerle PA, Rupec RA (1997) J Biol Chem 272:23435
73. Sugden PH, Fuller SJ, Michael A, Clerk A (1997) Biochem Soc Trans 25:S565
74. Klotz LO, Holbrook NJ, Sies H (2001) Curr Probl Dermatol 29:95
75. Mahns A, Melchheier I, Suschek CV, Sies H, Klotz LO (2003) Free Radic Res 37:391
76. Assefa Z, Garmyn M, Bouillon R, Merlevede W, Vandenheede JR, Agostinis P (1997) J Invest Dermatol 108:886
77. Rosette C, Karin M (1996) Science 274:1194
78. Brenneisen P, Wenk J, Klotz LO, Wlaschek M, Briviba K, Krieg T, Sies H, Scharffetter-Kochanek K (1998) J Biol Chem 273:5279
79. Sachsenmaier C, Radler-Pohl A, Zinck R, Nordheim A, Herrlich P, Rahmsdorf HJ (1994) Cell 78:963
80. Raingeaud J, Gupta S, Rogers JS, Dickens M, Han J, Ulevitch RJ, Davis RJ (1995) J Biol Chem 270:7420

The Handbook of Environmental Chemistry Vol. 2, Part O (2005): 219–234
DOI 10.1007/b101153

Mitochondrial Free Radical Production, Antioxidant Defenses and Cell Signaling

Enrique Cadenas · Alberto Boveris (✉)

[1] University of Southern California, Department of Molecular Pharmacology and Toxicology, School of Pharmacy, Los Angeles CA, USA
[2] University of Buenos Aires, Facultad de Farmacia y Bioquímica, Junin 956 C1113AAD, Buenos Aires, Argentina
aboveris@ffyb.uba.ar

Abstract Mitochondria were classically recognized as the organelles that produce the energy required to drive the endergonic processes of cell life, but now they are considered as the most important cellular source of free radicals, as the main target for free radical regulatory and toxic actions, and as the source of signaling molecules that command cell cycle, proliferation, and apoptosis. Mitochondrial production of the primary free radicals superoxide anion ($O_2^{\bullet -}$) and nitric oxide (NO), as well as of the termination products H_2O_2 (hydrogen peroxide) and peroxynitrite ($ONOO^-$), is described. A specialized network of intramitochondrial antioxidants consisting of the enzymes Mn-superoxide dismutase and glutathione peroxidase and of the reductants $NADH_2$, ubiquinol and reduced glutathione, appears operative in minimizing the potentially harmful effects of $O_2^{\bullet -}$, NO, H_2O_2 and $ONOO^-$. The effective operation of the intramitochondrial network of antioxidants is necessary to prevent the oxidant-driven mitochondrial damage that leads to mitochondrial dysfunction, a syndrome recognized in several diseases, inflammation, and aging. Deficiency in nutritional antioxidants and environmental metal toxicity are considered in relation to mitochondrial oxidative stress and dysfunction. Parkinson's and Alzheimer's diseases have been claimed to be associated with neuronal and systemic oxidative stress. There are reports on the beneficial effects of vitamins and antioxidants for the treatment of both diseases. Nitric oxide and H_2O_2 participate in the cell signaling that commands cell cycle, proliferation, and apoptosis. A mechanism involving mitogen-activated protein kinases is described. The role of mitochondria in apoptosis is well established through the mitochondrion-dependent pathway of cell death, which includes increased NO production, loss of membrane potential, appearance of dysfunctional mitochondria, cytochrome c release, and opening of the outer membrane voltage-dependent anion channel.

Keywords Mitochondria · Superoxide radical · Hydrogen peroxide · Nitric oxide · Peroxynitrite · Mn-SOD · Glutathione peroxidase · Nutritional antioxidants · Toxic metals

1 Mitochondrial Production of Free Radicals

Mitochondria were classically considered as the subcellular organelles of eukaryotic cells that produce the energy required to drive the endergonic biochemical processes of cell life. Such a concept is now complemented by the consideration of mitochondria as the most important cellular source of free radicals, as the main target for free radical regulatory and toxic actions, and as the source of signaling molecules that command cell cycle, proliferation, and apoptosis.

Mitochondria were recognized as source of the free radical superoxide anion ($O_2^{\bullet-}$) in simultaneous reports by Boveris and Cadenas [1] and by Dionisi et al. [2] showing that $O_2^{\bullet-}$ is the stoichiometric precursor of mitochondrial H_2O_2. The quantitative determination of the rates of $O_2^{\bullet-}$ and H_2O_2 production by subcellular organelles and structures indicated that in most mammalian organs mitochondria are the most important intracellular source of $O_2^{\bullet-}$ and H_2O_2. Liver, where the rates of $O_2^{\bullet-}$ and H_2O_2 production by isolated endoplasmic reticulum (microsomes) are comparable to the mitochondrial rates, seems to constitute the exception to the previous statement [3]. A few years ago, mitochondria were also recognized as sources of nitric oxide (NO) by two independent research groups [4, 5]. Although analysis of the relative contribution of mitochondria to the total cellular capacity of NO synthesis in mammalian organs has just started, preliminary data indicate that the mitochondrial contribution is far from being negligible. The mitochondrial contribution to cellular NO synthesis ranges from 20–25% in brain, to 30–40% in liver and kidney, and to 50–60% in heart. Interestingly, the reactive oxygen species that are continuously produced in the mitochondrial matrix show cross-reactivity in a series of collisional non-enzymatic reactions as illustrated in Fig. 1.

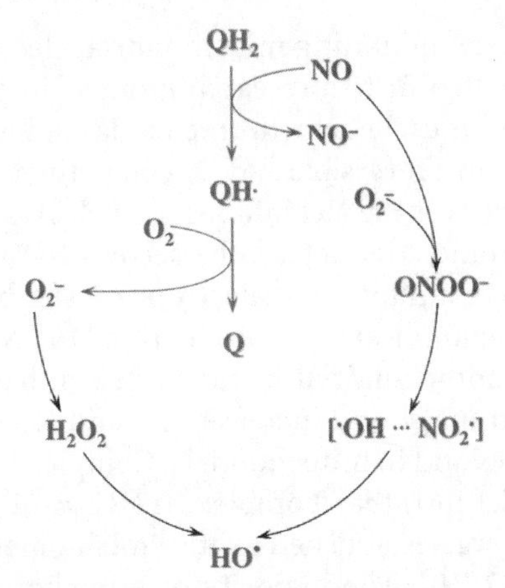

Fig. 1 Biochemical relationships of the reactive species in the mitochondrial matrix. QH_2 ubiquinol, *QH·* ubisemiquinone, *Q* ubiquinone, *NO* nitric oxide, NO^- nitroxyl anion, O_2 molecular oxygen, $O_2^{\cdot-}$ superoxide anion radical, $ONOO^-$ peroxynitrite, H_2O_2 hydrogen peroxide, *HO·* hydroxyl radical

1.1 Mitochondrial Production and Release of Superoxide Radical and Hydrogen Peroxide

Isolated mitochondria were observed as sources of hydrogen peroxide (H_2O_2) about 30 years ago [6, 7]. The mitochondrial production of H_2O_2 was characterized as a by-product of electron transfer by the auto-oxidation of components of the respiratory chain in a process called "electron leak". The component of the respiratory chain that is the quantitative main source of oxygen reduction is ubisemiquinone (UQH) [8–12]. The process of mitochondrial production of H_2O_2 is strongly modulated by the mitochondrial metabolic state and by the intramitochondrial concentration of nitric oxide (NO). In the case of rat heart and liver, mitochondrial H_2O_2 production accounts for about 0.5% of the physiological organ O_2 uptake [8]. The production of H_2O_2 in state 4 (the controlled or resting mitochondrial metabolic state with slow respiratory rate, excess of reducing substrate, and absence of ADP) is about 4–8 times higher than in state 3 (the active mitochondrial metabolic state, characterized by the fastest physiological rate of oxygen uptake and ATP synthesis, and with excess of reducing substrate and ADP). The marked difference in the rates of H_2O_2 production between both mitochondrial metabolic states indicates that the H_2O_2 generator is a component of the respiratory chain, markedly changing its redox state in the state 4/state 3 transition. The rates of H_2O_2 production of mitochondria isolated from mammalian organs are 0.3–0.8 nmol H_2O_2 min^{-1} mg^{-1} protein for state 4 and 0.05–0.15 nmol H_2O_2 $min^{-1}mg^{-1}$ protein for state 3 [3]. Ion movements through the inner mitochondrial membrane markedly change the rates of H_2O_2 production, indicating a membrane potential regulation of the auto-oxidation of UQH· [13].

The property of NO of inhibiting mitochondrial electron transfer was first recognized in 1994 by two British research groups [14, 15] that reported the inhibition of brain and muscle cytochrome oxidase (complex IV) activity by low NO concentrations in a reversible and O_2-competitive manner. More related to the scope of this review is the NO inhibition of electron transfer at complex III, ubiquinol-cytochrome c reductase, the second NO-sensitive point in the respiratory chain, where inhibition of electron transfer between cytochromes b and c enhances mitochondrial H_2O_2 production [16]. Nitric oxide, produced by NO donors or by mitochondrial nitric oxide synthase (mtNOS), inhibits complex III electron transfer and increases $O_2^{\bullet-}$ and H_2O_2 production in submitochondrial particles and in mitochondria. Complex IV is more sensitive to NO inhibition (IC_{50}=0.1 μM) than complex III (IC_{50}=0.2 μM).

Superoxide radical was established as the stoichiometric precursor of mitochondrial H_2O_2 [1, 2, 10]. The majority of mitochondrial $O_2^{\bullet-}$, 70–80%, is vectorially released to the mitochondrial matrix, but recently it was found that 20–30% is released to the mitochondrial intermembrane space [11].

Two main $O_2^{\bullet-}$ generating reactions have been described for the mitochondrial respiratory chain. In both cases, the intermediate semiquinone of two redox pairs of components of the respiratory chain, ubiquinol/ubiquinone and the $FMNH_2$/FMN coenzyme of NADH dehydrogenase, $UQH^{\bullet}$ and $FMNH^{\bullet}$, respectively, are collision and non-enzymatically auto-oxidized by molecular O_2 to yield $O_2^{\bullet-}$ [3]:

$$UQH^{\bullet} + O_2 \rightarrow UQ + H^+ + O_2^{\bullet-} \quad (1)$$

$$FMNH^{\bullet} + O_2 \rightarrow FMN + H^+ + O_2^{\bullet-} \quad (2)$$

The auto-oxidation of other components of the mitochondrial respiratory chain is thermodynamically possible, but the rates are not kinetically important on a quantitative basis.

1.2
Mitochondrial Production and Release of Nitric Oxide

In the first years of the 1990s, the extraordinary research activity on biological NO production identified three different nitric oxide synthases (NOS): neuronal NOS (nNOS or NOS-I), inducible or macrophague NOS (iNOS or NOS-II), and endothelial NOS (eNOS or NOS-III) [17]. Two research groups located in Zurich [4] and Los Angeles/Buenos Aires [5], produced a breakthrough in the knowledge of mitochondrial regulation with the report of NO production by rat liver mitochondria and mitochondrial fragments. The responsible enzyme, mitochondrial NOS (mtNOS), is a classic NOS in biochemical terms (reaction 3). The intramitochondrial concentrations of $NADPH_2$, arginine, O_2, and Ca^{2+} are in excess or in the range needed for enzymatic activity. Mitochondria and submitochondrial preparations yield rates of 0.25–0.90 nmol NO $min^{-1}mg^{-1}$ protein.

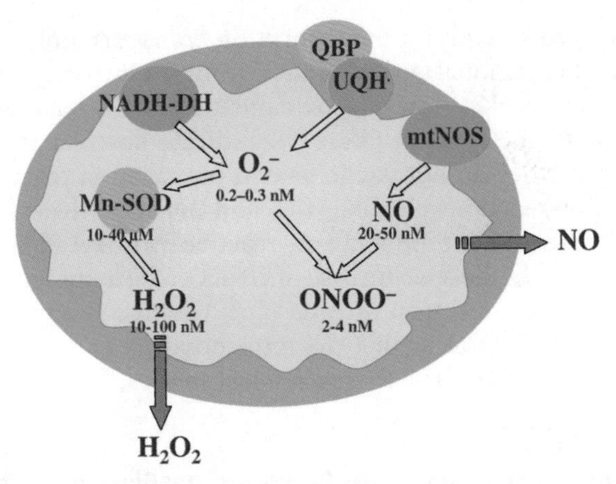

Fig. 2 Metabolism of superoxide radical and nitric oxide in the mitochondrial matrix. The numbers below the symbols indicate approximate steady state concentrations for mammalian organs under physiological conditions. The *arrows* reaching outside the mitochondrion indicate diffusion of H_2O_2 and NO to the cytosol. *QBP* ubiquinone binding protein, *NADH-DH* NADH dehydrogenase

$$\text{NADPH2} + \text{Arg} + O_2 \rightarrow \text{NADP} + H_2O + \text{Cit} + \text{NO} \quad (3)$$

Intramitochondrial steady state concentrations of NO were calculated as 20–50 nM NO [12] and a release of 29 nM NO was electrochemically measured after supplementation of a single mitochondrion with Ca^{2+} [18]. Under physiological conditions the tissues are oxygenated in the range of 20 μM O_2, with $[O_2]/[NO]$ ratios of 500–1000, which should competitively inhibit cytochrome oxidase by 16–26% [19].

The cellular conditions in which NO diffuses from mitochondria to cytosol, as well as the conditions in which NO diffuses from cytosol to mitochondria (called the mitochondria–cytosol NO cross-talk) constitute a key question for the command of the complex process of intracellular signaling.

The main mitochondrial pathways of $O_2^{\cdot -}$ and NO metabolism are shown in Fig. 2, which also gives the approximate steady state concentrations of both metabolites and related species in mammalian mitochondria.

2 Intramitochondrial Antioxidant Network

Aerobic organisms evolved a complex system of antioxidant defenses to minimize the oxidative damage resulting from the presence of oxygen. Antioxidant enzymes and low molecular weight antioxidants, such as reduced glutathione (GSH), comprise the biochemical constitutive antioxidant defense. In addition, mammals in general complement the antioxidant defenses with nutritional antioxidants, such as ascorbic acid, α-tocopherol, and β-carotene, and for such purpose developed systems for the absorption, transportation, and biochemical utilization of nutritional antioxidants. The intramitochondrial metabolites $O_2^{\bullet-}$, H_2O_2, NO, and $ONOO^-$ are pro-oxidants potentially leading to oxidative damage. Two of them, $O_2^{\bullet-}$ and NO, are free radicals; however, they are unreactive and sluggish, they do not participate in propagation reactions and only show termination reactions yielding H_2O_2 and $ONOO^-$. The latter two species are potentially harmful in producing the reactive hydroxyl radical ($HO^{\bullet}$) after homolytic scission. In the specific case of $ONOO^-$, the hemolytic scission also yields the free radical nitrogen dioxide ($O_2N^{\bullet}$), which is involved in protein nitration [20]. It is then clear that the mitochondrial matrix has to have an antioxidant network to control these chemical species.

2.1 Antioxidant Enzymes: Mn-Superoxide Dismutase and Glutathione Peroxidase

The main part of mitochondrial $O_2^{\bullet-}$ is vectorially released to the matrix, where it encounters specific intramitochondrial Mn-superoxide dismutase (Mn-SOD) [21–23] (reaction 4). Steady state concentrations of 0.2–0.3 nM $O_2^{\bullet-}$ were estimated for the mitochondrial matrix, with a content of 10–40 μM Mn-SOD reaction centers [24]. The $O_2^{\bullet-}$ released into the intermembrane space [11] reacts with cytochrome c, located on the P side of the inner membrane, and with the Cu, Zn-SOD of the intermembrane space [25]:

$$O_2^{\bullet-} + 2H^+ \rightarrow O_2 + H_2O_2 \quad (4)$$

$$H_2O_2 + 2GSH \rightarrow 2H_2O + GSSG \quad (5)$$

Glutathione peroxidase (reaction 5) is the unique mitochondrial enzyme that uses H_2O_2 in the mitochondria of most mammalian organs [26], the exception being heart, where a mitochondrial catalase has been described [27]. The activity of glutathione peroxidase and the concentration of GSH provide a maximal mitochondrial rate of H_2O_2 utilization of 2.4 μM s^{-1}, calculated from 240 $s^{-1} \times [H_2O_2]$ and a 10 nM steady state concentration of H_2O_2 [24]. This rate of H_2O_2 utilization, which accounts for about 60% of the rate of H_2O_2 production, appears to indicate both a role of H_2O_2 as messenger and a function of glutathione peroxidase in the reduction of mitochondrial hydroperoxides. Glutathione deficiency is associated with widespread mitochondrial dysfunction that leads to other types of cell damage [24, 28]. Glutathione peroxidase requires a continuous supply of GSH, since this reductant becomes oxidized during H_2O_2 and hydroperoxide reduction. A separate enzyme, $NADPH_2$-dependent glutathione reductase is located in the mitochondrial matrix and shows a high activity that keeps the GSH/GSSG couple in a reduced state [26]. Mitochondria are the result of the symbiotic co-evolution of primitive bacteria and eukaryotic cells, a process beginning around 2 billions years ago [29, 30]. Modern bacteria have a set of the classic antioxidant enzymes, Mn-SOD, catalase, and glutathione peroxidase [31–33], indeed the presence of Mn-SOD and catalase is the criterion for distinguishing aerobes, anaerobes, and microaerobes [33]. Then, either catalase was lost during mitochondrial evolution or modern bacteria gained catalase as a more efficient way to deal with H_2O_2.

2.2 Antioxidants of the Metabolism of Nitric Oxide and Peroxynitrite

Nitric oxide and $O_2^{\cdot-}$ metabolism in the mitochondrial matrix are linked by the very fast, diffusion limited, reaction between NO and $O_2^{\cdot-}$ to produce peroxynitrite ($ONOO^-$) (reaction 6 [34, 35]). This oxidative utilization of NO is the main (60–70%) pathway of NO metabolism but only a minor part (15%) of mitochondrial $O_2^{\cdot-}$ utilization, whereas the reductive utilization of NO by ubiquinol and cytochrome oxidase provides a minor (20%) pathway of NO catabolism [12].

$$NO + O_2^{\cdot-} \rightarrow ONOO^- \qquad (6)$$

Peroxynitrite is a powerful oxidant that, as a charged species, is poorly diffusible from the intramitochondrial space. When $ONOO^-$ is produced in excess, as in inflammation and in ischemia-reperfusion, there is tyrosine nitration of the mitochondrial proteins. Mitochondrial levels of 2–5 nM $ONOO^-$ have been estimated for the mitochondrial matrix under physiological conditions [12, 22], and levels above 20–30 nM are considered cytotoxic. The existence of a stable low $ONOO^-$ concentration is indicated by the detection of nitrotyrosine in normal mitochondria [36]. At high levels, $ONOO^-$ overwhelms the reducing reactions, and oxidation and nitration of lipids and proteins may impair mitochondrial function. The whole syndrome of mitochondrial dysfunction appears driven by excess NO and $ONOO^-$. This mitochondrial syndrome, observed in ischemia-reperfusion, inflammation, and aging [37–39] is characterized by decreased rates of state 3 respiration and ATP synthesis, decreased respiratory controls and membrane potential, increased rates of state 4 respiration, and increased mitochondrial size and fragility.

Protein oxidation and nitration may have a prime importance in the cell life cycle. The proteolytic enzymes that degrade modified proteins decline with age; this is then associated with a less efficient removal and an accumulation of oxidized proteins [40, 41].

There are effective reductants for matrix $ONOO^-$ due to their reaction constants and mitochondrial levels. In rat liver, 1.9 mM $NADH_2$, 1.3 mM UQH_2, and 1.5 mM GSH in the mitochondrial matrix (taken as 2 $\mu L\ mg^{-1}$ protein) account for 33, 46, and 21%, respectively, of the matrix capacity to reduce $ONOO^-$ [22]. Peroxynitrite readily reacts with CO_2 to yield the adduct $ONOOCO_2^-$ ($k = 6\times10^4\ M^{-1}\ s^{-1}$, [42, 43]) that also participates in oxidation and nitration reactions. Formation of the adduct decreases the intramitochondrial steady state level of $ONOO^-$ from 30 nM to 2 nM, without changing the type of product formed [23]. Thus, adduct formation can be considered as a detoxification pathway for mitochondrial $ONOO^-$ that takes advantage of the high CO_2

Table 1 Production and utilization rates and steady state concentrations of the reactive species $O_2^{\bullet-}$, H_2O_2, NO, and $ONOO^-$ in the mitochondrial matrix. Adapted from data of Poderoso et al. [12], Valdez et al. [22], Boveris and Cadenas [24], and Alvarez et al. [82]

Reactive species (nM)	Steady state concentration	Production and utilization rate ($\mu M\ s^{-1}$)	
$O_2^{\bullet-}$	0.2–0.3	d $[O_2^{\bullet-}]$/dt	8.2
H_2O_2	10–100	d $[H_2O_2]$/dt	4.1
NO	20–50	d [NO]/dt	2.3
$ONOO^-$	2–4	d $[ONOO^-]$/dt	0.2

concentration in mitochondria (1 mM). Finally, NO has been found to be an effective chain-breaker antioxidant by termination reactions with the free radical intermediates of lipoperoxidation [44].

Table 1 gives an integrated view of the steady state concentrations of the reactive species in the mitochondrial matrix, as well as of their production and utilization rates. In the steady state approach, both production and utilization rates are equalized to proceed with the differential equations that allow the estimation of the steady state concentrations.

3 Nutritional Antioxidants and Mitochondrial Function

The scientific community has been discussing for quite some time now the relationship between oxidative stress, defined as the imbalance between oxidant and antioxidants [45], and the health-disease status. An impressive amount of information available in the literature deals with the effects of the classic antioxidants, ascorbic acid, α-tocopherol, and β-carotene in a huge series of pathophysiological situations in experimental animals and humans. Concerning the effects of the classic antioxidants on mitochondrial function in situations of oxidative stress, the information is not so vast and most of the time it is not conclusive. However, substantial progress has been made in the description of the mitochondrial alterations in neurodegenerative diseases and in the α-tocopherol effects, both as prevention and as treatment [46]. We will briefly review some reports related to vitamin E and mitochondrial dysfunction in oxidative metabolic disorders and in the neurodegenerative Alzheimer's and Parkinson's diseases.

3.1 Vitamin E and Mitochondrial Oxidative Stress

The mitochondrial level of α-tocopherol (0.2–0.3 nmol mg^{-1} protein) is decreased in oxidative stress situations such as acute or chronic alcohol treatments [47, 48] and in aging [49]. Decreased levels of α-tocopherol in heart mitochondrial membranes were reported to be associated with increased rates of lipoperoxidation in vitro [50, 51].

3.2 Vitamin E and Parkinson's and Alzheimer's Diseases

The occurrence of mitochondrial damage and oxidative stress has been repeatedly found in the clinical setting and in experimental models of Parkinson's disease [52]. Treatment of Parkinson's disease patients with vitamin E has not provided definite positive results; however, a combination of antioxidants [46, 53], including vitamins such as niacin and riboflavin, is strongly recommended [46, 54]. Treatment of patients with Alzheimer's disease with vitamin E (2000 mg day^{-1}) has been reported to improve the patients "event-free-survival" (cognitive activity) [55]. It has also been claimed that an increased intake of vitamin E lowers the risk of Alzheimer's disease [56, 57].

4 Environmental Toxic Metals and Mitochondrial Oxidative Stress and Dysfunction

Many toxicity studies on contaminating metals show a cellular oxidative stress situation not mechanistically defined in terms of the subcellular localization of the initial damage. Growing evidence shows that metals such as Zn^{2+}, Pb^{2+}, and Cd^{2+} and ambient particulate matter may exert their toxic action involving mitochondrial oxidative stress and dysfunction [58].

4.1 Lead

The neurotoxic effect of Pb^{2+} has been associated with interference with the mitochondrial Na^{+}/Ca^{2+} anti-porter [59] and with the initiation of mitochondria-dependent apoptosis in rod photoreceptors [60]. Apparently, cytosolic Pb^{2+} competes with Ca^{2+}, leading to mitochondrial dysfunction.

4.2 Zinc

An increasing body of evidence suggests that Zn^{2+} induces neuronal death by impairing mitochondrial function through inhibition of electron transport at complex III [61], membrane depolarization [62, 63], and inhibition of α-ketoglutarate dehydrogenase activity [64]. Altogether, these results indicate that Zn^{2+} toxicity decreases mitochondrial energy production. It has been observed that Zn^{2+} also increases mitochondrial $O_2^{\bullet-}$ production [63], which may afford a mechanism for Zn^{2+}-triggered apoptosis.

4.3 Ambient Particulate Matter

These particles are complex mixtures of organic and inorganic components that increase the level of oxidative species in the lung and the heart after inhalation [65]. The time course of organ oxidative stress was followed by in situ lung and heart chemiluminescence and the phenomenon seems to involve mitochondrial participation [66].

5 Mitochondrial Signaling

The early recognition that both $O_2^{\bullet-}$ and H_2O_2 are able to initiate reactions harmful to cell and tissues is now complemented by the concept that both $O_2^{\bullet-}$ and H_2O_2 [8, 67, 68] are carefully regulated metabolites capable of signaling to the regulatory devices of the biochemical and genetic systems of the cell. Such a concept is also completed by the idea that inner membrane potential and selective permeability, with its system of uni-porters, anti-porters, and ion channels, constitute a regulatory device that signals to the sensitive proteins of the biochemical and genetic systems involved in the regulation of cell cycle and in the execution of apoptosis.

Nitric oxide was initially recognized as an intercellular messenger [17] and later as an intracellular regulator [14–16, 69]. At present, $O_2^{\bullet-}$, H_2O_2, and NO are considered part of an integrated system of mitochondrial signaling for cellular regulation. The redox regulation of gene expression and intercellular communication is just starting to be understood as a vital mechanism in health and disease.

The fine regulation by H_2O_2 of the cell cycle was advanced by Antunes and Cadenas [70] who showed that in Jurkat-T cells, steady state concentrations below 0.7 μM H_2O_2 place cells in a proliferative state, whereas at 1.0–3.0 μM H_2O_2 cells develop programmed cell death, and at levels higher that 3.0 μM H_2O_2 cells undergo necrosis. Hydrogen peroxide and NO share the chemical property of being uncharged and the biological property of being highly diffusible through biological membranes, at variance with $O_2^{\bullet-}$ and $ONOO^-$ that are charged and non-diffusible molecules. Consequently the first two species are suitable for cellular and intercellular signaling, and the two latter for cytotoxicity in confined spaces.

There is evidence that H_2O_2 and NO modulate mitogen-activated protein kinases (MAPK), the widespread integral components of intracellular phosphorylation and dephosphorylation signaling cascades involved in cell survival, proliferation, differentiation, and death [71]. The interactions are complex and seem to involve GSH and $ONOO^-$ [72,73]. Both H_2O_2 and NO diffusing from mitochondria to the cytosol appear to constitute a pleiotropic signal for a series of cellular processes, among them JNK signaling [74, 75]. Nitric oxide diffusing out of mitochondria may also inactivate JNK1 via S-nitrosylation [75]. The concept of narrow concentration ranges of the messenger molecules for different or opposite biological actions, as observed for H_2O_2 levels in the proliferation/apoptosis transition [70], is now considered for effector systems with two regulators, H_2O_2 and NO. This gives four different responses for the combination of two signals with two, low or high, levels each.

The role of mitochondria in apoptosis is well established; a specific mitochondria-dependent pathway of cell death is differentiated from the death-receptor mediated apoptosis [76]. Mitochondria-dependent apoptosis follows oxidative stress in a process that is modulated by MAPK signaling through JNK activation, associated with Bcl_2 down-regulation, cytochrome c release, and the downstream activation of caspases [77, 78]. A very early activation of mtNOS has been reported in thapsigargin-induced thymocyte apoptosis [79]. A cross talk between mitochondria and endoplasmic reticulum, in which Ca^{2+} and NO are the signals, occurs in the initial phase of apoptosis followed by the well-established steps of mitochondria-mediated apoptosis: fall in membrane potential, mitochondrial dysfunction, opening of the membrane transition pores (MTP), and cytochrome c release [76, 79, 80]. The central role of mitochondria in the cell death program is exerted during the first and reversible phase of apoptosis through the release of NO and H_2O_2 to activate cytosolic JNK that, attached to mitochondria, catalyzes phosphorylation of Bcl-2 and Bcl-XL and of intramitochondrial proteins [81]. The regulatory actions of $O_2^{\bullet-}$, NO, and H_2O_2 in early apoptosis are illustrated by Fig. 3, according to the data

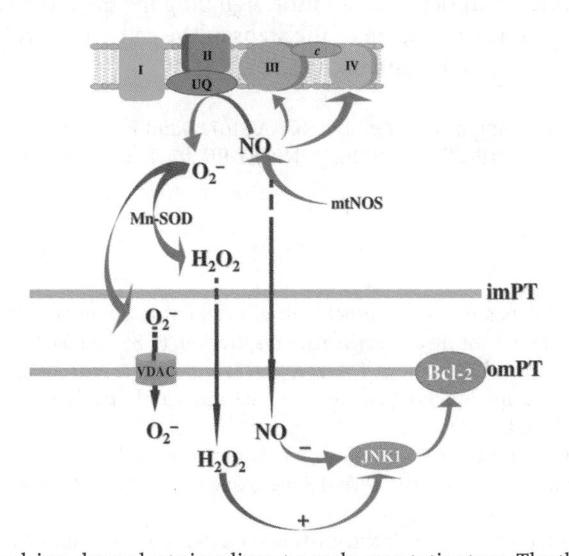

Fig. 3 Mitochondrion-dependent signaling at a early apoptotic stage. The three mayor actions of NO in mitochondria are indicated as reversible binding to complex IV, inhibition of electron transfer at complex III, and oxidation of ubiquinol yielding $O_2^{\bullet-}$. The fates of $O_2^{\bullet-}$, H_2O_2, and NO are indicated together with their release into cytosol (the former through VDAC). Modulation of JNK activity (stimulation by H_2O_2 and inhibition by NO) is indicated. JNK phosphorylates Bcl-2 and Bcl-XL

of Schroeter et al. [81]. The second and irreversible phase of apoptosis is initiated by JNK-mediated phosphorylations that open the membrane transition pore and release Ca^{2+} and cytochrome c to cytosol simultaneous to the opening of the voltage-dependent anion channel (VDAC) with release of $O_2^{\bullet -}$ to the cytosol. The next steps of the cell death program, such as apoptosome assembly and caspase activation, are independent of mitochondrial intervention.

6 Concluding Remarks

Mitochondria have been recognized as the cell powerhouses for half a century. In recent years it has become clear that these organelles are the most important physiological source of oxygen free radicals and at the same time a cellular target for free radical-mediated damage, as described in a series of diseases and in aging. Mitochondria evolved an antioxidant defense to minimize the oxidative damage due to the presence of oxygen. This defense consists of the enzymes Mn-SOD, glutathione peroxidase, and glutathione reductase, of low molecular weight antioxidants, such as GSH, and of the mitochondrial reductants $NADH_2$ and UQH_2. Mitochondrial oxidative stress by deficient antioxidants seems associated with the neurodegeneration as in Parkinson's and Alzheimer's diseases, and by metal-catalyzed free radical formation in toxicity by environmental metals. The role of mitochondria in the signaling for the biochemical regulation and the genetic expression of the steps of the cell cycle, proliferation, and apoptosis is starting to be understood.

Acknowledgements Supported by grants R01-AG16718 and R01-ES11342 from NIH to EC and by grants from ANPCYT 01-08710, CONICET PIP 02271 and UBACYT B 075 to AB.

References

1. Boveris A, Cadenas E (1975) Mitochondrial production of superoxide anions and its relationship to the antimycin-insensitive respiration. FEBS Lett 54:311–314
2. Dionisi O, Galeotti T, Terranova T, Azzi A (1975) Superoxide radicals and hydrogen peroxide formation in mitochondria from normal and neoplastic tissues. Biochim Biophys Acta 403:292–301
3. Boveris A, Cadenas E (1982) Production of superoxide radicals and hydrogen peroxide in mitochondria. In: Oberley LW (ed) Superoxide dismutase. CRC, Boca Raton, Florida, pp 15–30
4. Ghafourifar P, Richter C (1997) Nitric oxide synthase activity in mitochondria. FEBS Lett 418:291–295
5. Giulivi C, Poderoso JJ, Boveris A (1998) Production of nitric oxide by mitochondria. J Biol Chem 273:11038–11043
6. Boveris A, Chance B (1973) The mitochondrial generation of hydrogen peroxide. General properties and effect of hyperbaric oxygen. Biochem J 134:707–716

7. Boveris A, Oshino N, Chance B (1972) The cellular production of hydrogen peroxide. Biochem J 128:617–630
8. Boveris A, Cadenas E (2000) Mitochondrial production of hydrogen peroxide. Regulation by nitric oxide and the role of ubisemiquinone. IUBMB Life 50:1–6
9. Boveris A, Cadenas E, Stoppani AOM (1976) Role of ubiquinone in the mitochondrial generation of hydrogen peroxide. Biochem J 156:435–444
10. Cadenas E, Boveris A, Ragan CI, Stoppani AOM (1977) Production of superoxide radicals and hydrogen peroxide by NADH-ubiquinone reductase and ubiquinol-cytochrome c reductase from beef heart mitochondria. Arch Biochem Biophys 180:248–257
11. Han D, Williams E, Cadenas E (2001) Mitochondrial respiratory chain-dependent generation of superoxide anion and its release to the intermebrane space. Biochem J 353: 411–416
12. Poderoso JJ, Lisdero CL, Schöpfer F, Riobó NA, Carreras MC, Cadenas E, Boveris A (1999) The regulation of mitochondrial oxygen uptake by redox reactions involving nitric oxide and ubiquinol. J Biol Chem 274:37709–37716
13. Cadenas E, Boveris A (1980) Enhancement of hydrogen peroxide formation by protophores and ionophores in antimycin-supplemented mitochondria. Biochem J 188:31–37
14. Brown GC, Cooper CE (1994) Nanomolar concentrations of nitric oxide reversibly inhibit synaptosomal respiration by competing with oxygen at cytochrome oxidase. FEBS Lett 356:295–298
15. Cleeter MWJ, Cooper JM, Darley-Usmar VM, Moncada S, Shapira AHV (1994) Reversible inhibition of cytochrome c oxidase, the terminal enzyme of the mitochondrial respiratory chain. Implications for neurodegenerative diseases. FEBS Lett 345:50–54
16. Poderoso JJ, Carreras MC, Lisdero CL, Riobó NA, Schöpfer F, Boveris A (1996) Nitric oxide inhibits electron transfer and increases superoxide radical production in rat heart mitochondria and submitochondrial particles. Arch Biochem Biophys 328:85–92
17. Ignarro LJ (2000) In: Ignarro LJ (ed) Nitric oxide: biology and pathobiology. Academic Press, New York
18. Kanai AJ, Pearce LL, Clemens PR, Birder LA, Van Bibber MM, Choi SY, de Groat WC, Peterson J (2001) Identification of a neuronal nitric oxide synthase in isolated cardiac mitochondria using electrochemical detection. Proc Natl Acad Sci USA 98:14126–14131
19. Boveris A, Costa LE, Poderoso JJ, Cadenas E (2000) Regulation of mitochondrial respiration by oxygen and nitric oxide. Ann NY Acad Sci 899:121–135
20. Radi R, Cassina A, Hodara R, Quijano C, Castro L (2002) Peroxynitrite reactions and formation in mitochondria. Free Radic Biol Med 33:1451–1464
21. Tyler DD (1975) Polarographic assay and intracellular distribution of superoxide dismutase in rat liver. Biochem J 147:493–504
22. Valdez LB, Alvarez S, Lores-Arnaiz S, Schöpfer F, Carreras MC, Poderoso JJ, Boveris A (2000) Reactions of peroxynitrite in the mitochondrial matriz. Free Radic Biol Med 29:349–356
23. Weisinger RA, Fridovich I (1973) Mitochondrial superoxide dismutase. Site of synthesis and intramitochondrial localization. J Biol Chem 248:4793–4796
24. Boveris A, Cadenas E (1997) Cellular sources and steady-state levels of reactive oxygen species. In: Biadasz-Clerch L, Massaro DJ (eds) Oxygen, gene expression and cellular function. Dekker, New York, pp 1–25
25. Okado-Matsumoto A, Fridovich I (2001) Subcellular distribution of superoxide dismutases (SOD) in rat liver: Cu,Zn-SOD in mitochondria. J Biol Chem 276:38388–38393
26. Chance B, Seis H, Boveris A (1979) Hyperoxide metabolism in mammalian organs. Physiol Rev 59(3):527–605

27. Radi R, Turrens JF, Chang LY, Bush KM, Crapo JD, Freeman BA (1991) Detection of catalase in rat heart mitochondria. J Biol Chem 266:22028–22034
28. Meister A (1995) Mitochondrial changes associated with glutathione deficiency. Biochim Biophys Acta 1271:35–42
29. Fridovich I (1974) Evidence for the symbiotic origin of mitochondria. Life Sci 14:819–826
30. Lane N (2002) Oxygen. The molecule that made the world. Oxford University Press, New York, USA, pp 50–51
31. Kohrl J, Brigelius-Fole R, Bock A, Gartner R, Meyer O, Flohe L (2000) Selenium in biology: facts and medical perspectives. Biol Chem 381:849–864
32. McCord JM, Keele BB, Fridovich I (1973) An enzyme-based theory of obligate anaerobiosis: the physiological function of superoxide dismutase. Proc Natl Acad Sci USA 68: 1024–1027
33. Georgiou G (2002) How to flip the (redox) swith. Cell 111:607–610
34. Beckman JS (1990) Ischaemic injury mediator. Nature 345:27–28
35. Kissner R, Nauser T, Bugnon P, Lye PG, Koppenol WH (1997) Formation and properties of peroxynitrite as studied by laser flash photolysis, high pressure stopped-flow technique, and pulse radiolysis. Chem Res Toxicol 10:1285–1292
36. Lisdero CL, Carreras MC, Meuleman A, Melani M, Aubier M, Boczkovski J, Poderoso JJ (2004) The mitochondrial interplay of ubiquinol and nitric oxide in endotoxemia. Methods Enzymol 382:67–81
37. Beckman KB, Ames BN (1998) The free radical theory of aging matures. Physiol Rev 78:547–581
38. Cutrin JC, Boveris A, Zingaro B, Corvetti G, Poli G (2000) In situ determination by surface chemiluminescence of temporal relationships between evolving warm ischemia-reperfusion injury in rat liver and phagocyte activation and recruitment. Hepatology 31:622–632
39. Gonzalez-Flecha B, Cutrin JC, Boveris A (1993) Time course and mechanisms of oxidative stress and tissue damage in rat liver subjected to in vivo ischemia-reperfusion. J Clin Invest 91:456–464
40. Buchczyk DP, Grune T, Sies H, Klotz LO (2003) Modifications of glyceraldehydes-3-phosphate dehydrogenase induced by increasing concentrations of peroxynitrite: early recognition by the proteasome. Biol Chem 384:237–241
41. Grune T, Merker K, Sandig G, Davies KJ (2003) Selective degradation of oxidatively modified protein substrates by the proteasome. Biochem Biohys Res Commun 305:709–718
42. Bonini MG, Radi R, Ferrer-Sueta G, Ferreira AM, Augusto O (1999) Direct EPR detection of the carbonate radical anion produced from peroxynitrite and carbon dioxide. J Biol Chem 274:10802–10806
43. Lymar SV, Hurst JK (1995) J Am Chem Soc 117:8867–8868
44. Rubbo H, Radi R, Anselmi D, Kirk M, Barnes S, Butler J, Eiserich JP, Freeman BA (2000) Nitric oxide reaction with lipid peroxyl radicals spares α-tocopherol during lipid peroxidation. Greater oxidant protection from the pair nitric oxide/α-tocopherol than α-tocopherol/ascorbate. J Biol Chem 275:10812–10818
45. Sies H (1985) Oxidative stress: introductory remarks. In: Sies H (ed) Oxidative stress. Academic Press, USA, pp 1–7
46. Marriage B, Cladinin MT, Glerum DM (2003) Nutritional cofactor treatment in mitochondrial disorders. J Am Diet Assoc 103:1029–1038
47. Bjorneboe GE, Bjorneboe A, Hagen BF, Morland J, Drevon CA (1987) Reduced hepatic α-tocopherol content after long-term administration of ethanol to rats. Biochim Biophys Acta 918:236–241

48. Frank O, Luisada-Oper A, Sorrell MF, Zetterman R, Baker H (1976) Effects of a single intoxicating dose of ethanol on the vitamin profile of organelles in rat liver and brain. J Nutr 106:606–614
49. Grinna LS (1976) Effect of dietary α-tocopherol on liver microsomes and mitochondria of aging rats. J Nutr 106:918–929
50. Ernster L, Forsmark P, Nordenbrand K (1992) The mode of action of lipid-soluble antioxidants in biological membranes. Relationship between the effects of ubiquinol and vitamin E as inhibitors of lipid peroxidation in submitochondrial particles. J Nutr Sci Vitaminol (Tokyo) Sp. Issue, pp 548–551
51. Scholz RW, Minicucci LA, Reddy CC (1997) Effects of vitamin E and selenium on antioxidant defense in rat heart. Biochem Mol Biol Int 42:997–1006
52. Fahn S, Cohen G (1992) The oxidative stress hypothesis in Parkinson's disease. Evidence supporting it. Neurology 47:S161–S170
53. Prasad KN, Cole WC, Kumar B (1999) Multiple antioxidants in the prevention and treatment of Parkinson's disease. J Am Coll Nutr 18:413–423
54. Coimbra CC, Junqueira VBC (2002) High doses of riboflavin and the elimination of red meat promote the recovery of some motor functions in Parkinson's disease patients. Braz J Med Biol Res 36:1409–1417
55. Sano M, Ernesto C, Thomas RG et al. (1997) A controlled trial of selegiline, α-tocopherol, or both as treatment for Alzheimer's disease. The Alzheimer's Disease Cooperative Study. N Engl J Med. 336:1216–1222
56. Engelhart MJ, Geerlings MI, Ruitenberg A et al. (2002) Dietary intake of antioxidants and risk of Alzheimer disease. J Am Med Assoc 287:3223–3229
57. Morris MC, Evans DA, Bienias JL et al. (2002) Dietary intake of antioxidant nutrients and the risk of incident Alzheimer disease in a biracial community study. J Am Med Assoc 287:3230–3237
58. Oteiza P, Mackenzie GG, Verstraeten SV (2004) Metals in neurodegeneration: involvement of oxidants and oxidant-sensitive transcription factors. Mol Aspects Med 25:103–115
59. Silbergeld EK, Adler HS (1978) Subcellular mechanisms of lead neurotoxicity. Brain Res 148:451–467
60. He L, Perkins GA, Poblenz AT, Harris JB, Hung M, Ellisman MH, Fox DA (2003) Bcl-xL overexpression blocks bax-mediated mitochondrial contact site formation and apoptosis in rod photoreceptors of lead-exposed mice. Proc Natl Acad Sci USA 100:1022–1027
61. Lorusso M, Cocco T, Sardanelli AM, Minuto M, Bonomi F, Papa S (1991) Interaction of Zn^{2+} with the bovine-heart mitochondrial bc1 complex. Eur J Biochem 197:555–561
62. Dineley KE, Votyakova TV, Reynolds IJ (2003) Zinc inhibition of cellular energy production: implications for mitochondria and neurodegeneration. J Neurochem 85:563–570
63. Sensi SL, Yin HZ, Carriedo SG, Rao SS, Weiss JH (1999) Preferential Zn^{2+} influx through Ca^{2+}-permeable AMPA/kainate channels triggers prolonged mitochondrial superoxide production. Proc Natl Acad Sci USA 96:2414–2419
64. Brown AM, Kristal BS, Effron MS, Shestopalov AI, Ullucci PA, Sheu KF, Blass JP, Cooper AJ (2000) Zn^{2+} inhibits alpha-ketoglutarate-stimulated mitochondrial respiration and the isolated alpha-ketoglutarate dehydrogenase complex. J Biol Chem 275:13441–13447
65. Gurgueira SA, Lawrence J, Coull B, Krishna-Murti GG, Gonzalez-Flecha B (2002) Rapid increases in the steady state concentration of reactive oxygen species in the lungs and the heart after particulate air pollution inhalation. Environ Health Perspect 110:749–755
66. Gonzalez-Flecha B (2004) Oxidant mechanisms in response to ambient air particles. Mol Aspects Med 25:169–182
67. Cross CE, Halliwell B, Borisch HT, Pryor WA, Ames BN, Saul RL, McCord JM, Harman D (1987) Oxygen radicals and human disease. Ann Intern Med 107:526–545

68. McCord JM (2000) The evolution of free radicals and oxidative stress. Am J Med 108: 652–659
69. Brown GC (2003) NO says yes to mitochondria. Science 299:838–839
70. Antunes F, Cadenas E (2001) Cellular titration of apoptosis with steady state concentrations of H_2O_2: submicromolar levels of H_2O_2 induce apoptosis through Fenton chemistry independent of the cellular thiol state. Free Radic Biol Med 30:1008–1018
71. Boyd CS, Cadenas E (2002) Nitric oxide and cell signaling pathways in mitochondrial-dependent apoptosis. Biol Chem 383:411–423
72. Go YM, Patel RP, Maland MC, Park H, Beckman JS, Darley-Usmar VM, Jo H (1999) Evidence for peroxynitrite as a signaling molecule in flow-dependent activation of c-Jun NH_2 terminal kinase. Am J Physiol Heart Circ Physiol 277:H1647–H1653
73. Lander HM, Jacovina AT, Davis RJ, Tauras JM (1996) Differential activation of mitogen-activated protein kinases by nitric oxide-related species. Proc Natl Acad Sci USA 271: 19705–19709
74. Martindale JL, Holbrook NJ (2002) Cellular responses to oxidative stress: signaling for suicide and survival. J Cell Physiol 192:1–15
75. Park HS, Park E, Kim MS, Ahn K, Kim IY, Choi EJ (2000) Selenite inhibits the c-Jun N-terminal kinase/stress activated kinase (JNK/SAPK) through a thiol redox mechanism. J Biol Chem 275:2527–2531
76. Kroemer G, Dallaporta B, Resche-Rigon M (1998) The mitochondrial death/life regulators in apoptosis and necrosis. Annu Rev Physiol 60:619–642
77. Tounier C, Hess P, Yang DD, Xu J, Turner TK, Nimnual A, Bar-Sagi D, Jones SN, Flavell RA, Davis RJ (2000) Requirement of JNK for stress-induced activation of the cytochrome c-mediated death pathway. Science 288:870–874
78. Whitfield J, Neame SJ, Paquet L, Bernard O, Ham J (2001) Dominant-negative c-Jun promotes neuronal survival by reducing BIM expression and inhibiting cytochrome c release. Neuron 29:629–643
79. Bustamante J, Bersier G, Aron-Badin R, Cymeryng C, Parodi A, Boveris A (2002) Sequential NO production by mitochondria and endoplasmic reticulum during induced apoptosis. Nitric Oxide 3:333–341
80. Ott M, Robertson JD, Gogvadze V, Zhivotovsky B, Orrenius S (2002) Cytochrome c release from mitochondria proceeds by a two-step process. Proc Natl Acad Sci USA 99:1259–1263
81. Schroeter H, Boyd CS, Ahmed R, Spencer JPE, Duncan RF, Rice-Evans C, Cadenas E (2003) JNK-mediated modulation of brain mitochondria function: new target proteins for JNK-signaling in mitochondria-dependent apoptosis. Biochem J 372:359–369
82. Alvarez S, Valdez LB, Zaobornyj T, Boveris A (2003) Oxygen dependence of mitochondrial nitric oxide synthase activity. Biochem Biophys Res Commun 305:771–775

Subject Index